Carbon Sequestration Potential of Agroforestry Systems

Advances in Agroforestry

Volume 8

Series Editor:
P.K.R. Nair
School of Forest Resources and Conservation,
University of Florida, Gainesville, Florida, U.S.A.

Aims and Scope

Agroforestry, the purposeful growing of trees and crops in interacting combinations, began to attain prominence in the late 1970s, when the international scientific community embraced its potentials in the tropics and recognized it as a practice in search of science. During the 1990s, the relevance of agroforestry for solving problems related to deterioration of family farms, increased soil erosion, surface and ground water pollution, and decreased biodiversity was recognized in the industrialized nations too. Thus, agroforestry is now receiving increasing attention as a sustainable land-management option the world over because of its ecological, economic, and social attributes. Consequently, the knowledge-base of agroforestry is being expanded at a rapid rate as illustrated by the increasing number and quality of scientific publications of various forms on different aspects of agroforestry.

Making full and efficient use of this upsurge in scientific agroforestry is both a challenge and an opportunity to the agroforestry scientific community. In order to help prepare themselves better for facing the challenge and seizing the opportunity, agroforestry scientists need access to synthesized information on multi-dimensional aspects of scientific agroforesty.

The aim of this new book-series, *Advances in Agroforestry*, is to offer state-of-the art synthesis of research results and evaluations relating to different aspects of agroforestry. Its scope is broad enough to encompass any and all aspects of agroforestry research and development. Contributions are welcome as well as solicited from competent authors on any aspect of agroforestry. Volumes in the series will consist of reference books, subject-specific monographs, peer-reviewed publications out of conferences, comprehensive evaluations of specific projects, and other book-length compilations of scientific and professional merit and relevance to the science and practice of agroforestry worldwide.

For further volumes:
http://www.springer.com/series/6904

B. Mohan Kumar • P.K. Ramachandran Nair
Editors

Carbon Sequestration Potential of Agroforestry Systems

Opportunities and Challenges

Springer

Editors
B. Mohan Kumar
Department of Silviculture and Agroforestry
College of Forestry
Kerala Agricultural University
KAU P.O.
Thrissur, Kerala 680 656,
India
bmkumar.kau@gmail.com

P.K. Ramachandran Nair
School of Forest Resources and Conservation
University of Florida
Newins-Ziegler Hall 351
Gainesville, FL 32611-0410
USA
pknair@ufl.edu

ISSN 1875-1199
e-ISSN 1875-1202
ISBN 978-94-007-1629-2
e-ISBN 978-94-007-1630-8
DOI 10.1007/978-94-007-1630-8
Springer Dordrecht Heidelberg London New York

Library of Congress Control Number: 2011934684

Printed on acid-free paper

Springer is part of Springer Science+Business Media (www.springer.com)

Preface

Global climate change caused by rising levels of carbon dioxide (CO_2) and other greenhouse gases is recognized as a serious environmental issue of the twenty-first century. The role of land use systems in stabilizing the CO_2 levels and increasing the carbon (C) sink potential has attracted considerable scientific attention in the recent past, especially after the Kyoto Protocol to the United Nations Framework Convention on Climate Change (UNFCCC). The Kyoto Protocol recognizes the role of afforestation, reforestation, and natural regeneration of forests in increasing the C storage capacity of terrestrial ecosystems. The post-Kyoto Protocol discussions on climate change are also heavily oriented towards an agenda on mitigating the rising atmospheric CO_2 levels through C sequestration in terrestrial vegetation systems.

Although the pristine natural forest ecosystems represent the largest vegetation and soil C sinks, a considerable extent of this has already been lost especially in the less developed and developing countries of the world. It is unlikely that these degraded and deforested sites will be returned to natural forest cover. The need for transforming some of the lower biomass land uses (such as arable croplands and fallows) to carbon-rich tree based systems such as plantation forests and agroforestry therefore assumes significance. Agroforestry systems (AFS) spread over one billion ha in diverse ecoregions around the world have a special relevance in this respect. These woody perennial-based land use systems have relatively high capacities for capturing and storing atmospheric CO_2 in vegetation, soils, and biomass products.

According to the Intergovernmental Panel on Climate Change, AFS offer important opportunities of creating synergies between both adaptation and mitigation actions with a technical mitigation potential of 1.1–2.2 Pg C in terrestrial ecosystems over the next 50 years. Additionally, 630 million ha of unproductive croplands and grasslands could be converted to agroforestry representing a C sequestration potential of 0.586 Tg C/yr by 2040 (1 Tg = 1 million tons). The total C storage in the aboveground and belowground biomass in an AFS is generally much higher than that in land use without trees (i.e. tree-less croplands) under comparable conditions. Various agroforestry practices such as alley cropping, silvopasture, riparian buffers, parklands, forest farming, homegardens, and woodlots, and other similar land use patterns have thus raised considerable expectations as a C sequestration strategy in both industrialized and developing countries.

Estimates of aboveground C sequestration potential (CSP) for AFS vary considerably. As can be expected, the CSP values are a direct manifestation of the ecological production potential of the system, depending on a number of factors including site characteristics, land use types, species involved, stand age, and management practices. In most cases, however, baseline information is either non-existent or is only anecdotal; besides, the methodologies used to derive such estimates often lack the required rigor.

Although C sequestration is a focal theme of discussion in most agroforestry and climate conferences, publications on C sequestration in agroforestry are scattered. Indeed, comprehensive publications focused on agroforestry and its C sequestration potentials are rare. This book is an attempt to address that deficiency. The book originated from a technical session "Carbon sequestration in Agroforestry" at the 2nd World Congress of Agroforestry, August 2009, Nairobi, Kenya (http://www.worldagroforestry.org/wca2009/), which featured 42 presentations (oral+ poster) on the topic. Out of the several manuscripts that originated from these presentations, six were selected following peer-review. Additionally, 10 chapters were organized as contributions from technical experts on the subject, some of which were based on presentations at the XXIII IUFRO (International Union of Forest Research Organizations) World Congress, August 2010, Seoul, South Korea. Five of these 16 chapters are research articles and are presented in the conventional research-publication format. Others deal with either case studies or provide regional overviews, and focus on the current trends in carbon sequestration research. The 16 chapters are organized into three broad sections: Measurement and Estimation, Agrobiodiversity and Tree Management, and Policy and Socioeconomic Aspects. Together they represent a cross section of the opportunities and challenges in current research and emerging issues in harnessing C sequestration potential of AFS.

The tedious task of putting together such a book would not have been possible without the unflinching cooperation and unfailing support of a number of collaborators. First of all, we thank the chapter authors who showed the highest level of commitment and professionalism in coping with repeated requests for revisions and improvement following rigorous peer review of their manuscripts. The reviewers (as per the attached list) did a splendid job of providing insightful comments and valuable suggestions, often at very short notice, which helped enhance the professional quality of the chapters. We also thank the publishers and/or other copyright holders of the original publications for permission to reproduce some of the tables and figures in some chapters as indicated in the respective places. Once again, we sincerely thank all the authors, reviewers, and others who directly or indirectly supported and cooperated with us in bringing out this publication.

Thrissur, Kerala, India
Gainesville, FL, USA
February 2011

B. Mohan Kumar
P.K. Ramachandran Nair

Contents

Contributors

André Aasrud Carbon Finance Unit, The World Bank, 1818 H St, NW, Washington, DC 20433, USA, andre.aasrud@iea.org

Michael G. Andreu School of Forest Resources and Conservation, Institute of Food and Agricultural Sciences, University of Florida, P.O. Box 110410, Gainesville, FL 32611–0410, USA, mandreu@ufl.edu

André Rodrigues de Aquino Carbon Finance Unit, The World Bank, 1818 H St, NW, Washington, DC 20433, USA, adeaquino@worldbank.org

Jens B. Aune Department of International Environment and Development Studies, Noragric, Norwegian University of Life Sciences, P.O. Box 5003, 1432 Aas, Norway, jens.aune@umb.no

Tracy Beedy World Agroforestry Centre (ICRAF), Lilongwe, Malawi, t.beedy@cgiar.org

Christian Böhm Soil Protection and Recultivation, Brandenburg University of Technology, Konrad-Wachsmann-Allee 6, D-03046 Cottbus, Germany, boehmc@tu-cottbus.de

Shushan Ghirmai Brakas Department of International Environment and Development Studies, Noragric, Norwegian University of Life Sciences, P.O. Box 5003, 1432 Aas, Norway, shushanfree@yahoo.com

Chetphong Butthep Office of International Affairs, National Research Council of Thailand, 196 Phaholyothin Road, Chatuchak, Bangkok 10900, Thailand, butthep@msu.edu

S.A.O. Chamshama Faculty of Forestry and Nature Conservation, Department of Forest Biology, Sokoine University of Agriculture, P.O. Box 3010, Morogoro, Tanzania, chamstz@yahoo.com

Johannes Dietz World Agroforestry Centre (ICRAF), Nairobi, Kenya, j.dietz@cgiar.org

Francis Dube Department of Silviculture, Faculty of Forest Sciences, University of Concepción, Victoria 631, Casilla 160-C, Concepción, Chile, fdube@udec.cl

Miguel Espinosa Department of Silviculture, Faculty of Forest Sciences, University of Concepción, Victoria 631, Casilla 160-C, Concepción, Chile, mespinos@udec.cl

Elsa Esquivel Cooperativa Ambio, Cuitlahuac 30, San Cristóbal de las Casas, Chiapas 29290, México, elsaesquivelb@hotmail.com

Dirk Freese Soil Protection and Recultivation, Brandenburg University of Technology, Konrad-Wachsmann-Allee 6, D-03046 Cottbus, Germany, freese@tu-cottbus.de

A.C. Gama-Rodrigues Soil Laboratory, Norte Fluminense State University, Campos dos Goytacazes, RJ 28013-602, Brazil, tonygama@ufl.edu

E.F. Gama-Rodrigues Soil Laboratory, Norte Fluminense State University, Campos dos Goytacazes, RJ 28013-602, Brazil, emanuela@uenf.br

Rasmo Garcia Animal Science Department, Federal University of Viçosa, P.H. Rolfs Avenue, MG 36570-000, Viçosa, Brazil

Goats and Sheep, EMBRAPA, Sobral, Ceara State, Brazil, rgarcia@ufv.br

Andrew M. Gordon School of Environmental Sciences, University of Guelph, Guelph, ON N1G 2W1, Canada, agordon@uoguelph.ca

Leticia Guimarães School of Public Policy, University of Maryland, 2101 Van Munching Hall, College Park, MD 20742, USA, leticiagsguimaraes@gmail.com

Terry Hills Conservation International, Cairns, Australia, t.hills@conservation.org

Reinhard F. Hüttl Soil Protection and Recultivation, Brandenburg University of Technology, Konrad-Wachsmann-Allee 6, D-03046 Cottbus, Germany

Helmholtz Centre, Potsdam – GFZ German Research Centre for Geosciences, Telegrafenberg, G 320, D-14473 Potsdam, Germany, Reinhard.Huettl@gfz-potsdam.de

M.E. Isaac Department of Physical and Environmental Sciences, University of Toronto, Scarborough, 1265 Military Trail, Toronto, ON M1C 1A4, Canada, marney.isaac@utoronto.ca

Shibu Jose 203 ABNR, The Center for Agroforestry, School of Natural Resources, University of Missouri, Columbia, MO 65211, USA, joses@missouri.edu

A.A. Kimaro Department of Soil Science, University of Saskatchewan, 51 Campus Drive, Saskatoon, SK S7N 5A8, Canada, anthony.kimaro@usask.ca

Usa Klinhom Mahasarakham University, Tambon Kamriang, Kantarawichai District, Mahasarakham 44150, Thailand, usa_klinhom@yahoo.com

B. Mohan Kumar Department of Silviculture and Agroforestry, College of Forestry, Kerala Agricultural University, KAU P.O., Thrissur, Kerala 680 656, India, bmkumar.kau@gmail.com

T.K. Kunhamu Department of Silviculture and Agroforestry, College of Forestry, Kerala Agricultural University, KAU P.O., Thrissur, Kerala 680 656, India, kunhamutk@yahoo.com

Teerawong Laosuwan Mahasarakham University, Tambon Kamriang, Kantarawichai District, Mahasarakham 44150, Thailand, Teerawong@msu.ac.th

Eike Luedeling World Agroforestry Centre (ICRAF), Nairobi, Kenya, e.luedeling@cgiar.org

Eduardo da Silva Matos Soil Protection and Recultivation, Brandenburg University of Technology, Konrad-Wachsmann-Allee 6, D-03046 Cottbus, Germany, eduardo.matos@embrapa.br

Federico Morales PROIMMSE, Universidad Nacional Autónoma de México, Calle Cuauhtemoc, San Cristobal de las Casas, Chiapas 29290, México, fmorales@servidor.unam.mx

M.R. Mosquera-Losada Crop Production Department, Escola Politécnica Superior, Universidad de Santiago de Compostela, Campus de Lugo, Lugo 27002, Spain, mrosa.mosquera.losada@usc.es

P.K. Ramachandran Nair Center for Subtropical Agroforestry, School of Forest Resources and Conservation, University of Florida, P.O. Box 110410, Gainesville, FL 32611–0410, USA, pknair@ufl.edu

Vimala D. Nair Soil and Water Science Department, University of Florida, P.O. Box 110510, Gainesville, FL 32611, USA, vdn@ufl.edu

Charlie Navanugraha Mahasarakham University, Tambon Kamriang, Kantarawichai District, Mahasarakham 44150, Thailand, encnv@hotmail.com

Stephanie Paladino El Colegio de la Frontera Sur, ECOSUR, Carretera Panamericana y Periférico Sur s/n, San Cristóbal de las Casas, Chiapas 29200, México, macypal@gmail.com

Ansgar Quinkenstein Soil Protection and Recultivation, Brandenburg University of Technology, Konrad-Wachsmann-Allee 6, D-03046 Cottbus, Germany, quinkenstein@tu-cottbus.de

A. Rigueiro-Rodríguez Crop Production Department, Escola Politécnica Superior, Universidad de Santiago de Compostela, Campus de Lugo, Lugo 27002, Spain, antonio.rigueiro@usc.es

Celia Ruiz-De-Oña-Plaza PROIMMSE, Universidad Nacional Autónoma de México, Calle Cuauhtemoc, San Cristobal de las Casas, Chiapas 29290, México, celia.ecosur@gmail.com

Subhrajit K. Saha Ashoka Trust for Research in Ecology and the Environment (ATREE), Bangalore, India, subhrajit_s@yahoo.com

Jay H. Samek Department of Forestry, Michigan State University, East Lansing, MI 48823, USA, samekjay@msu.edu

S. Samuel Department of Silviculture and Agroforestry, College of Forestry, Kerala Agricultural University, KAU P.O., Thrissur, Kerala 680 656, India, samuel.sijo@gmail.com

Götz Schroth Mars Incorporated, Santarém, Pará, Brazil

Federal University of Western Pará, Santarém, Pará, Brazil, goetz.schroth@gmail.com

Gudeta Sileshi World Agroforestry Centre (ICRAF), Lilongwe, Malawi, Sweldesemayat@cgiar.org

David L. Skole Department of Forestry, Michigan State University, East Lansing, MI 48823, USA, skole@msu.edu

Maria do Socorro Souza da Mota Federal University of Western Pará, Santarém, Pará, Brazil, msmota13@ig.com.br

Lorena Soto-Pinto El Colegio de la Frontera Sur, ECOSUR, Carretera Panamericana y Periférico Sur s/n, San Cristóbal de las Casas, Chiapas 29290, México, lsoto@ecosur.mx

Taylor V. Stein School of Forest Resources and Conservation, Institute of Food and Agricultural Sciences, University of Florida, P.O. Box 110410, Gainesville, FL 32611–0410, USA, tstein@ufl.edu

Neal B. Stolpe Department of Soils and Natural Resources, Faculty of Agronomy, University of Concepción, Vicente Méndez 595, Casilla 537, Chillan, Chile, nstolpe@udec.cl

Naresh V. Thevathasan School of Environmental Sciences, University of Guelph, Guelph, ON N1G 2W1, Canada, nthevath@uoguelph.ca

Rafael G. Tonucci Animal Science Department, Federal University of Viçosa, P.H. Rolfs Avenue, MG 36570-000, Viçosa, Brazil

Goats and Sheep, EMBRAPA, Sobral, Ceara State, Brazil, rgtonucci@gmail.com

Ranjith P. Udawatta Department of Soil, Environmental and Atmospheric Sciences, School of Natural Resources, University of Missouri, 203 ABNR, Columbia, MO 65211, USA

The Center for Agroforestry, School of Natural Resources, University of Missouri, Columbia, MO 65211, USA, UdawattaR@missouri.edu

Pornchai Uttaruk Mahasarakham University, Tambon Kamriang, Kantarawichai District, Mahasarakham 44150, Thailand, pornchai.u@msu.ac.th

Iwan Wijayanto Conservation International-Indonesia, Jakarta, Indonesia, i.wijayanto@conservation.org

Candra Wirawan Arief Conservation International-Indonesia, Jakarta, Indonesia, candra.arief@gmail.com

Erick Zagal Department of Soils and Natural Resources, Faculty of Agronomy, University of Concepción, Vicente Méndez 595, Casilla 537, Chillan, Chile, ezagal@udec.cl

Yatziri Zepeda Conservation International-Mexico, Tuxtla Gutierrez, México, yatzirizepeda@gmail.com

Reviewers

Aasrud, André, International Energy Agency, Climate Change Unit - Energy Efficiency and Environment Division, 9 rue de la Fédération, Paris, France. <andre.aasrud@iea.org>.

Allen, Harriet, University of Cambridge, Cambridge, UK. <hda1@cam.ac.uk>.

Aune, Jens B., Department of International Environment and Development Studies, Noragric, Norwegian University of Life Sciences, Aas, Norway. <jens.aune@umb.no>.

Balooni Kulbhooshan, Indian Institute of Management, Kozhikode, India. <kbalooni@iimk.ac.in>.

Fraisse, C., Agriculture and Biological Engineering, University of Florida, Gainesville, FL, USA. <cfraisse@ufl.edu>.

Gockowski, James, Sustainable Tree Crops Program, International Institute of Tropical Agriculture, Accra, Ghana. <j.gockowski@cgiar.org>.

Henry, M., Facoltà di Agraria, Università degli Studi della Tuscia, Via Camillo de Lellis, Viterbo, Italy. <henry@unitus.it>.

Howlett, D., School of Forest Resources and Conservation, University of Florida, Gainesville, FL, USA. <davhowlett@yahoo.com>.

Isenhart, Thomas M., NREM, Iowa State University, Ames, Iowa, USA. <isenhat@iastate.edu>.

Kaonga, Martin L., A Rocha International, Compass House, Vision Park, Chivers Way, Histon, Cambridge, UK. <martin.kaonga@arocha.org>.

Kimaro A.A., Department of Soil Science, University of Saskatchewan, 51 Campus Drive, Saskatoon, SK, Canada. <anthony.kimaro@usask.ca>.

Kjosavik, Darley Jose, Department of International Environment and Development Studies, Norwegian University of Life Sciences, Aas, Norway. <darley.kjosavik@umb.no>.

Kristjanson, Patti, CGIAR/ESSP Program on Climate Change, Agriculture & Food Security (CCAFS), World Agroforestry Centre, Nairobi Kenya. <P.Kristjanson@cgiar.org>.

McAdam Jim, Crops Grassland and Ecology Branch, Agri Food and Biosciences Institute and Queens University Belfast, Belfast, UK. <jim.mcadam@yahoo.co.uk>.

McNeely, Jeffrey A., World Conservation Union, Gland, Switzerland. <jam@iucn.org>.

Mosquera-Losada, M.R., Crop Production Department. Escola Politécnica Superior. Universidad de Santiago de Compostela, Campus de Lugo, Spain. <mrosa.mosquera.losada@usc.es>.

Nair, V.D., Soil and Water Science Department, University of Florida, Gainesville, Florida, USA. <vdn@ufl.edu>.

Nairan Felix de Barros, Soil Science Department, Federal University of Viçosa, Viçosa, MG, Brazil. <nfbarros@ufv.br>.

Rahman, S.A., Department of Sociology, University of Rajshahi, Rajshahi, Bangladesh.

Rao, K.S., Department of Botany, University of Delhi, Delhi, India. <sumonsociology @yahoo.com>.

Rayment, Mark, School of Environment, Natural Resources and Geography, Bangor University, Wales, UK. <m.rayment@bangor.ac.uk>.

Rivest, David, Département de Phytologie, Université Laval, Québec, Canada. <david.rivest.1@ulaval.ca>.

Saha, S., Ashoka Trust for Research in Ecology and the Environment (ATREE), Bangalore, India. <subhrajit_s@yahoo.com>.

Sathyapalan, Jyothis, Centre for Economic and Social Studies, Hyderabad, India. <jyothis.cess@gmail.com>.

Sayer, Jeffrey, Cook University. Cairns Queensland, Australia. <jeffrey.sayer@jcu.edu.au>.

Schroth, G., Federal University of Western Pará, Santarém, Pará, Brazil. <goetz.schroth@effem.com>

Showalter, J.M., Soil and Water Science Department, University of Florida, Gainesville, Florida, USA. <juliashowalter@gmail.com>.

Sileshi, G.W., World Agroforestry Centre (ICRAF), Lilongwe, Malawi. <sgwelde@yahoo.com>.

Snelder, Denyse, Institute of Environmental Sciences (CML), Faculty of Science, Centrum voor Milieuwetenschappen, University of Leiden, Einsteinweg, Leiden, The Netherlands. <Snelder@cml.leidenuniv.nl>.

Somarribo, E. CATIE, Turrialba, Costa Rica. <esomarri@catie.ac.cr>.

Steiner, Christoph, University of Georgia, Athens, USA. <christoph.steiner@biochar.org>.

Takimoto, Asako, Environment & Energy Group, Bureau for Development Policy, UNDP, New York, USA. <asako.takimoto@undp.org>.

Teklehaimanot, Zewge, School of Environment, Natural Resources and Geography, Bangor University, Wales, UK. < afs032@bangor.ac.uk>.

Tipper, Richard, Ecometrica, Edinburgh, UK. <richard.tipper@ecometrica.co.uk>.

Upadhyay, Thakur Prasad, Faculty of Forestry and the Forest Environment, Lakehead University, Thunder Bay, ON, Canada. <tpupadhy@lakeheadu.ca>.

Verchot, L.V., Center for International Forestry Research, Bogor, Indonesia. <L.verchot@cgiar.org>.

Wang, G., Department of Environmental Studies, Western Washington University, Bellingham, WA, USA. <Grace.wang@wwu.edu>.

Whalen K. Joann, Department of Natural Resources Sciences, Macdonald Campus, McGill University, Quebec, Canada. <Joann.whalen@mcgill.ca>.

Zamora, D., University of Minnesota, St. Paul, MN, USA. <zamor015@umn.edu>.

Part I
Measurement and Estimation

Methodological Challenges in Estimating Carbon Sequestration Potential of Agroforestry Systems

P.K. Ramachandran Nair

Abstract The methods used to estimate carbon sequestration in agroforestry systems (AFS) vary widely. Consequently, there is enormous inconsistency in the available datasets. Moreover, the estimations entail several assumptions, some of which are erroneous. A serious one is that C in the biomass and soil are equated to sequestered C. The amount of C stored in root biomass is also subject to widely variable estimations. Large-scale global models that are based on extrapolation of field measurements from sample plots, as used for C sequestration estimates in forestry, are thus likely to result in serious under- or overestimations of total C stock. These methodological problems that are common to most land use systems are of a higher order of magnitude in AFS compared with agricultural systems because of the integrated nature of AFS and the lack of rigorous data on the area under the practice. While there are no easy, fast, and pragmatic solution to these complex issues in the short term, agroforestry researchers could, at the very minimum, include accurate description of the methods and procedures they use while reporting results. That will help researchers at large to examine the datasets even at a later time and possibly incorporate the reported results in larger databases and help agroforestry earn its deserving place in mainstream efforts. Missing the opportunity to capitalize on the environmental services of agroforestry for the lack of rigorous research and consistent procedures for data reporting will be a serious setback to the development of agroforestry.

Keywords Allometric equations • Biomass determination • Global carbon models • Design and sampling problems • Soil bulk density

P.K.R. Nair (✉)
Center for Subtropical Agroforestry, School of Forest Resources and Conservation, University of Florida, P.O. Box 110410, Gainesville, FL 32611–0410, USA
e-mail: pknair@ufl.edu

B.M. Kumar and P.K.R. Nair (eds.), *Carbon Sequestration Potential of Agroforestry Systems: Opportunities and Challenges*, Advances in Agroforestry 8,
DOI 10.1007/978-94-007-1630-8_1,

Introduction

Agroforestry practices are said to be characterized by four "I" words: intentional, intensive, integrated, and interactive (Gold and Garrett 2009). Perhaps another one could be added: imprecise. This is not said in a pejorative sense. It only reflects the lack of precision in dealing with issues concerning agroforestry. Starting with the definition, agroforestry is not entirely precise or definitive in many of its attributes. In fact, that is just the "nature of the beast": various attributes of integrated and interactive land use systems that are practiced in concert with nature and environment in accordance with the local socio-cultural norms and traditions cannot be expected to be measured in quantitative terms with 100% precision and accuracy because of the multiplicity of factors involved and their complex interactions. This lack of precision may not be a serious problem in managing the systems because they are location-specific and their management is less dependent on machinery than in the case of commercial agriculture and forestry systems. However, when it comes to quantifying their attributes to lay the foundations for future scientific developments, accurate measurements are important. Thus, measurement of the perceived benefits and advantages of agroforestry is essential; but it is a challenge, indeed a serious one. We are faced with such a serious challenge in our efforts to estimate carbon (C) sequestration in agroforestry systems (AFS).

The role of land use systems in capturing atmospheric carbon dioxide (CO_2) and storing the C in plant parts and soil became an important area of research during the past decade. Agroforestry attracted special attention as a C sequestration strategy following its recognition as a C sequestration activity under the afforestation and reforestation (A & R) activities of the Kyoto Protocol. This was in recognition of the perceived advantages of the large volume of aboveground biomass (AGB) and deep root systems of trees in accomplishing that task. Consequently a large number of estimates and reports on C sequestration potential of various agroforestry systems under different ecological regions have become available since the mid-1990s starting with the reports of Dixon et al. (1994), Schroeder (1994), and others. Most of these available reports on C sequestration in AFS are estimates of C stocks: how much C is, or potentially could be, accumulated and stored in above- and belowground compartments of AFS under different conditions of ecology and management. The estimates range from 0.29 to 15.21 Mg ha^{-1} year^{-1} aboveground, and 30–300 Mg C ha^{-1} up to 1-m depth in the soil (Nair et al. 2010).

Collecting (or estimating) such C stock data is important in itself for feeding into massive global datasets such as those of the IPCC (Intergovernmental Panel on Climate Change: www.ipcc.ch, accessed 13 February 2011) and for other planning and developmental purposes. The methods and procedures adopted in collecting such datasets have to be consistent and standardized, so that development plans for the future are based on rigorous databases of unquestionable value. Therefore, we have the responsibility of stepping up our norms, criteria, and standards for reporting C sequestration data in AFS. With that in mind, this chapter aims to bring together, first of all, some basic concepts of C sequestration and then identify some of the common mistakes and pitfalls in C sequestration studies in AFS and ways to avoid them. Developing a uniform or standardized set of procedures is a long and arduous task; that is not even attempted here; the hope, however, is that this effort will stimulate some thinking in organizing future efforts in that direction. While raising these issues, it is recognized that most of them deserve considerable discussion. That, however, is beyond the scope of this paper. Furthermore, all the supporting literature, based on which many statements are made in an abstract manner in the text, is not cited for reason on brevity.

Carbon Sequestration

During the past two decades, there has been a veritable explosion of the literature on C sequestration. Internet search engines and abstracting services are virtually flooded with all sorts of literature on all aspects of the process. Unfortunately, considerable variations exist among different user groups about the concept of C sequestration and the term is not used or understood uniformly in different contexts. This has led to serious difficulties in consolidating and synthesizing available reports and publications according to a uniform pattern and set of norms.

The United Nations Framework Convention on Climate Change (UNFCCC) defines carbon sequestration as the process of removing C from the atmosphere and depositing it in a reservoir. It entails the transfer of atmospheric CO_2, and its secure storage in long-lived pools (UNFCCC 2007). From the agroforestry point of view, C sequestration primarily involves the uptake of atmospheric CO_2 during photosynthesis and the transfer of fixed C into vegetation, detritus, and soil pools for "secure" (i.e. long-term) storage (Nair et al. 2010). It occurs in two major segments of the AFS: aboveground and belowground. Each can be partitioned into sub-segments: the former into specific plant parts (stem, leaves, etc., of trees and herbaceous components), and the latter into living biomass such as roots and other belowground plant parts, soil organisms, and C stored in various soil horizons. The total amount sequestered in each compartment differs greatly depending on a number of factors including the ecoregion, the type of system (and the nature of components and age of perennials such as trees), site quality, and previous land use. On average, the aboveground parts and the soil (including roots and other living biomass) are estimated to hold roughly one-thirds and two-thirds, respectively, of the total C stored in tree-based land use systems (Lal 2010). Based on the notion that tree incorporation in croplands and pastures would result in greater net C storage above- and belowground (Palm et al. 2004; Haile et al. 2008), AFS are believed to have a higher potential to sequester C than pastures or field crops growing under similar ecological conditions (Roshetko et al. 2002; Kirby and Potvin 2007).

Measurement of Carbon Sequestration in Agroforestry Systems

Aboveground (Vegetation)

Aboveground measurements of C stock (and, by implication, C sequestration) are direct derivatives of aboveground biomass (AGB) measurements/estimates, assuming that 50% of the biomass is made up by C. The AGB is often derived by summing up the amount of harvested and standing biomass, and the measurements are relatively straight-forward compared to those of the belowground compartment. Estimation of tree biomass by whole-tree harvesting is an old approach: it consists of cutting down sample trees, separating various parts (stem, leaves, inflorescence, etc.), digging out and washing the roots, determining their dry weights from samples of each part, and adding them up to get the total biomass. After dividing up the harvested representative trees into their various components (branches, dead branches, branchlets, leaves, roots and fine roots), and determining their dry weight, the C content in each is measured. Using the data, allometric equations are developed as regression models with the measured variables such as diameter at breast height (DBH), total tree height or commercial bole height, and sometimes wood density, as the independent variables and total dry weight as the dependent variable. The destructive method of determining tree biomass, though comparatively accurate, is extremely time- and labor-intensive, especially for large trees. It is often used to validate other, less invasive and costly methods, such as the estimation of C stock using nondestructive in-situ measurements and remote sensing. Such allometric equations developed based on biophysical properties of trees and validated by occasional measurements of destructive sampling are widely used in forestry for estimating standing volumes of forests. With increasing understanding about the role of forests in sequestering C, various allometric equations have been developed for different forest types (Brown 1997; FAO 2004; Pearson et al. 2005; Chave et al. 2005; Basuki et al. 2009; Fernández-Núñez et al. 2010).

Efforts in developing allometric equations for agroforestry situations have generally been slow and researchers trying to use this approach are forced to use broad approximations. For example, for estimating the standing tree biomass in the parkland AFS in the Sahel where species-specific allometric equations were not available for the region, Takimoto et al. (2008) followed the UNFCCC (2006) recommendation to use the Brown (1997) general equations for parkland trees. In other cases, more simple analyses were used for large-scale estimations. Dixon et al. (1993) made estimations by measuring the volume of stem wood and multiplying it with species-specific wood density; that number was then multiplied by 1.6 to get an estimation of whole-tree biomass; C content was assumed as 50% of the estimated whole-tree biomass, and root biomass was excluded. This rough estimation was then used for more extensive estimations of global forest biomass. More recently, databases for tree characteristics such as wood density for agroforestry species http://www.worldagroforestrycentre.org/sea/Products/AFDbases/WD/ (accessed 13 February 2011) developed at the World Agroforestry Centre (www.cgiar-icraf.org, accessed 13 February 2011) are being used in such allometric calculations.

As Kumar et al. (1998) noted following their efforts to develop allometric equations for some common agroforestry tree species in Kerala, India, such equations vary greatly with species, age, wood density, bole shape, and other factors, and could lead to excessive inaccuracies. Besides, such determinations can be difficult for smallholder agroforestry plots that comprise much of the agroforestry in developing countries. These systems involve a multitude of plants of varying growth habits yielding diverse economic products, and the species are planted and their products harvested, mostly for household consumption, throughout the year. Variations in tree management can be another issue: trees in AFS may be pruned depending on management practices or may have different growth forms due to differences in spacing compared to natural (forest) systems. Furthermore, no two agroforestry plots are similar: each may be unique in terms of plant composition, planting arrangements, and stand densities. Thus, determination of biomass production from indigenous AFS is a challenging task, and makes extrapolation from one system to others very difficult.

Belowground (Soils)

The determination of belowground organic C dynamics in AFS is crucial for understanding the impact of the system on C sequestration, but it is difficult – more difficult than that for aboveground C. Organic C occurs in soils in a number of different forms including living root and hyphal biomass, microbial biomass, and as soil organic matter (SOM) in labile and more recalcitrant forms. Difficulties of separating these different forms and their complex interactions make measurement, estimation and prediction of soil C sequestration a daunting task. The most common method of estimating the amount of C sequestered in soils is based on soil analysis, whereby the C content in a sample of soil is determined (mass per unit mass of soil, such as g C per 100 g soil) and expressed usually in megagrams ($Mg = 10^6$ g or tons) per hectare.

Soil organic C (SOC) is often measured on a whole soil basis. The Walkley-Black procedure that used to be employed extensively in the past, and is perhaps used even now in some places, is no longer recommended because of concerns about the accuracy of determination (in view of the correction factor that is usually applied, leading to over- or underestimations) as well as environmental concerns due to the impact of the use of potassium dichromate (Kimble et al. 2001). Currently, many studies measure SOC by quantifying the amount of CO_2 produced through heating in a furnace. Other studies measure the change in weight of the sample after heating. However, the temperature used can vary; it needs to be standardized for accurate comparison of different studies. The presence of carbonates and charcoal in the soil can also skew results (Kimble et al. 2001). These measurements of C on a whole soil basis give information about total concentrations, but other analytical procedures are needed to determine details of the form and recalcitrance of the stored C as well as where it is stored. In order to gain a better understanding of such details of C sequestration in soils, attention has focused on the study of soil aggregates (Nair et al. 2010).

Since the majority of SOC is found in soil aggregates, understanding the structure and cycling of these aggregates will give us a better understanding of how C is entering, moving through, and leaving the soil, and thus the ability to predict future levels based on inputs and current conditions. By knowing the factors that are likely to influence aggregate formation and stability, we can predict the factors to be taken into consideration, and thus be able to better develop and adopt new agricultural and land management practices to optimize C sequestration both immediately and for the long term. Soil aggregate analyses, however, have not yet become a step in routine soil C determination.

Belowground Living Biomass

In addition to SOM, belowground biomass is a major C pool (Nadelhoffer and Raich 1992). However, belowground biomass is difficult to measure. The root-to-shoot ratio is therefore commonly used to estimate below ground living biomass. The ratios differ considerably among species (e.g., higher in palms than in dicot trees) and across ecological regions (e.g., higher in cold than in warm climates). In the absence of measured values, many researchers assume that the belowground biomass constitutes a defined portion of the aboveground biomass and the values so assumed range from 25% to 40% depending on such factors as nature of the plant and its root system and ecological conditions.

Modeling

In order to understand global carbon cycling, models that incorporate rates of terrestrial C cycling are used. Such models are based on a set of assumptions that are formed from our understanding of ecological processes including tree growth, and decomposition processes in the soil. The CENTURY and RothC models are the most widely used soil C models. The former models the cycling of C and other elements (phosphorus, nitrogen and sulfur) and their interactions, focusing specifically on the effects of species type and management practices such as tillage to model agricultural systems. It accounts for agricultural systems, forests, or savannas but not for integrated tree-crop systems such as agroforestry; adding agroforestry could be interesting and important to this model in order to improve its C sequestration estimates in global soils. The RothC model (Rothamsted model), based on the long-term experiments studying organic matter on the Rothamsted sites in England, takes into consideration organic pools in terms of how labile they are. Although the parameters of the model are comparatively simple, the model may not be quite appropriate for predictions of tropical agroforestry sites.

Numerous mathematical models have been developed to predict the response of SOM to agricultural practices at various scales, from soil profile or small plot scales to larger spatial extents, especially in response to the demand for national inventories of soil C sequestration potential (Viaud et al. 2010). Discussing such models, Nair et al. (2010) have noted that difficulties in obtaining information that is essential for the models could limit the applicability of the models to many tropical AFS. In general, models used in agroforestry research are developed for natural ecosystems and planted forests or agricultural systems; they rely on assumptions that are not fully relevant to AFS, and are often hard to incorporate into larger ecosystem models.

Global Estimates: Seeing the World for Trees and Forests

In the wake of increasing global initiatives and agreements such as REDD+(Reducing Emissions from Deforestation and Forest Degradation: www.un-redd.org: accessed 7 February 2011), various massive efforts are under way to assess the extent and health of the Earth's forests and other ecosystems. For example, ALERTS [Automated Land-change Evaluation, Reporting, and Tracking System: www.planetaryskin.org (accessed 7 February 2011), a unit of the Planetary Skin Institute (PSI), a not-for-profit organization set up jointly by NASA (the US National Aeronautics and Space Administration) and Cisco Systems, a large computing firm] is a decision support system – and one of several such tools – that has been launched in collaboration with several national agencies around the world to assess the actual weight of the world forest biomass and how much C they are storing. To calculate this, tree data such as DBH measured from sample plots are combined with images from NASA's 'super' cameras and satellites to estimate the plant biomass and therefore C in an area. As better ways of measurements and monitoring become more available, it will be possible to arrive at more accurate figures on amounts of CO_2 released from deforestation and forest degradation, used up in photosynthesis, and stored in "long-lived" above- and belowground compartments of ecosystems. They appear massive and impressive; nevertheless, their application in the short term and to small and scattered agroforestry plots sound uncertain. Furthermore, the accuracy and reliability of all these efforts depend on field measurements and calibration.

Methodological Challenges

As can be seen from the above, the methods and procedures adopted in collecting or estimating the data are quite inconsistent and are often incomparable and inconclusive. They vary widely in details of all aspects such as sampling, analytical methods, computations, data interpretation and presentation. This can greatly affect the conclusions made when comparing the differences under various management practices, soils, environments, social conditions, etc. Obviously, these problems and challenges have to be addressed; but that is not an easy or simple task. As a preliminary effort in that direction, let us analyze the major types of challenges and examine what, if anything at all, can be done until proper procedures are developed, tested, and put in place. But, first, the concept of C sequestration itself needs to be examined and understood.

The Concept of Carbon Sequestration

An important part of the UNFCCC definition of C sequestration is the secure storage C (CO_2) that is removed from the atmosphere in long-lived pools. There is considerable ambiguity in the understanding of this concept, especially when it comes to "long-lived" pools. The literature on C sequestration in land use systems, especially AFS, is not clear on this. Most reports equate C stock to C sequestration. Most such determinations are simple computations, in which aboveground biomass is estimated from arbitrarily chosen or general allometric equations; belowground biomass is considered as a fraction, usually 30%, of AGB, and 50% of the total biomass is taken as C stock (and sequestered C). Some reports do not specify if belowground biomass is factored into the estimations. In the case of soil, the C content as determined by soil analysis (which is then extrapolated to a region or country with or without the aid of remotely sensed or otherwise computed data) is expressed as C stock (=sequestered C).

Erroneous Assumptions

Estimations and computations of C stock in AFS as described above are approximations. Depending on the procedures used, the estimates may have deficiencies and inadequacies arising from both the assumptions used and the procedures adopted. Some of the commonly used assumptions and the errors involved in them are listed below:

- Carbon content in biomass is 50%. Often it is less than that.
- All biomass represents sequestered C. All biomass does not end up in "long-lived" pools. The foliage that falls on ground decomposes rapidly and releases CO_2 back to the atmosphere. The fraction of the biomass that can be considered as sequestered C is variable depending on a number of factors including the species, plant part, and ecological conditions.
- All C in soil represents sequestered C. Recent additions to organic C in surface soil through litterfall and external additions are subject to rapid decomposition and release of CO_2 with only a small percentage of it getting transformed to stable C in "long-lived" pools. If C stocks increase through time, that is a form of sequestration because the total amount is greater. These and other issues imply that there are some complexities to quantifying C sequestration and how it relates to C stocks.

- Carbon stock is the same as C sequestration: C sequestration is a rate process involving the time factor (e.g., Mg C ha^{-1} $year^{-1}$), whereas C stock (Mg ha^{-1}) does not have the time factor.
- Growth form of trees has little to do with root biomass. Differences in growth form of trees and management practices can lead to under- or over-estimations of root biomass.
- Amount of C sequestered is generally uniform for a given agroforestry practice. High levels of spatial heterogeneity exist among similar types of agroforestry practices at different locations such that extrapolation between one AFS and another or even from one area of an agroforestry farm to another can be misleading.

Operational Inadequacies and Inaccuracies

The procedures for collecting and processing plant- and soil samples for nutrient analyses and productivity measurements are well established; the lack of such procedures is not the issue in the context of this discussion; the "devil is in the details." The problem about the lack of rigorous allometric equations for estimating biomass has already been presented. The uncertainty arising from the lack of uniform methods for describing area under agroforestry (Nair et al. 2009; Udawatta and Jose 2011) is another difficulty in gauging the importance of agroforestry in carbon sequestration. While some progress has been made in resolving this puzzle in the tropical arena (thanks to the ICRAF-sponsored study, which, using geospatial analysis of remote sensing derived global datasets at 1 km resolution, reported the area under agroforestry as about one billion hectares of agricultural lands worldwide: Zomer et al. 2009), no such progress seems to have been made in assessing the area under agroforestry in the industrialized world. Additionally, a few of the common challenges, primarily in soil-related estimates, are considered briefly here.

Sampling depth: A major issue that lacks uniformity is soil sampling depth. Most soil studies are limited to the surface soils to 20 or 30 cm depth. The importance of sampling beyond the surface soil cannot be overemphasized while studying tree-based systems such as agroforestry, not only because tree roots extend to deeper soil horizons, but also because of the role of subsoil in long-term stabilization of C. The lack of uniformity in breaking points between soil-horizon depths is another procedural problem: results of a C study in the 0–5 cm surface horizon cannot realistically be compared with those of 0–50 cm study.

Sample preparation (sieve size): The 2 mm sieve that is almost universally used for preparing soils for laboratory analyses is also an issue to be considered. The fractions more than 2 mm in size (retained in the sieve) are often discarded; but they may constitute a sizeable amount of the soil and may contain some C (Howlett et al. 2011).

Pseudoreplication: Repeated sampling from the same contiguous experimental unit without true replicates of treatments is an issue that comes up often in sampling procedures in agroforestry field research. The purpose of replication is to reduce random or stochastic error and increase the precision of comparisons. Therefore, if true replicates are not used, the treatments cannot be statistically compared. While the results from such studies may still be valid, statistical comparisons between treatments may be invalid and the treatments cannot be declared as statistically different or not. While this is unquestionable in the statistical sense, the concept of replication needs to be taken into consideration in these discussions. For example, C stock is estimated based on samples drawn from existing field plots rather than replicated field experiments as in many ecological studies where pre-existing conditions are used for research. The question may arise as to what constitutes true replication in the case of treatments that extend over several hundred hectares of land as in some commercial agroforestry operations such as the silvopastoral systems in Brazil (Tonucci et al. 2011). Some argue that when a treatment occupies such a large area, randomly distributed sampling plots that are replicated within the "contiguous" unit itself but are quite far (200 m or more) from each other can be taken as having fulfilled the concept of replication. In such studies, spatial interspersion of replications together with the use of a systematic design is used to alleviate possible pseudoreplication problems (Stamps and Linit 1999; Peichl et al. 2006; Dube et al. 2011; Tonucci et al. 2011). Forestry researchers have used composite samples drawn from large experimental units as replicates considering the land use systems as fixed effect treatments (Lugo et al. 1990). In the statistical sense, a fixed effect model means that inferences are restricted to the treatments in the study; the results cannot be used to make conclusions about other agroforestry systems. The fixed effects model also applies to the so-called "repeated measures."

Repeated measures: These refer to measurements made in time or space on the same subject or experimental unit, such as a tree or a plot. For example, in agroforestry experiments, we may draw soil samples from depth increments from the same sites, or at defined horizontal distances from trees or transects. In experimental designs, measurements are made statistically independent by randomly assigning treatments to the experimental units. However, when time and space are considered as treatments, they cannot be randomly assigned; the depth/distance increments are treated as repeated measures rather than as independent measurements (Moser et al. 1990; Stern et al. 2004). The non-randomized nature of repeated measures designs often results in the violation of the assumptions necessary for valid univariate analysis. However, statistical procedures are available to address the limitations imposed by the model. In certain instances, standard univariate approaches, such as ANOVA (analysis of variance) with randomized block or split-plot models can be applied and valid tests of hypothesis obtained (Moser et al. 1990). In the case of soil depths at the same site, they could be stratified and each soil depth considered independently treating each site as a replication.

Chronosequence studies: Although some studies carry out chronosequences to see the change in C, these are few and not well standardized. Since changes in C stock is unlikely to be linear through time, understanding the nature of the curve of C storage over time is important to understand the periods when most C is being sequestered.

In addition, it is difficult to know if the residence time of C that is sequestered initially in a system differs from that of C that is sequestered later. Are the cycles that the initial C and later C additions go through the same? A large number of many such questions need to be answered for realistically assessing the impact of agroforestry and other management practices on C sequestration.

Calculations and Reporting of Results

The most common inconsistency in reporting C stock and C sequestration data in AFS from different locations is related to soil. Soil C stock is conventionally expressed in mass per area such as Mg C ha^{-1}. These data are derived by multiplying the analytical data, which is usually in mass per unit mass of soil (g C 100 g $soil^{-1}$) with the soil's bulk density (BD), which is expressed in mass per volume of soil (g cm^{-3} or Mg m^{-3}), and with soil (sampling) depth. There is an anomaly in this conversion because the BD value involves a volume measure whereas the C stock value is expressed in an area measure (ha). To overcome this, C stock reported in Mg ha^{-1} is assumed to be for 1 cm thickness (depth) of the soil unless the depth is otherwise specified. Thus, when the C stock to, say 40 cm or 100 cm depth is reported, that depth should be mentioned. Unfortunately, many reports on soil C stock in AFS, either do not report such details, or do not follow any uniform norm about the depth (for example, Table 3, Nair et al. 2010). This can lead to confusion and speculation when the data are compiled or compared. Based on the results accrued so far from AFS research (Nair et al. 2010), it seems fair to stipulate that soil C stock in AFS should be reported to at least 1 m depth.

The soil BD is an important factor in these computations, but is not reported in many research papers on soil C sequestration in AFS. Consider two soils, soil A and soil B, both with the same C concentration of 2 g C 100 g $soil^{-1}$, but with different BD values, 1.0 and 1.2 Mg m^{-3}, respectively. The total soil C stock to 1 m depth in the two soils will be as follows:

$$\text{Soil A}: 2.0\ \text{g}\ 100\text{g}^{-1} \times 1.0\ \text{Mg m}^{-3} \times 1\ \text{m} \ = 200\ \text{Mg ha}^{-1}$$
$$\text{Soil B}: 2.0\ \text{g}\ 100\text{g}^{-1} \times 1.2\ \text{Mg m}^{-3} \times 1\ \text{m} \ = 240\ \text{Mg ha}^{-1}$$

[Note that the units of ha (=10,000 m^2) and 1 m depth are used in the calculation.]

Thus, soil B will have 20% more C stock than soil A to the same depth although both soils have the same C concentration (It is a different matter if both soils have same C concentrations throughout the 1-m depth). The point here is that while estimating C stock to 1 m depth factoring in BD values, soil B consisted of 20% more soil mass than soil A. Such differences are often overlooked while compiling regional and global datasets based on "standard" values of soil C stock (Mg C ha^{-1}). Therefore, the influence of soil bulk density on measured C stocks is particularly important when comparing land use treatments that result in different BD values, as may be the case with AFS compared with annual crops or pastures. The problem is compounded when soil depth, to which the value reported is related, is not specified. These highlight the importance of reporting soil BD data and soil depth on the one hand, and the need for exercising caution while using reported values of soil C sequestration on the other.

Another issue is the "one-size fits all" approach to computations of regional and global statistics. Currently, most policy documents and projections including major ones such as the IPCC reports have the tendency to assign a single, uniform value or sets of narrow-range values, for C stock and C sequestration potential of AFS irrespective of their site conditions and system characteristics. For example, the IPCC estimated a global value of 630 million hectares of unproductive croplands and grasslands that could be converted to agroforestry and could potentially sequester 1.43 and 2.15 Tg (10^{12}g = megatons) of CO_2 annually by 2010 and 2040, respectively (IPCC 2000). Several other such estimates are also available (for example, MIT, 2010, Mission 2013, Carbon sequestration, Massachusetts Institute of Technology: http://igutek.scripts.mit.edu/terrascope/index.php?page=Agroforestry, accessed 13 February 2011). It is important that the variability among soils to store C is factored into such global estimates and projections.

Conclusions

Carbon sequestration in land use systems is a rather loosely defined concept. Several methodological challenges, arising from difficulties related to sampling, analysis, computations, and interpretation make its measurement a difficult task. These difficulties are of a higher magnitude in the case of AFS because often the systems involve complex multispecies combinations and the measurements are made from pre-existing sites rather than randomized and replicated experiments. There is no easy, fast, and pragmatic solution to these issues in the short term. In the circumstances what can the common researcher do? The author's recommendation is that before setting out to undertake the study, the researcher should think through the problems they might encounter while reporting the results. While reporting results, they should describe accurately how the data were collected, analyzed, and managed. That means, explaining unambiguously how the samples were drawn, estimations were made and computations were calculated for extrapolation to broader scale such as Mg ha^{-1}. Such a clear presentation of the results will make it possible for researchers at large to understand and decide whether, how, and to what extent to incorporate the reported results in larger databases, and help agroforestry earn its deserving place in the mainstream of such efforts. Mistakes might be made; but that is better than not doing anything for fear of making mistakes. In this era of rapid progression of science and efforts to understand and quantify the underexploited ecosystem services, agroforestry researchers have to position themselves to ensure that agroforestry is not left behind in these global efforts, because, only what gets measured gets recognized and managed.

Acknowledgments The author thankfully acknowledges the critical comments and suggestions on the manuscript received from BM Kumar, M-R Moequera-Losada, VD Nair, G Schroth, and JM Showalter.

References

Basuki TM, van Laake PE, Skidmore AK, Hussin YA (2009) Allometric equations for estimating the above-ground biomass in tropical lowland dipterocarp forests. For Ecol Manag 257:1684–1694

Brown S (1997) Estimating biomass and biomass change of tropical forests: a primer. FAO Forestry paper 134, Rome, 55 p

Chave J, Andalo A, Brown S, Cairns MA, Chambers JQ, Eamus D, Fölster H, Fromard F, Higuchi N, Kira T, Lescure J-P, Nelson BW, Ogawa H, Puig H, Riéra B, Yamakura T (2005) Tree allometry and improved estimation of carbon stocks and balance in tropical forests. Oceologia 145:87–99

Dixon RK, Winjum JK, Schroeder PE (1993) Conservation and sequestration of carbon: the potential of forest and agroforest management practices. Global Environ Chang 2:159–173

Dixon RK, Wimjum JK, Lee JJ, Schroeder PE (1994) Integrated systems: assessing of promising agroforest and alternative land use practices to enhance carbon conservation and sequestration. Clim Chn 30:1–23

Dube F, Thevathasan NV, Zagal E, Gordon AM, Stolpe NB, Espinosa M (2011) Carbon sequestration potential of silvopastoral and other land use systems in the Chilean Patagonia. In: Kumar BM, Nair PKR (eds.) Carbon sequestration potential of agroforestry systems: opportunities and challenges. Springer, Dordrecht, pp 101–127

FAO (2004) Assessing carbon stocks and modelling win–win scenarios of carbon sequestration through land-use changes. FAO, Rome. Available at http://www.fao.org/docrep/007/y5490e/y5490e00.htm. Accessed 13 Feb 2011

Fernández-Núñez E, Rigueiro-Rodríguez A, Mosquera-Losadaa MR (2010) Carbon allocation dynamics one decade after afforestation with *Pinus radiata* D. Don and *Betula alba* L. under two stand densities in NW Spain. Ecol Eng 36:876–890

Gold MA, Garrett HE (2009) Agroforestry nomenclature, concepts, and practices. In: Garrett HE (ed.) North American Agroforestry: an integrated science and practice, 2nd edn. American Society of Agronomy, Madison, pp 45–55

Haile SG, Nair PKR, Nair VD (2008) Carbon storage of different soil-size fractions in Florida silvopastoral systems. J Environ Qual 37:1789–1797

Howlett DS, Mosquera-Losada MR, Nair PKR, Nair VD, Rigueiro-Rodríguez A (2011) Soil C storage in silvopastoral systems and a treeless pasture in northwestern Spain. J Environ Qual 40:784–790. doi:10.2134/jeq2010.0145

IPCC (Intergovernmental Panel on Climate Change) (2000) In: Watson RT, Noble IR, Bolin B, Ravindranath NH, Verardo DJ, Dokken DJ (eds.) Land use, land-use change, and forestry. A special report of the IPCC. Cambridge University Press, Cambridge, 375 p

Kimble JM, Lal R, Follett RF (2001) Methods for assessing soil C pools. In: Lal R, Kimble JM, Follett RF, Stewart BA (eds.) Assessment methods for soil carbon. Lewis Publishers, Boca Raton, pp 3–12

Kirby KR, Potvin C (2007) Variation in carbon storage among tree species: implications for the management of a small-scale carbon sink project. For Ecol Manag 246:208–221

Kumar BM, George SJ, Jamaludheen V, Suresh TK (1998) Comparison of biomass production, tree allometry and nutrient use efficiency of multipurpose trees grown in wood lot and silvopastoral experiments in Kerala, India. For Ecol Manag 112:145–163

Lal R (2010) Managing soils and ecosystems for mitigating anthropogenic carbon emissions and advancing global food security. Bioscience 60:708–721

Lugo AE, Cuevas E, Sanchez MJ (1990) Nutrients and mass in litter and soil of ten tropical tree plantations. Plant Soil 125:263–280

Moser EB, Saxton AM, Pezeshki SR (1990) Repeated measures analysis of variance: application to tree research. Can J For Res 20:524–535

Nadelhoffer KJ, Raich JW (1992) Fine root production estimates and belowground carbon allocation in forest ecosystems. Ecology 73:1139–1147

Nair PKR, Kumar BM, Nair VD (2009) Agroforestry as a strategy for carbon sequestration. J Plant Nutr Soil Sci 172:10–23

Nair PKR, Nair VD, Kumar BM, Showalter JM (2010) Carbon sequestration in agroforestry systems. Adv Agron 108:237–307

Palm C, Tomich T, Van Noordwijk M, Vosti S, Alegre J, Gockowski J, Verchot L (2004) Mitigating GHG emissions in the humid tropics: case studies from the Alternatives to Slash-and-Burn Program (ASB). Environ Dev Sust 6:145–162

Pearson T, Walker S, Brown S (2005) Sourcebook for land use, land use change and forestry project. Winrock International and the BioCarbon Fund of the World Bank, Washington DC, 64 p. Available at http://www.winrock.org/ecosystems/files/winrock-biocarbon_fund_sourcebook-compressed.pdf. Accessed 13 Feb 2011

Peichl M, Thevathasan NV, Gordon AM, Huss J, Abohassan R (2006) Carbon sequestration potentials in temperate tree-based intercropping systems, southern Ontario, Canada. Agroforest Syst 66:243–257

Roshetko JM, Delaney M, Hairiah K, Purnomosidhi P (2002) Carbon stocks in Indonesian homegarden systems: can smallholder systems be targeted for increased carbon storage? Am J Altern Agric 17:138–148

Schroeder P (1994) Carbon storage benefits of agroforestry systems. Agroforest Syst 27:89–97

Stamps WT, Linit MJ (1999) The problem of experimental design in temperate agroforestry. Agroforest Syst 44:187–196

Stern RD, Coe R, Allan EF, Dale IC (2004) Good statistical practice for natural resources research. CABI, Wallingford, UK, 388p

Takimoto A, Nair PKR, Nair VD (2008) Carbon stock and sequestration potential of traditional and improved agroforestry systems in the West African Sahel. Agric Ecosyst Environ 125:159–166

Tonucci RG, Garcia R, Nair PKR, Nair VD, Bernardino FS (2011) Soil carbon storage in silvopasture and related land-use systems in the Brazilian cerrado. J Environ Qual 40:833–841. doi:10.2134/jeq2010.0162

Udawatta RP, Jose S (2011) Carbon sequestration potential of agroforestry practices in temperate North America. In: Kumar BM, Nair PKR (eds.) Carbon sequestration potential of agroforestry Systems: opportunities and challenges. Springer, Dordrecht, pp 17–42

UNFCCC (2006) Revised simplified baseline and monitoring methodologies for selected small-scale afforestation and reforestation project activities under the clean development mechanism. Bonn. Available at http://cdm.unfccc.int/UserManagement/FileStorage/CDMWF_AM_A3II6AX6KGW5GBB7M6AI98UD3W59X4. Accessed 7 Feb 2011

UNFCCC (2007) Report of the conference of parties on its thirteenth session, Bali, Indonesia. United nations framework convention on climate change, Geneva

Viaud V, Angers D, Walter C (2010) Toward landscape-scale modeling of soil organic matter dynamics in agroecosystems. Soil Sci Soc Am J 74:1847–1860

Zomer RJ, Trabucco A, Coe R, Place F (2009). Trees on farm: an analysis of global extent and geographical patterns of agroforestry. ICRAF Working paper no. 89. World Agroforestry Centre, Nairobi. Available at http://www.worldagroforestry.org/downloads/publications/PDFs/WP16263.PDF. Accessed 13 Feb 2011

Carbon Sequestration Potential of Agroforestry Practices in Temperate North America

Ranjith P. Udawatta and Shibu Jose

Abstract Agroforestry, an ecologically and environmentally sustainable land use, offers great promise to sequester carbon (C). The objectives of this chapter are to (1) provide a review of C sequestration opportunities available under various agroforestry practices in temperate North America, and (2) estimate C sequestration potential by agroforestry in the US. Since accurate land area under agroforestry was not available, the potential C sequestration was estimated based on several assumptions about the area under different agroforestry practices in the US: 1.69 million ha under riparian buffer, 17.9 million ha (10% of total cropland) under alley cropping, and 78 million ha under silvopasture (23.7 million ha or 10% of pasture land and 54 million ha of grazed forests). Based on these, we estimated C sequestration potential for riparian buffers, alley cropping, and silvopasture in the US as 4.7, 60.9, and 474 Tg C year^{-1}, respectively. Establishment of windbreaks to protect cropland and farmstead could sequester another 8.79 Tg C year^{-1}. Thus, the potential for C sequestration under agroforestry systems in the US is estimated as 548.4 Tg year^{-1}. The C sequestered by agroforestry could help offset current US emission rate of 1,600 Tg C year^{-1} from burning fossil fuel (coal, oil, and gas) by 34%. These preliminary estimates indicate the important role of agroforestry as a promising CO_2 mitigation strategy in the US, and possibly in other parts of North America. The analysis also reveals the need for long-term C sequestration research in all

R.P. Udawatta (✉)
Department of Soil, Environmental and Atmospheric Sciences, School of Natural Resources, University of Missouri, 203 ABNR, Columbia, MO 65211, USA

The Center for Agroforestry, School of Natural Resources, University of Missouri, Columbia, MO 65211, USA
e-mail: UdawattaR@missouri.edu

S. Jose
203 ABNR, The Center for Agroforestry, School of Natural Resources, University of Missouri, Columbia, MO 65211, USA
e-mail: joses@missouri.edu

B.M. Kumar and P.K.R. Nair (eds.), *Carbon Sequestration Potential of Agroforestry Systems: Opportunities and Challenges*, Advances in Agroforestry 8,
DOI 10.1007/978-94-007-1630-8_2, © Springer Science+Business Media B.V. 2011

regions and for all agroforestry practices, establishment of standardized protocols for C quantification and monitoring, inventory of agroforestry practices, development of models to understand long-term C sequestration, and development of agroforestry design criteria for optimum C sequestration for all regions.

Keywords Alley cropping • Riparian buffers • Silvopasture • Windbreaks

Introduction

Rising levels of atmospheric carbon dioxide (CO_2) and associated global warming have moved to the center stage of climate change discussion in the past two decades. While many dispute the global warming hypothesis, projected doubling of atmospheric CO_2 by the latter half of the Twenty-first century raises concerns for everyone. Significant reductions in the atmospheric CO_2 concentrations can only be achieved with substantial additional costs and major changes in living standards. Therefore, adoption of CO_2 reduction strategies are widely debated, not well received, and not agreed upon by all nations. The world needs carbon (C) sequestration techniques that provide social, environmental, and economic benefits while reducing atmospheric CO_2 concentration.

Management of agricultural systems to sequester C has been accepted as a partial solution to climate change (Morgan et al. 2010). Establishing and maintaining perennial vegetation to enhance C sequestration is less costly compared to most other techniques, and these practices have minimal environmental and health risks. Perennial vegetation is more efficient than annual vegetation as it allocates a higher percentage of C to below-ground and often extends the growing season (Morgan et al. 2010), therefore enhancing C sequestration potential of agricultural systems even further (Lal et al. 1999; Watson et al. 2000; Oelbermann et al. 2006a; Jose 2009).

Agroforestry, as a system that combines trees and/or shrubs (perennial) with agronomic crops (annual or perennial), offers great promise to sequester C both above- and below-ground. Agroforestry practices have been approved as a strategy for soil C sequestration under afforestation and reforestation programs and also under the Clean Development Mechanisms of the Kyoto Protocol (Watson et al. 2000; IPCC 2007; Smith et al. 2007). Adoption of agroforestry practices has greater potential to increase C sequestration of predominantly agriculture dominated landscapes than monocrop agriculture (Lee and Jose 2003; Nair and Nair 2003; Nair et al. 2009; Schoeneberger 2009; Morgan et al. 2010). Within agroforestry systems (AFS), C can be stored in above- and below-ground biomass, soil, and living and dead organisms. The quantity and quality of residue supplied by trees/shrubs/grass in agroforestry systems enhance soil C concentration (Oelbermann et al. 2006b). In addition, C stored by trees could stay in soils or as wood products for extended periods of time. If agroforestry systems are managed sustainably, C can be retained in these systems for centuries (Dixon 1995). The amount of C stored on a site is a balance between long-term fluxes. However, the net C gain depends on the C content of the previous system that the agroforestry practice replaces (Schroeder 1994).

The enhanced C sequestration concept is based on efficient use of resources by the structurally and functionally more diverse and complex plant communities in agroforestry systems compared to sole crop or grass systems (Sanchez 2000; Sharrow and Ismail 2004; Thevathasan and Gordon 2004; Steinbeiss et al. 2008; Marquez et al. 1999). Agroforestry practices accumulate more C than forests and pastures because they have both forest and grassland sequestration and storage patterns active (Schroeder 1994; Kort and Turnock 1999; Sharrow and Ismail 2004). For example, an alley cropping system with pine trees and pasture grass could efficiently utilize light energy at different canopy levels compared to a monocropping system. Species in agroforestry systems often have different physiological needs for particular resources in certain amounts, at certain times, and possess different structural or functional means to obtain them (Jose et al. 2004). The utilization of the environment by species includes three main components: space, resources, and time. Any species utilizing the same exact combination of these resources as another will be in direct competition which could lead to a reduction in C sequestration. However, if one species differs in utilization of even one of the components, for example light saturation of C3 vs. C4 plants, C sequestration will be enhanced.

Although agroforestry has come of age during the past three decades and scientific data has expanded, our understanding of C storage and dynamics in AFS is still minimal (Nair et al. 2010). Similarly, a complete picture of C distributions in AFS in the North American Continent is lacking in the literature, thus restricting development of suitable mitigation strategies to enhance C sequestration associated with establishment of agroforestry practices on the agricultural landscape (Udawatta and Godsey 2010). Reliable estimates of soil C sequestration are essential for development of management plans related to climate change (Watson et al. 2000). This is especially important in AFS due to their complex nature, differences among practices, climatic conditions, and site factors. Well designed long-term research is needed to fill the knowledge gap so that appropriate agroforestry systems could be developed to maximize C sequestration benefits (Reed 2007). The objectives of this chapter are to (1) provide a review of C sequestration opportunities available under various agroforestry practices in temperate North America, and (2) synthesize available data and estimate C sequestration potential by agroforestry in the US.

Data Collection and Analysis

A literature search was conducted to identify peer-reviewed papers and government documents pertaining to agroforestry related C sequestration in five major temperate agroforestry practices namely; riparian buffers, alley cropping, silvopasture, windbreaks, and forest farming (Table 1). Scientific conclusions on C storage and sequestration as influenced by management practice and other factors were included in the analyses. Studies on C sequestration were categorized by practice (Table 1). Forest farming was not included since sufficient information was not available for an in depth review. When C concentrations were not provided, biomass was assumed

Table 1 Five main agroforestry practices in temperate North America

Practice	Predominant region and distribution	Function
Riparian and upland buffers	All regions	Ameliorate non point source pollution Protect watersheds and stream banks
Silvopasture	West and Southeast; all regions	Economic diversification Improve animal health Create wildlife habitat Fire protection Timber management
Alley cropping	Midwest; all regions	Increase bio diversity Increase income
Windbreaks	Great plains; all regions	Protect crop, animal, and structures Enhance crop and animal production Control erosion Distribute snowfall
Forest farming	All regions	Diversify income

Source: Gold and Garrett (2009). Reproduced with permission

to contain 50% C. Although literature from both the US and Canada were reviewed, the combined data set was used to estimate overall C sequestration potential of agroforestry practices in the US only.

Riparian Buffers

Riparian areas have many definitions which vary with the intended function and geographic region, but are generally defined as a complex terrestrial assemblage of plants and other organisms adjacent to an aquatic environment (Table 1). These include the transition zone between upland and aquatic habitats such as wetlands, streams, rivers, lakes, and bays. They are linear in shape and characterized by laterally flowing water that rises and falls at least once within a growing season (Lowrance et al. 1985; Welsh 1991).

Riparian systems store C in above- and below-ground biomass of the vegetation and in soils. Biomass accrual varies by region, plant composition, soil, climate, age, and management. The diverse species mixture of riparian buffers helps enhance C sequestration potential spatially and temporally compared to monocropping systems. The different functional groups such as trees and grasses in these systems colonize and capture both the above- and below-ground resources more effectively than the row crop agriculture.

In general, C sequestration potential and storage are greater in the above-ground portion of riparian buffer systems compared to row crops or upland forests. In riparian systems, tree density and basal area are often greater than or equal to those of upland

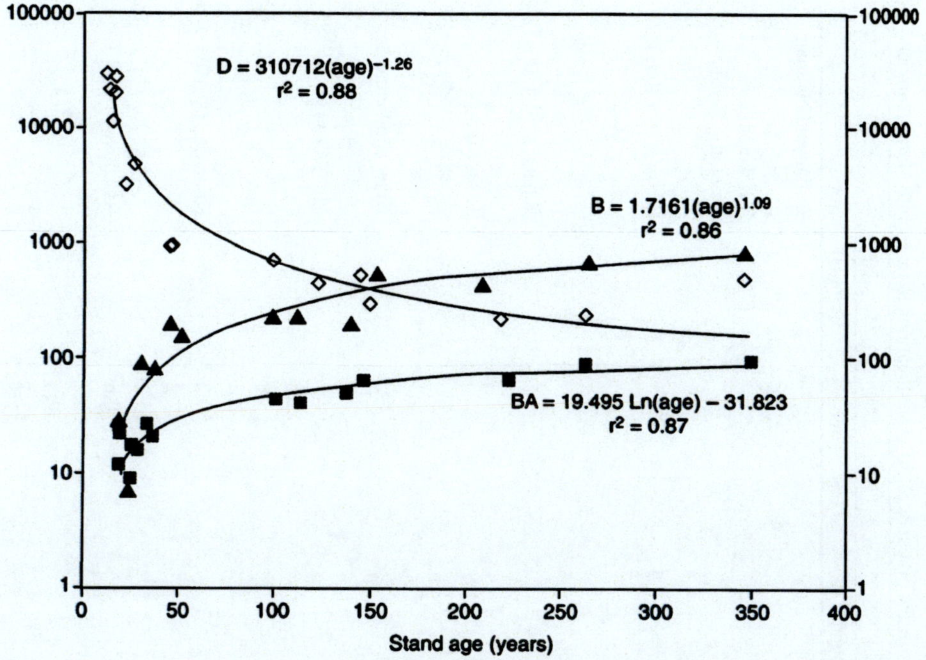

Fig. 1 Changes in stem density (D, stems ha^{-1}), biomass (B, Mg ha^{-1}), and basal area (BA, m^2 ha^{-1}) of a riparian forest buffer with age in Washington, USA. Y axis in logarithmic scale for stems ha^{-1}, Mg ha^{-1}, and m^2 ha^{-1} (Source: Naiman et al. 2005. Reproduced with permission)

forests due to prevailing favorable growth conditions. Above-ground C of a mature riparian forest ranged from 50 to 150 Mg ha^{-1} (Naiman et al. 2005). As riparian systems mature, the above- and below-ground biomass of the understory and overstory vegetation increase, giving an overall increase in the system level C stock. According to Naiman et al. (2005), stem biomass accrual of a riparian forest buffer can be determined by stand age (stem biomass kg ha^{-1} = 1.7161* age$^{1.09}$; $r^2 = 0.86$). Stem biomass accrual and thereby C stock increased at a diminishing rate for stands < 150 year and reached a plateau after 250 year (Fig. 1). Biomass accumulation pattern of another riparian system in Washington, USA, showed similar patterns with an increase in C from 9 to 271 Mg ha^{-1} as the system matured (age ~ 250 year). Almost 90% of the stem density and biomass accumulation occurred during the first 20–40 years (Table 2; Balian and Naiman 2005).

Similar observations were made by Boggs and Weaver (1994) and Harner and Stanfoord (2003) in Montana, USA. In their study, willow (*Salix* spp.) and cottonwood (*Populus deltoides* Bortr. ex Marsh.) riparian buffers developed into a mature system (~ 60 year) where above-ground C increased from 0.5 to 97 Mg ha^{-1} while stem density decreased from > 10,000 to < 1,300 stems ha^{-1}. Tufekcioglu et al. (2003) observed four and eight times greater above-ground C for poplar areas (~20 Mg ha^{-1}) of the riparian buffer in Iowa compared to 5 and 2.5 Mg C ha^{-1} for switchgrass

Table 2 Biomass (above and below), soil, and microbial carbon stocks of various agroforestry practices at different locations in temperate North America

Agroforestry practice	Location	Age (years)	Species/treatment	C (Mg ha^{-1})				Source
				Above-ground[a]	Below-ground	Soil	Microbial	
Riparian buffers	Washington, USA	~250	N/A	9–271				Balian and Naiman (2005)
	Iowa, USA		Poplar	20				Tufekcioglu et al. (2003)
			Switchgrass	5				
			Cool season grasses	2.5				
	South Carolina, USA	2	N/A	<7.5	2.5			Giese et al. (2003)
		8		17.5	3.7			
		12		17.5	5			
		60		106	5.5			
	Northeast Ontario, Canada	95		29.3–269.1				Hazlett et al. (2005)
	Iowa, USA	6	Poplar+	35	6			Tufekcioglu et al. (1999)
			Switchgrass+		9			
			Cool season grass		7			
	New York, USA			0.25–14.4 mean 6.6				Kiley and Schneider (2005)
	Iowa, USA	6	Poplar			2.4		Marquez et al. (1999)
		6	Switchgrass			1.8		
		6	Cool season grass			1.8		
		1	Soybean			0.4		
	Iowa, USA	7–17	Riparian buffer			50.2		Kim et al. (2010)
			Warm season grass			47.2		
			Cool season grass			55.3		
	Iowa, USA	16–26	Riparian buffer			70.8		As above
			Warm season grass			56.2		
			Cool season grass			57.8		
			Corn-soybean			57.1		

Alley cropping	Missouri, USA	5	Pin oak	0.03			Udawatta et al. (2005)
			Bur oak	0.01			
			Swamp white oak	0.015			
	Georgia, USA	1	Mimosa tree mulch + grain sorghum + winter wheat	2.5			Rhoades et al. (1998)
	Guelph, Ontario, Canada	15	Poplar intercrop	96.5	3.0		Peichl et al. (2006)
			Spruce intercrop	75.3	2.5		
			Barley sole crop	68.5	2.4%		
	St. Remi, Quebec, Canada.	8	Tree-based conventional systems		77.1 43.5		Bambrick et al. (2010)
	Guelph, Ontario, Canada	21	Poplar		57		As above
			Norway spruce conventional systems		51 51		
	Florida, USA	3	Pecan orchard			1.2%	Lee and Jose (2003)
		3	Pecan + cotton			1.9%; 0.38	
		47	Pecan orchard			4.3%	
		47	Pecan + cotton			3.4%; 0.78	
Silvopasture	Oregon, USA		Pastures	0	102.5		Sharrow and Ismail (2004)
			Agroforestry	12.24	95.9		
			Plantation	6.95	91.94		
	Oregon, USA		*Understory*				As above
			Pastures	1.0			
			Agroforestry	1.17			
			Plantation	2.23			
	Florida, USA		Pasture		1033		Haile et al. (2010)
			Center of alley		1376		
			Between tree row		1318		
Windbreak	Nebraska, USA	35	Windbreak		39.94		Sauer et al. (2007)
			Crop field		36.23		

[a]Assumed 50% C in the biomass to estimate C when C concentration was not provided

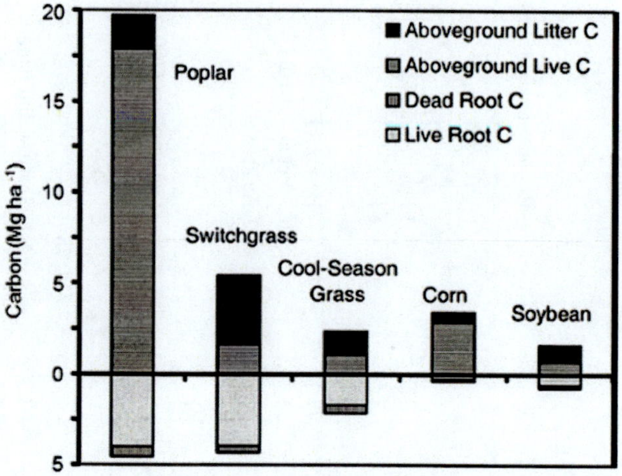

Fig. 2 Litter and root carbon distributions in a riparian system with trees, grass, and crops in Iowa, USA (Source: Tufekcioglu et al. 2003. Reproduced with permission)

(*Panicum virgatum* L.) and cool season grass areas, respectively (Fig. 2). Adjacent corn (*Zea mays* L.) and soybean [*Glycine max* L. (Merr.)] areas had 3.0 and 1.3 Mg ha^{-1} above-ground C, respectively. Giese et al. (2003) reported 106 Mg ha^{-1} C in a 60 year-old riparian buffer compared to <7.5, 17.5, and 17.5 Mg ha^{-1} in 2, 8, and 12 year-old buffers, respectively in South Carolina (Table 2). The total amount of C (including roots, herbs, and shrubs) stored in the mature riparian forest buffer in this study was four times that of the younger stands. Studying C storage in riparian (0–5 m from the water body) versus upslope forested area (60–75 m from the water body) in northeastern Ontario, Canada, Hazlett et al. (2005) observed 3% more C in the riparian zones (Table 2).

The aforementioned studies provide a realistic estimate of above-ground C stock of mature riparian buffer systems in temperate North America. If the system is maintained, these values may reflect the sequestration potential at maturity and would allow estimates of annual accrual rates. From these data, we estimate, for mature riparian buffers, an average above-ground C stock of 123 Mg C ha^{-1} for a 50 year cutting cycle. The estimated average above-ground C sequestration potential is 2.46 Mg C ha^{-1} $year^{-1}$ (Table 3). In Canada, C accruals of 29–269 Mg ha^{-1} were reported for riparian buffers (Hazlett et al. 2005). Using published data (n=4), Schroeder (1994) estimated 63 Mg C ha^{-1} above-ground storage for temperate zone riparian buffers with a 30 year cutting cycle. Our mean estimate of 123 Mg C ha^{-1} is twice the value estimated by Schroeder (1994). According to Hoover and Heath (2011), above-ground C stock for forest stands could range from 74 to 106 Mg ha^{-1} with a mean of 90 Mg ha^{-1}. Riparian areas are generally highly productive and therefore the value could be greater than for a typical forested site.

Roots of the riparian buffers also sequester significant quantities of C belowground and this C is retained in the soil C pool as roots decay. Studying root densities in riparian-crop transects in Iowa, Tufekcioglu et al. (1999) found significantly greater root biomass in the riparian vegetation compared to the row-crop areas (Fig. 3). On average, poplar (*P. deltoides* x *nigra* DN-177), switchgrass, and cool season grass root C during the study were 3, 4.5, and 3.5 Mg C ha^{-1}, respectively (Table 2; Fig. 3).

Table 3 Estimated C sequestration potential in above-ground and below-ground vegetation parts and soil for major agroforestry practices in temperate North America

Practice		C Stock[a] (Mg C ha^{-1}) Minimum	Maximum	Mean	C sequestration rate[b] (Mg C ha^{-1} $year^{-1}$)
Riparian buffers	Above-ground	7.5	269	123	2.6
	Below-ground	2.0	14.4	4.6	
	Soil	1.8	5.5	3.6	
Alley cropping	Above-ground	0.05	96.5	26.8	3.4
	Soil	0.05	25	6.9	
Silvopasture	Above-ground	1.17	12.2	4.9	6.1
	Soil	1.03	1.38	1.21	
Windbreaks[c]	Above-ground	0.68	105		
	Soil		23.1		6.4
	Hybrid poplar[d]			367	0.73
	White spruce[d]			186	

[a]This analysis used published data for the United States and Canada as reported in Table 2. If not given, we assumed 50% C in the above- and below-ground biomass to estimate C stocks
[b]Harvest age of 50 year was assumed for riparian buffers. Harvest age of 20 year and tree density of 40 tree ha^{-1} were assumed to estimate annual C accrual rates for windbreaks on cropland
[c]C Stock in windbreaks s expressed as Mg C km^{-1}
[d]Mean C stock for hybrid poplar and white spruce are in kg $tree^{-1}$

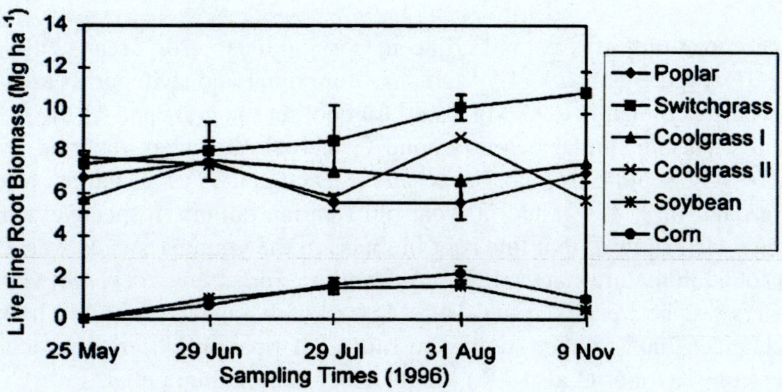

Fig. 3 Fine root biomass of trees, grasses, and crops in a riparian-row crop continuum in Central Iowa, USA (Source: Tufekcioglu et al. 1999. Reproduced with permission)

The riparian vegetation consisting of trees and grasses also had more fine (0–2 mm dia.) and medium (2–5 mm dia.) sized roots in the surface soil and throughout the 165 cm soil profile. Coarse and medium roots of poplar trees in the riparian zone extended beyond 165 cm depth while no crop roots were found below 125 cm. Four years after establishment, root biomass and thereby below-ground C were significantly greater in the riparian zone than the row crop areas (Marquez et al. 1999).

Another study, also from Iowa, demonstrated the potential to sequester greater quantities of C in soils under riparian buffers compared to row crops. Below-ground C in the tree and switchgrass areas of the riparian buffers was significantly greater

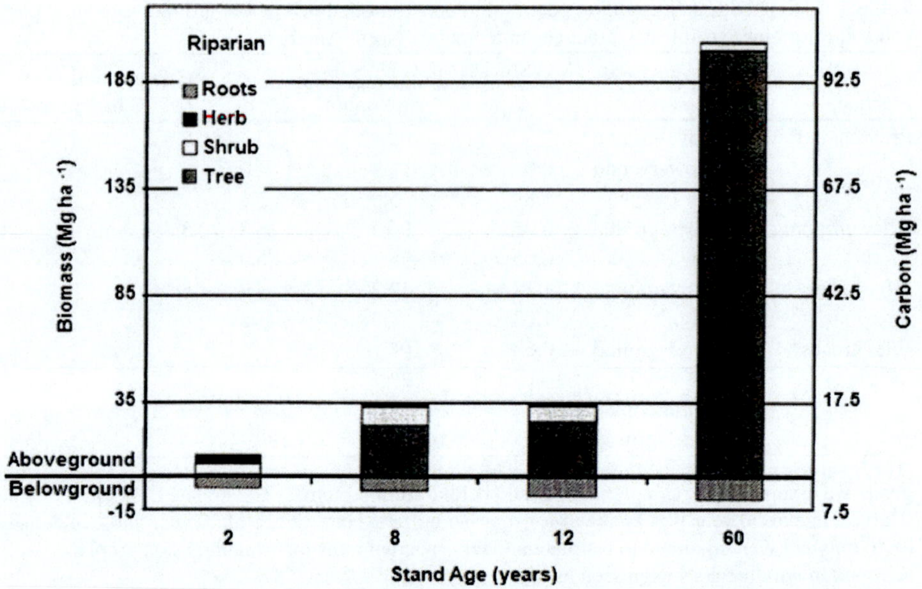

Fig. 4 Above- and below-ground biomass and carbon of 2-, 8-, 12-, and 60-year-old riparian stands in South Carolina, USA (Source: Giese et al. 2003. Reproduced with permission)

than accompanying grasses and adjacent corn-soybean crop areas (Tufekcioglu et al. 2003). Figure 2 shows >4.5 Mg C ha^{-1} for poplar and switchgrass areas of the riparian buffer compared to <2 Mg C ha^{-1} for cool season grass and <1 Mg C ha^{-1} for corn and soybean. Similar below-ground C accrual rates were reported by Giese et al. (2003) in South Carolina. The results showed 2.5, 3.7, 5.0, and 5.5 Mg C ha^{-1} below-ground in 2, 8, 12, and 60 year-old riparian buffers, respectively (Fig. 4). This study also showed that fine root biomass in the younger stands was 25–50% of that found in mature stands. In the Adirondack Park, New York, root C of riparian buffers was between 0.25 and 14.5 Mg ha^{-1} with a mean of 6.6 Mg ha^{-1} (Kiley and Schneider 2005). Other studies on root C of riparian buffers reported values ranging from 1 (Jones et al. 1996) to 3 Mg ha^{-1} (Tufekcioglu et al. 1999). Greater root mass of mature riparian stands indicate the importance of long-term management of riparian buffers for enhanced C sequestration.

The aforementioned results were used to estimate the below-ground C sequestration potential of riparian buffers (Table 3). Root biomass and C accumulation follows an asymptote as in above-ground biomass and C accumulation with an early increase in accumulation rates and a plateau as the system matures (Giese et al. 2003). We estimated a mean C sequestration of 0.09 Mg C ha^{-1} year^{-1} in below-ground tissues for riparian buffer systems for a 50 year harvest cycle.

In addition to the C sequestered in roots, riparian soils store C in soil organic matter (SOM). The SOM, which contains about 50% C, is greater in mature riparian stands compared to monocropped agroecosystems or younger riparian buffers (Giese et al. 2003). A riparian system consisting of poplar trees and grasses established in Central Iowa (Marquez et al. 1999) in 1990 showed significantly greater total soil C

Table 4 Soil carbon and soil organic matter percentage in 2-, 8-, 12-, and 60-year-old riparian buffer systems in South Carolina, USA

Stand age (year)	Soil carbon (%)	Soil organic matter (%)
2	4.2	12.3
8	4.7	12.9
12	4.0	10.7
60	11.4	30.3

Source: Giese et al. (2003). Reproduced with permission

concentrations in 1991 and 1996 than the nearby crop areas. In the poplar and switchgrass zones, soil organic carbon (SOC) accrued at the rate of 1.2 and 0.9 Mg C ha^{-1} $year^{-1}$. The results of this study introduce a promising and important observation: changes in soil C can occur in a relatively short period of time and therefore establishment of riparian buffers with the appropriate vegetation combination may help sequester soil C in a short period at a low cost. Giese et al. (2003) observed that soil C content was 2.6 times greater in the 60 year-old buffer in South Carolina compared to 2, 8, and 12 year-old riparian buffers (Table 4). Kim et al. (2010) studied riparian buffer soils to a 15 cm depth in Iowa and showed a SOC increase of 50–71 Mg ha^{-1} in 7 years, representing a 29% increase (Table 2). Warm season and cool season grass buffers contained 47 and 56 Mg C ha^{-1} in the sampled 15 cm soil depth.

The litter material in the riparian zones, either from the riparian vegetation or flooding, also contributes to soil C sequestration. The litter is approximately 47% of the above-ground biomass production (Piedade et al. 2001). The litter production rate is inversely related to the latitude of the riparian systems (Benfield 1997). In general, riparian buffers with infrequent flooding produce 5.5 Mg ha^{-1} litter material (Piedade et al. 2001); however, this varies by vegetation type (Tufekcioglu et al. 1999). Riparian buffers that are frequently or permanently flooded produce less litter than infrequently flooded buffers (Conner et al. 1981; Piedade et al. 2001).

One of the factors determining net soil C sequestration is soil respiration. In Iowa, annual soil respiration rates within a riparian buffer and adjacent crop field varied between 7.4 and 12.2 Mg C ha^{-1} $year^{-1}$ (Tufekcioglu et al. 2001) with the highest in the streamside cool season grass buffer and the lowest in the corn areas. Annual respiration rates were 11.5, 11.4, 10.3, and 7.5 Mg C ha^{-1} $year^{-1}$ for crop side cool season grass, poplar buffer, switchgrass, and soybean, respectively. Although the perennial vegetation in the buffer areas (trees or grasses), had significantly greater respiration rates compared to the annual crops, trees leaf out and grasses begin to grow before the crop is established and, thus increasing the overall C sequestration potential of the system (Marquez et al. 1999).

Management Implications

As the above literature reveals, riparian buffer systems have tremendous potential to sequester C in above- and below-ground plant parts and in the soil compared to monocropping systems in temperate North America. These systems sequester C in a relatively short period of time and the sequestration rates are high during the early stages of development. Root C deposited in the deeper horizons of the soil profile could remain for extended periods, and thereby contributing to long-term soil organic C storage. Proper management operations such as maintenance of suitable buffer width along water bodies, proper species selection, and removal of older trees following best management practices (BMPs) would further enhance C sequestration potential of riparian systems.

Management agencies at both national and state levels have determined appropriate buffer dimensions for protection of water bodies. For example, the Massachusetts Department of Conservation and Recreation specifies that riparian buffer width should increase by 12 m for every 10% increment of slope greater than 10%. In general, buffer widths between 15 and 100 m have been suggested by various local, state, and federal agencies for water quality improvements, stream bank stabilization, and to reduce sediment and nutrient losses (NRCS 2007). Buffers of 91 m width have been proposed for levee protection and to stop flooding and other damages (Dwyer et al. 1997). The composition and the width of the riparian buffer system also vary and much wider buffers are required for the removal of soluble nutrients compared to stabilizing stream banks (Schultz et al. 2009). Multi-species riparian buffers could consist of fast and slow growing trees adjacent to the water body and shrubs and grasses between the trees and upland areas. Properly designed riparian buffers not only sequester C, but improve water quality, wildlife habitat, aesthetic value, economic returns, and land value (Qiu and Prato 2001; Schultz et al. 2009).

The total river and stream length in the US is approximately 5.65 million km (3.533 million miles; USEPA 2010). Lakes and estuaries occupy 16.8 million ha and 22.7 million ha, respectively. The nationwide NRCS goal was to establish 3.2 million km (two million miles) of conservation buffers by 2002 (http://www.nrcs.usda.gov/feature/buffers/BufrsPub.html#InitiativeBuff_7Anchor, accessed 15 January 2011). Documented goals or committed riparian buffer lengths vary by state. For example, Chesapeake Bay agreement for riparian buffers fulfilled their 960 km riparian buffer target for 2010 (http://www.unl.edu/nac/insideagroforestry.htm, accessed 24 January 2011). If a 30 m wide riparian buffer is established along both sides of 5% of total river length it would occupy 1.69 million ha. Using a conservative estimate of 2.6 Mg C ha^{-1} $year^{-1}$ accrual rate (Table 3), the potential C sequestration by riparian buffers along rivers in the US could be as high as 4.7 Tg C $year^{-1}$. This approximation ignored smaller and/or intermittent streams that were not part of the total river length as well as other water bodies that do not have a measurable perimeter for the estimation of buffer length. Some of these water bodies currently have riparian buffers established for various ecological and environmental reasons. Other water bodies that have disconnected buffers or no buffers offer a greater C sequestration potential through establishment of riparian buffers.

The 4.7 Tg $year^{-1}$ C sequestration potential estimated by this analysis is significantly greater than the 1.5 Tg C $year^{-1}$ estimated by Nair and Nair (2003). This difference is due to the area considered for the sequestration and the values used to estimate the sequestration potential. Nair and Nair (2003) used 30 m wide forested riparian buffer zone on one fourth of the 3.2 million km conservation buffers committed by the USDA in 2002 for their estimate.

Alley Cropping

Alley cropping has received increased attention in temperate North America in recent years. These systems could include widely spaced single or multi-species tree, grass, and/or shrub rows with agronomic crops or pasture grass grown in the alleys (Table 1). The selection of companion perennial vegetation depends on landowner objectives and site suitability for a particular species. Expected benefits include improvements in environmental quality, economic returns, C sequestration, and wildlife benefits. In these systems, spatial heterogeneity exists in C stocks and sequestration due to tree and crop row configuration, differences in C input into the soil, decomposition rate, previous management, and associated soil micro fauna (Udawatta et al. 2008, 2009; Bambrick et al. 2010).

Only a few studies have examined above-ground biomass accumulation in alley cropping practices. In a 5 year-old alley cropping system in northeast Missouri (Udawatta et al. 2005), pin oak (*Quercus palustris* Muenchh) had twice the above-ground C of bur oak (*Quercus macrocarpa* Michx.) and swamp white oak (*Quercus bicolor* Willd.) (Table 2). The system sequestered 0.05 Mg C ha^{-1} in 5 years. The lower biomass accumulation of the site was attributed to persistent deer browsing during the initial 3 years of the study. Another study in Georgia, with *Albizia julibrissin* (L.) Benth. (mimosa) and grain sorghum (*Sorghum bicolor* (L.) Moench) during summer and wheat (*Triticum aestivum* L.) grown over winter, reported 50 times greater C than the Missouri study (Rhoades et al. 1998). The estimated tree density was 2,400 ha^{-1} (0.5 m spacing within rows and 4 m spacing between rows). The C input from pruning of leaves and twigs (second year at 1 m height) were 1.42 and 1.08 Mg ha^{-1} year^{-1}, respectively. In Southern Ontario, Canada, Peichl et al. (2006) showed that 13 year-old poplar and spruce alley cropping, and barley monocrop contained 96.5, 75.3, and 68.5 Mg C ha^{-1} (Table 2). In central Missouri, Pallardy et al. (2003) reported a biomass accumulation of 2.7 and 13 Mg ha^{-1} for first and second year harvests of poplar clones (1.3 and 6.5 Mg C ha^{-1}, assuming 50% C in the biomass).

Based on the limited data we estimate the above-ground C stock in alley cropping system as 26.8 Mg C ha^{-1} (Table 3). This is 4.6 times lower compared to the C stocks of riparian buffers. It should be noted that the alley cropping systems reviewed in this analysis are much younger (1–13 year-old) compared to the riparian buffers (2–250 year-old). We estimate that alley cropping has an average above-ground C sequestration potential of 2.7 Mg ha^{-1}.

In general 40–50% of C sequestered by trees is believed to be below-ground (Turnock 2001). However, this value changes by species and location because height growth, assimilation rates, litterfall, and root turnover differ by species. In an alley cropping practice in southern Ontario, Norway spruce (*Picea abies* L.) sequestered twice as much C as poplar in a 13 year-old study (Peichl et al. 2006). Although the above-ground C stocks of poplars and spruce were almost the same (85% and 82%, respectively), spruce had 63% of the C in branches and needles that provided greater quantities of litter material and thereby greater potential to add C to the soil pool. In addition, the spruce branches and needle were lignin-rich compared to poplar leaves and decomposed slowly in the soil.

In alley cropping, differences in SOC do not occur in a short period of time and in some situations, the SOC decreased with time (Thevathasan and Gordon 2004; Oelbermann et al. 2006a, b; Bambrick et al. 2010). According to Young (1997), tropical environments require at least 10 years to observe significant differences in SOC of alley cropping systems compared to monocropping systems. A longer timeframe is required to detect changes in the SOC content of these systems in the temperate zone due to colder climatic conditions and low C inputs (Peichl et al. 2006; Oelbermann et al. 2006a, b).

Studying 4, 8, and 21 year-old tree-based oat (*Avena sativa* L.)- maize-maize rotational alley cropping systems in Quebec, Canada, Bambrick et al. (2010) observed that differences in SOC among systems were not significant. However, spatial variation in SOC was obvious. The SOC concentrations were significantly greater at 0.75 m distance from the tree row than at 5 and 11 m. Also, 8 and 21 year-old sites showed significantly greater SOC concentration (77% and 12%, respectively) in the tree-based system than the conventional oat-maize rotations. The authors concluded that these systems required at least 6 years to sequester significantly more C in the soil than the conventional agricultural systems under existing soil and climatic conditions. Other studies, however, speculate that it would take at least 10 years to accrue significantly measurable differences in soil C between alley cropping and monocrop systems (Peichl et al. 2006; Oelbermann et al. 2006a, b).

The spatial variation in SOC in alley cropping systems results from the distinct spatial pattern of above-ground biomass and litterfall. Initially, more litter material tends to accumulate near the tree base (Bambrick et al. 2010). However, SOC concentration became non-significant with distance from trees as trees mature and spread roots and branches evenly. For example, Thevathasan and Gordon (2004) observed significant litter accumulation closer to the tree row and decreasing amounts away from the trees in a 6 year-old poplar-barley (*Hordeum vulgare* L.) alley cropping system in Ontario, Canada. The SOC content was 35% higher near the tree base and this effect extended up to 4 m in the crop alley when the system was 8 year-old. With time, crop alleys also showed increased SOC due to evenly distributed leaf biomass. The spatial variation in root biomass in alley cropping could also contribute to the SOC distribution. Jose et al. (2001) observed significantly greater root biomass in the black walnut (*Juglans nigra* L.) and red oak (*Quercus rubra* L.) tree rows compared to maize alleys in Indiana, indicating greater C stocks in the tree rows. Red oak root biomass was 2.1 and 1.8 times greater than the maize root biomass at the tree base and 1.1 m from the base. Black walnut had 1.1 and 1.37 times more roots, at those distances respectively, than maize. Trees had fewer roots at distances greater than 2.3 m from the tree row.

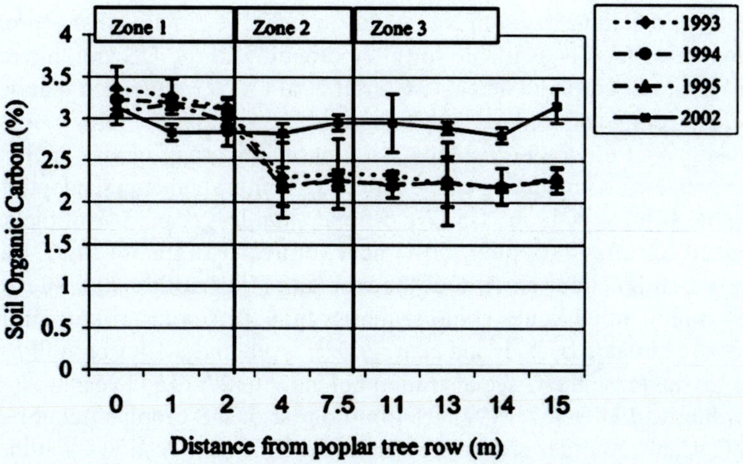

Fig. 5 Soil organic carbon concentrations in tree rows of 6- (1993), 7- (1994), 8- (1995), and 15- (2002)-year-old intercropped agroforestry practice in southern Ontario, Canada (Source: Thevathasan and Gordon 1997. Reproduced with permission)

Overall, soil C sequestration potential is much greater in alley cropping than in monocropping agronomic systems. For example, C inputs through litterfall on a poplar-spruce alley cropping with wheat-soybean-maize rotation were 0.6 and 0.95 Mg C ha^{-1} in the 11th and 12th years in Guelph, Ontario, Canada (Oelbermann 2002). In a 6 year-old hybrid poplar site (111 trees ha^{-1}) in Canada, Thevathasan and Gordon (1997) reported 1.07 Mg C ha^{-1} contributed by litterfall (Fig. 5). In the same study, hybrid poplar leaves and branches had C stocks of 1.3 and 5.5 Mg C ha^{-1} when trees were 13 year-old (Peichl et al. 2006). After 13 years the tree component of the system added 14 Mg C ha^{-1} in addition to the 25 Mg C ha^{-1} added by litter and fine roots (Thevathasan and Gordon 2004). The total C sequestration was therefore 39 Mg C ha^{-1}. The authors estimated that the system had immobilized 156 Mg ha^{-1} CO_2 or 43 Mg C ha^{-1} by age 13 and the system could potentially sequester significantly more C at the end of a 40 year harvest cycle.

One of the aspects neglected in soil C quantification in agroforestry is microbial C. In a pecan (*Carya illinoinensis* (Wangenh.) K. Koch)-cotton (*Gossypium hirsutum* L.) alley cropping system in Florida, Lee and Jose (2003) found significantly greater microbial biomass in a 47 year-old system compared to a 3 year-old system. Soils in the mature pecan system had 1.75 Mg C ha^{-1} (398 mg C kg^{-1} soil) as opposed to 0.38 Mg C ha^{-1} (88 mg C kg^{-1}) in the 3 year-old system (bulk density was assumed to be 1.25 g cm^{-3}). The highest SOM (4.3%) was also observed in the older alley cropping system and the authors attributed these differences to roots, leaves, branches, and other components from older pecan trees, as well as accrued, decomposing litter.

According to the USDA NASS (2008) and USDA NRCS (2003), cropland in the US is about 179 million ha, which includes approximately 16 million ha of idle land.

Montagnini and Nair (2004) and Nair and Nair (2003) estimated that approximately 80 million ha of land is available for alley cropping in the US and this represents 44.7% of the total cropland acreage. Garrett et al (2009) suggested that 40 million ha of highly erodible nonfederal croplands could be suitable for alley cropping. This represents 22% of the total croplands. Although alley cropping has the potential to sequester greater C compared to conventional agricultural practices, adoption of alley cropping has been slow in the US. We estimate that less than 10% of the croplands will be used for alley cropping in the near future. Using a 3.4 Mg C ha^{-1} $year^{-1}$ C sequestration potential on 10% of the cropland (17.9 million ha), alley cropping practices in the United States could sequester 60.9 Tg C $year^{-1}$. If 80 million ha of cropland, as estimated by Nair and Nair (2003), is put under alley cropping, it would significantly increase the C sequestration potential to 272 Tg C $year^{-1}$.

According to Lal et al. (1999), 154 million ha of US cropland could sequester 73.8 Tg C $year^{-1}$. Another estimate by Nair and Nair (2003) shows that the 80 million ha of erodible and marginal agricultural land available for alley cropping in the US could potentially sequester 73.8 Tg C $year^{-1}$ in above- and below-ground biomass. The C sequestration potential for US cropland and alley cropping, if expressed per ha, would be 0.479 (Lal et al. 1999) and 0.922 (Nair and Nair 2003) Mg C ha^{-1} $year^{-1}$, respectively. These estimates fall within the range (0.1–1 Mg C ha^{-1} $year^{-1}$) reported for improved agricultural management practices without incorporating perennial vegetation such as grasses, trees, and shrubs on cropland (CAST 2004). Our estimated C sequestration potential for alley cropping (3.4 Mg C ha^{-1} $year^{-1}$) is 7 times and 3.6 times greater than the estimates of Lal et al. (1999) and Nair and Nair (2003), respectively. However, the higher estimate is reasonable with the incorporation of trees as illustrated in Tables 2 and 3.

Silvopastoral Systems

Silvopasture is an agroforestry practice that intentionally integrates trees, forage crops, and livestock into a structural and functional system for optimization of benefits from planned biophysical interactions (Table 1). It is the most common form of agroforestry in North America (Clason and Sharrow 2000; Nair et al. 2008; Sharrow et al. 2009). In the US, approximately one-fifth of the forests or 54 million ha are grazed by livestock (Lubowski et al. 2006; Sharrow et al. 2009). In many regions, grazing also occurs either on marginal lands or as a secondary activity on high yielding timber lands. In temperate North America, silvopastoral systems have a great potential to sequester C due not only to high biological productivity, but also to the availability of larger areas under grazing management (Haile et al. 2008; Sharrow et al. 2009).

Conversion of pastureland to silvopasture has the potential to enhance rooting depth and distribution, quantity, and quality of organic matter input and thereby C sequestration potential (Haile et al. 2008). These systems could outperform C sequestration of either forest or pastures as they have both forest and grassland

Table 5 Soil organic carbon and nitrogen for the grazed pasture (GP), agroforestry buffer (AgB), grass buffer (GB), and row crop (RC) management treatments in Missouri, USA

Treatment	Soil organic carbon (% mass basis)	Total soil nitrogen (% mass basis)
GP	1.8[a]	0.20[a]
AgB	1.7[a]	0.20[a]
GB	1.7[a]	0.19[a]
RC	1.2[b]	0.13[b]

Source: Paudel et al. (2011). Reproduced with permission
Values with the same superscript within a column are not significantly different at $p \leq 0.05$

Table 6 Root dry weights and carbon to a 1-m soil depth in agroforestry (trees+grass; AgB), grass buffer (grass only; GB), rotationally grazed (RG), and continuously grazed (CG) treatments in a silvopasture practice in Missouri, USA

Treatment	Root dry weight (g 100 cm^{-3} soil)	C (g 100 cm^{-3} soil)
AgB	0.381	0.19
GB	0.475	0.23
RG	0.140	0.07
CG	0.074	0.04

Source: Kumar et al. (2010). Reproduced with permission

mechanisms of C capture that can maximize C sequestration both above- and below-ground. In general, trees store about 50–60% of the C in the above-ground biomass whereas pasture grasses store only 10% above-ground, the rest being allocated below-ground (Houghton and Hackler 2000; Sharrow and Ismail 2004). The greater potential to sequester C by silvopasture was illustrated by Sharrow and Ismail (2004) in their comparison of Douglas fir (*Pseudotsuga menziesii* (Mirb.) Franco)-cool season grass silvopastoral system with pasture and Douglas fir plantation in Oregon. These authors observed that the silvopastoral system sequestered an additional 0.74 Mg C ha^{-1} year^{-1} and 0.52 Mg C ha^{-1} year^{-1} than the plantation and pasture, respectively (Table 2). Individual trees in the silvopastoral systems grew faster than in conventional forests on the same site, allowing silvopastoral trees to store more C. The total amount of C stored in above- and below-ground biomass and soil was 5.8 and 8.2 Mg C ha^{-1} greater in silvopasture than pasture and Douglas fir plantation.

Roots of the perennial vegetation in silvopastoral systems shifts C deeper into the soil profile, compared to conventional pastures or row crops. Paudel et al. (2011) observed significantly greater percentages of C in soils under a cottonwood (*P. deltoides* Bortr. ex Marsh.) and grass silvopasture compared to maize-soybean rotation in Missouri (Table 5). In the same study area, Kumar et al. (2010) observed significantly greater root mass in the 1 m soil profile in tree-grass areas than the pasture grass (Table 6), clearly indicating the potential to deposit C deeper in the soil profile in silvopasture compared to pastures.

The spatial distribution of C, both above- and below-ground, can vary depending on the design of the silvopastoral systems and management practices. Soil organic

C derived from the tree component was significantly greater near the trees in a slash pine (*Pinus elliottii* Englem) and bahiagrass (*Paspalum notatum* Fluegge) silvopasture compared to open pasture areas in Florida (Haile et al. 2010). SOC contents were 1,033, 1,376, and 1,318 Mg ha^{-1} to a 1.25 m depth in open pasture, center of the pasture alley, and in-between trees in tree row, respectively (Table 2). Only the surface soil had more C derived from grasses. The SOC concentrations in open pastures were 94 and 26 Mg ha^{-1} for 0–75 and 75–125 cm depths, respectively. The silvopastoral system had 556 and 105 Mg ha^{-1} of SOC for the same depths, indicating the contribution of tree roots to the SOC pool, not only in the upper soil but also in the deeper soil profile.

Another factor that is not accounted for in many studies of silvopastoral systems is the amount of grass consumed by the grazing animals and the C deposited on soil via manure deposits. For example, sheep consumed a total of 30.5 Mg ha^{-1} forage in pastures and 22 Mg ha^{-1} of forage in silvopasture and deposited 9 and 7 Mg ha^{-1} manure in those two respective systems in the previously cited study in Oregon (Sharrow and Ismail 2004).

Strategies to enhance C sequestration in silvopaosture may include selection of complementary tree, shrub, and pasture grasses with optimal biomass accrual, deep rooting habits, and greater below-ground C accumulation potential. Proper maintenance of stocking rate, rotational grazing, and fertilizer application may also help enhance C sequestration. For example, Lee and Dodson (1996) estimated that conversion of 3.6 million ha marginal pasture lands in south central United States to silvopasture with pines could sequester 5.6 Tg C year^{-1} for the first 25 years and 1.1 Tg C year^{-1} for the subsequent 25 years. If this land is left for pasture, the sequestration would be 0.3 Tg C year^{-1}.

Although silvopasture remains the most common form of agroforestry in temperate North America, the precise land area under silvopasture is still unknown. Nair and Nair (2003) estimated the land available for silvopasture as 70 million ha. Pasture and grazed forestland areas in the United States are 237 and 54 million ha (www.ers.usda.gov/Data/MajorLandUses, accessed 24 December 2010), respectively. These land areas could be intensively managed for additional C sequestration.

The amount of SOC accrual in pasture lands ranged from 0.07 to 1.4 Mg C ha^{-1} year^{-1} (Franzluebbers 2005; Derner and Schuman 2007). According to Nair and Nair (2003), the C sequestration potential of silvopasture varies from a low of 1.8, medium 2.3, to a high of 3.3 Mg C yr^{-1}. Based on the data presented in the above sections, silvopastoral systems appear to sequester 6.1 Mg C ha^{-1} year^{-1} (Table 3). Using a sequestration potential of 6.1 Mg C ha^{-1} year^{-1} on 10% marginal pasture land (23.7 million ha) and 54 million ha of forests, the total C sequestration potential for silvopastoral lands in the Unites States could be as high as 474 Tg C year^{-1}. According to Montagnini and Nair (2004) and Nair and Nair (2003), 70 million ha of silvopasture in the US could store 9 Tg C year^{-1}. The value estimated in this analysis is 53 times greater than the previous estimate. We have used nearly the same acreage (77.7 million ha), but a much higher sequestration rate based on our literature review.

Windbreak

Windbreaks are designed with one or more rows of trees or shrubs planted across crop or grazing areas to reduce wind speed and enhance microclimate for crop and/or animal production (Table 1). Windbreaks have been used throughout history to protect homes, structures, livestock, and crops, control wind erosion and blowing snow, provide habitat for wildlife, improve landscape, and for odor mitigation (Brandle et al. 2004, 2009). Windbreaks are also used to reduce evaporation loss of water from soil and leaf surfaces (Brandle et al. 2009). The groundcover under the windbreak may also help reduce wind erosion and soil detachment by rain drops.

Like other agroforestry practices, windbreaks also offer great promise for C sequestration (Schoeneberger 2009). In addition to C sequestered by trees, windbreaks provide additional C sequestration due to improved crop and livestock production and energy savings (Kort and Turnock 1999). Indirectly, windbreaks reduce fuel use for heating and thereby reduce CO_2 emissions. Although shelterbelts and windbreaks have been planted in the Great Plains of the US since the 1930s, C sequestration in these systems have not been evaluated and there is a need for such estimates to determine the C sequestration capacity of these systems (Sauer et al. 2007).

The limited literature demonstrates the importance of species selection in maximizing the C sequestration potential of windbreaks or shelterbelts (Table 7). For example, hybrid poplar sequestered 367 kg C tree^{-1} in above- and below-ground compared to 110 kg C tree^{-1} in green ash (Kort and Turnock 1999). The above-ground C storage by single row conifer, hardwood, and shrubs for a windbreak in Nebraska was 9.14, 5.41, and 0.68 t km^{-1}, respectively (Brandle et al. 1992).

Table 7 Above- and below-ground biomass and carbon for shelterbelt trees commonly used in Saskatchewan, Canada

Vegetation type		Above-ground biomass	Below-ground biomass	Total C
		(kg tree^{-1})		
Deciduous	Green ash	161.8	64.7	110
	Manitoba maple	178.6	71.4	120
	Hybrid poplar	544.3	217.7	367
	Siberian elm	201.9	80.8	140
Conifers	White spruce	286.9	86.1	186
	Scot pine	164.1	49.2	107
	Colorado spruce	202.2	60.7	131
Shrubs	Choke cherry	402.6	201.3	302
	Villosa lilac	334.6	167.3	251
	Buffaloa berry	312.0	156	234
	Caranga	516.0	258	387
	Seabuckthorn	213.0	106	160

Source: Kort and Turnock (1999). Reproduced with permission

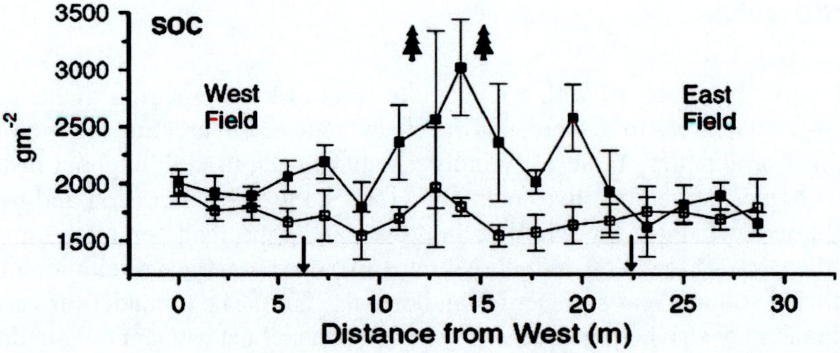

Fig. 6 Mean SOC values by position for 0–7.5 cm (closed squares) and 7.5–10 cm (open squares) across a shelterbelt crop transect in Nebraska, USA. Trees and arrows denote position of tree rows and extent of cultivation, respectively (Source: Sauer et al. 2007. Reproduced with permission)

In Saskatchewan, Canada, 17–90 year-old single row shelterbelts contained 24–41, 105, and 11 Mg C km^{-1} in conifer, poplar, and shrub shelterbelts, respectively (Kort and Turnock 1999). They also reported that above-ground C sequestration by hybrid poplar windbreaks in the US and Canada varied from <1 Mg C km^{-1} to >100 Mg C km^{-1}.

Windbreaks also contribute to the SOC pool albeit at a limited spatial scale on the landscape. In Nebraska, SOC concentration under a shelterbelt (3.04%) in the top soil (0–7.5 cm depth) was 55% more than that in the adjacent crop field (1.96%: Sauer et al. 2007). The shelterbelt treatment also contained 12% more SOC in the 7.5–15 cm depth compared to the crop field (Fig. 6). Overall, during a 35 year period, soils of 0–15 cm depth contained 3.71 Mg more SOC ha^{-1} in the shelterbelt than the cultivated region, which represents an annual sequestration of 0.11 Mg ha^{-1}. The authors attributed the increased SOC in the shelterbelt to absence of soil disturbance, increased inputs by litter, reduced erosion, and deposition of windblown material.

Nair and Nair (2003) estimated 85 million ha under windbreaks and sequestration potential of 4 Tg C $year^{-1}$. According to Brandle et al. (1992), 94 million ha of cropland in the North Central region need windbreaks to reduce damages. Another set of windbreaks are required to protect homes and roads. If 5% of the cropland, 120 million trees for protection of farmstead, and two million conifers for road protection are planted, these three categories would sequester 215, 13, 0.175 Tg C within 20 years or 11.4 Tg C $year^{-1}$ (Brandle et al. 1992).

Based on C stocks in individual trees (Table 7) and considering a 20 year harvest cycle for 120 million hybrid polar trees and two million white spruce trees, windbreaks could potentially sequester 2.2 Tg C $year^{-1}$ and 0.02 Tg C $year^{-1}$, respectively. The 5% (8.95 million ha) cropland with hybrid poplar could potentially sequester 131 Tg C or 6.56 Tg C $year^{-1}$. In this calculation, we considered 40 hybrid poplar trees for two rows of windbreaks per ha and 367 kg C $tree^{-1}$ in a 20 year harvest cycle. Thus, the total C sequestration potential estimated for windbreaks is 8.79 Tg C $year^{-1}$.

Limitations, Implications, and Future Directions

Although above-ground biomass data are available for many tree and shrub species for forest stands, the literature lacks such information for integrated agroforestry systems. Since agroforestry trees are often open grown or grown in linear configurations, the growth patterns and hence C sequestration potential could vary from conventional plantations or forest stands. There is a need for data on above- and below-ground biomass and C for trees and shrubs under agroforestry practices for all regions. Specifically, such data are needed for stems, branches, bark, leaves, litter, nuts, roots and any material that is not removed from the site in order to estimate accurate C sequestration potential of agroforestry practices. Below-ground data such as root biomass, dynamics, and morphology are an integral part C sequestration in agroforestry. Soil C data are currently available mostly for the upper 10–35 cm soil profile. Some additional parameters such as bulk density, moisture %, rock volume %, and actual sampling depth are required to express C concentration and stock. Quantitative information on CO_2 and methane emission may provide data to refine estimates of net C sequestration. Sampling intensity, time, and age at which samples were collected affect the final estimate and such information should be included in the data sets as well.

Standardized experimental procedures and data gathering protocols for all regions are required so that data can be compared among regions. This also permits development of widely acceptable conclusions for larger geographic areas. Remote sensing and satellite data need to be used to accurately estimate C stocks and sequestration by agroforestry practices at larger spatial and temporal scales.

Trees and shrubs sequester C over longer periods than annual crops. In general, harvest cycle vary from 10 to 80 years for tree species commonly used in agroforestry systems. Research focus needs to be changed to understand long-term benefits of these multi-species systems. Since agroforestry practices with trees take two to three decades to mature, tree growth models under agroforestry practices are needed to estimate C sequestration. Complex models for tree growth with crop, pasture, and/or livestock may be simulated to understand long-term benefits and also to scale-up for larger regions. Models need data for initial calibration and validation and therefore research plots are required for all regions before models are simulated and specific conclusions are drawn regarding long-term effects. As explained earlier, tree growth or biomass equations for open grown agroforestry tree species need to be developed so that biomass and C can be estimated non-destructively.

Major statistical inventory systems (USDA Forest Service and NRCS) do not collect agroforestry statistics (Morgan et al. 2010). Therefore, updated and representative statistics are not available for agroforestry practices. A national inventory system may be developed to collect agroforestry statistics, including land area under specific practices.

Data should be used to develop agroforestry design criteria for all regions and practices that optimize C sequestration, environmental benefits, and economic returns. Agroforestry designs should include perennial vegetation with desirable characteristics such as greater C sequestration, greater below-ground C allocation, and other complementary effects for optimal C accrual. Intensive and improved management techniques may be implemented in concert with genetically improved species for fast growth and greater resource use efficiency (e.g. higher fertilizer use efficiency). Agroforestry practices with perennial vegetation could be designed to protect and enhance C sequestration on sensitive landscape locations such as highly vulnerable areas for nonpoint source losses and steep slopes. Improved agroforestry designs that are strategically placed on agricultural landscapes will eventually allow development of suitable mitigation strategies to enhance C sequestration.

Conclusions

There are several limitations in the data sets used for this analysis. Lack of accurate estimates of C sequestration for all regions and systems and land area under each agroforestry practice can introduce errors in the calculations. However, our estimate clearly indicates possible net gains in C sequestration that could be used to promote agroforestry as a promising CO_2 mitigation strategy in the US and potentially in other parts of North America. There are four main land use categories that can be considered as the most suitable for agroforestry in North America: degraded or non productive land, permanent agriculture and pasture land, forest land, and disconnected narrow riparian corridors. As the literature reveals, incorporation of agroforestry by introducing improved plant stock and implementing improved and intensive management techniques, C sequestration could be enhanced on this land base in a short period of time.

Since agroforestry was not inventoried by the major natural resources inventories, our estimates of C sequestration were based on several assumptions. A coarse approximation was made with limited data by multiplying the C sequestration in each system by the land area. A 4.7 Tg C year^{-1} C sequestration potential for riparian buffers was based on a 30 m wide buffer along both sides of 5% of total river length that would occupy 1.69 million ha. The estimated area was multiplied by 2.6 Mg C ha^{-1} year^{-1} accrual rate. The estimated potential value could be much higher if we had the buffer data for all water bodies. For alley cropping, we used 10% of the cropland and sequestration value of 3.4 Mg C ha^{-1} year^{-1}. The cropland in the US has the potential to sequester 60.9 Tg C year^{-1} through alley cropping. Using a sequestration potential of 6.1 Mg C ha^{-1} year^{-1} on 10% pasture land (23.7 million ha) and 54 million ha of forests, the total C sequestration potential for silvopastoral lands in the US could be as high as 474 Tg C year^{-1}. Windbreaks that protect cropland, farmstead, and roads could sequester 8.79 Tg C year^{-1}. The total potential C sequestration by agroforestry in the US is therefore 548.4 Tg year^{-1}. This could offset the current US CO_2 emissions (1,600 Tg C year^{-1} from burning fossil fuel such as coal, oil, and gas) by 34%.

Finally, we draw the following conclusions: (1) Agroforestry is a promising practice to sequester C (548.4 Tg year^{-1} in the US alone) while providing numerous environmental, economical, and social benefits in temperate North America (2) Rigorous, long-term C sequestration research in all regions and all agroforestry practices is required to develop accurate estimates and to develop policies and guidelines to recommend agroforestry practices that satisfy landowner expectations, (3) A standardized protocol is required for sampling, sample analysis, and data handling so that all available C data can be used to simulate models to examine long-term effects and to scale-up for larger landscapes, (4) An inventory of agroforestry practices is essential not only to accurately estimate C sequestration potential, but to quantify the economic and environmental impact of agroforestry, and (5) Future research should focus on developing design criteria for appropriate configuration, species selection, and planting density for various agroforestry practices to optimize C sequestration.

Acknowledgements The authors express their sincere gratitude to Dr. P.K.R. Nair and three anonymous reviewers for their comments on an earlier version of the manuscript. We also thank Yarrow Titus for her help during the development of this chapter.

References

Balian E, Naiman RJ (2005) Abundance and production of dominant riparian trees in the lowland floodplain of the Queets River, Washington. Ecosystems 8:841–861

Bambrick AD, Whalen JK, Bradley RL, Cogliastro A, Gordon AM, Olivier A, Thevathasan NV (2010) Spatial heterogeneity of soil organic carbon in tree-based intercropping systems in Quebec and Ontario, Canada. Agroforest Syst 79:343–353

Benfield EF (1997) Comparison of litterfall inputs to streams. J North Am Benthol Soc 16:104–108

Boggs K, Weaver T (1994) Changes in vegetation and nutrient pools during riparian succession. Wetlands 14:98–109

Brandle JR, Wardle TD, Bratton GF (1992) Opportunities to increase tree planting in shelterbelts and the potential impacts on carbon storage and conservation. In: Sampson RN, Hair D (eds.) Forest and global change, Vol 1. Opportunities for Increasing Forest Cover. American Forests, Washington, DC, pp 157–176

Brandle JR, Hodges L, Zhou XH (2004) Windbreaks in North American agricultural systems. Agroforest Syst 61:65–78

Brandle JR, Hodges L, Tyndall J, Sudmeyer RA (2009) Windbreak practices. In: Garrett HE (ed.) North American Agroforestry, an integrated science and practice, 2nd edn. American Society of Agronomy, Madison, pp 75–104

CAST (Council for Agricultural Science and Technology) (2004) Climate change and greenhouse gas mitigation: challenges and opportunities for agriculture. Ames, Ames

Clason TR, Sharrow SH (2000) Silvopastural practices. In: Garrett HE, Rietveld WJ, Fisher RF (eds.) North American Agroforestry: An integrated science and practice. American Society of Agronomy, Madison, pp 119–147

Conner WH, Gosselink JG, Parrondo RT (1981) Comparison of vegetation of three Louisiana swamp sites with different flooding regimes. Am J Bot 68:320–331

Derner JD, Schuman GE (2007) Carbon sequestration and rangelands: a synthesis of land management and precipitation effects. J Soil Water Conserv 62:77–85

Dixon RK (1995) Agroforestry systems: sources or sinks of greenhouse gases? Agroforest Syst 31:99–116

Dwyer JP, Wallace DC, Larsen RD (1997) Value of woody river corridors in levee protection along the Missouri River in 1993. J Am Water Res Assoc 33(2):481–489

Franzluebbers AJ (2005) Soil organic carbon sequestration and agricultural greenhouse gas emission in the southeastern U.S. Soil Till Res 83:120–147

Garrett HE, McGraw RL, Walter WD (2009) Alley cropping practices. In: Garrett HE (ed.) North American Agroforestry: An integrated science and practice, 2nd edn. American Society of Agronomy, Madison, pp 133–162

Giese LAB, Aust WM, Kolka RK, Trettin CC (2003) Biomass and carbon pools of disturbed riparian forests. For Ecol Manag 180:493–508

Gold MA, Garrett HE (2009) Agroforestry nomenclature, concepts, and practices. In: Garrett HE (ed.) North American Agroforestry: An integrated science and practice, 2nd edn. American Society of Agronomy, Madison, pp 45–55

Haile SG, Nair PKR, Nair VD (2008) Carbon storage of different soil-size fractions in Florida silvopastoral systems. J Environ Qual 37:1789–1797

Haile SG, Nair VD, Nair PKR (2010) Contribution of trees to carbon storage in soils of silvopastoral systems in Florida, USA. Glob Change Biol 16:427–438

Harner MJ, Stanfoord JA (2003) Differences in cottonwood growth between a losing and a gaining reach of an alluvial floodplain. Ecology 84:1453–1458

Hazlett PW, Gordon AM, Sibley PK, Buttle JM (2005) Stand carbon stocks and soil carbon and nitrogen storage for riparian and upland forests of boreal lakes in northeastern Ontario. For Ecol Manag 219:56–68

Hoover CM, Heath LS (2011) Potential gains in carbon storage on productive forestlands in northeastern USA through stocking management. Ecol Appl. doi:10.1890/10-0046.1

Houghton RA, Hackler JL (2000) Changes in terrestrial carbon storage in the United States. 1. The role of agriculture and forestry. Glob Ecol Biogeogr 9:125–144

IPCC (2007) Intergovernmental panel on climate change 2007. Synthesis report. Available at http://www.ipcc.ch/pdf/assessment-report/ar4/syr/ar4_syr.pdf. Accessed 25 Jan 2011

Jones RH, Lockaby BG, Somers GL (1996) Effects of micro-topography and disturbance on fine root dynamics in wetland forests of low-order stream floodplains. Am Midl Nat 136:57–71

Jose S (2009) Agroforestry for ecosystem services and environmental benefits: an overview. Agroforest Syst 76:1–10

Jose S, Gillespie AR, Seifert JR, Pope PE (2001) Comparison of minirhizotron and soil core methods for quantifying root biomass in a temperate alley cropping system. Agroforest Syst 52:161–168

Jose S, Gillespie AR, Pallardy SG (2004) Interspecific interactions in temperate agroforestry. Agroforest Syst 61:237–255

Kiley DK, Schneider RL (2005) Riparian roots through time, space and disturbance. Plant Soil 269:259–272

Kim D, Isenhart TM, Parkin TB, Schultz RC, Loynachan TE (2010) Methane flux in cropland and adjacent riparian buffers with different vegetation covers. J Environ Qual 39:97–105

Kort J, Turnock R (1999) Carbon reservoir and biomass in Canadian prairie shelterbelts. Agroforest Syst 44:175–186

Kumar S, Udawatta RP, Anderson SH (2010) Root length density and carbon content of agroforestry and grass buffers under grazed pasture systems in a Hapludalf. Agroforest Syst 80:85–96

Lal R, Kimble JM, Follett RF, Cole CV (1999) The potential for US crop land to sequester carbon and mitigate the greenhouse effect. Lewis Publ, Boca Raton, pp 55–82

Lee JL, Dodson R (1996) Potential carbon sequestration by afforestation of pasture in south-central United States. Agroclimatology 88:381–384

Lee KH, Jose S (2003) Soil respiration and microbial biomass in a pecan-cotton alley cropping system in southern USA. Agroforest Syst 58:45–54

Lowrance R, Leonard R, Sheridan J (1985) Managing riparian ecosystems to control nonpoint source pollution. J Soil Water Conserv 40:87–91

Lubowski RN, Vesterby M, Bucholtz S, Baez A, Roberts MJ (2006) Major uses of land in the United States, 2002. USDA economic research service. Econ Inf Bull EIB 14:54 p

Marquez CO, Cambardella CA, Isenhart TM, Schultz RC (1999) Assessing soils quality in a riparian buffer by testing organic matter fractions in central Iowa, USA. Agroforest Syst 44:133–140

Montagnini F, Nair PKR (2004) Carbon sequestration: an underexploited environmental benefit of agroforestry systems. Agroforest Syst 61:281–295

Morgan JA, Follett RF, Allen LH, Grosso SD, Derner JD, Dijkstra F, Franzluebbers A, Fry R, Paustian K, Schoeneberger MM (2010) Carbon sequestration in agricultural land of the United States. J Soil Water Conserv 65:6A–13A

Naiman RJ, Decamps H, McClain ME (2005) Structural patterns. In: Naiman RJ, Decamps H, McClain ME (eds.) Riparia: ecology, conservation, and management of streamside communities. Elsevier, New York, pp 79–123

Nair PKR, Nair VD (2003) Carbon storage in North American agroforestry systems. In: Kimble JM, Heath LS, Birdsey RA, Lal R (eds.) The potential of U.S. Forest soils to sequester carbon and mitigate the greenhouse effect. CRC Press, Boca Raton, pp 333–346

Nair PKR, Gordon AM, Mosquera-Losada MR (2008) Agroforestry. In: Jorgenson SE, Faith BD (eds.) Encyclopedia of ecology. Elsevier, Oxford, pp 101–110

Nair PKR, Kumar BM, Nair VD (2009) Agroforestry as a strategy for carbon sequestration. J Plant Nutr Soil Sci 172:10–23

Nair PKR, Nair VD, Kumar BM, Showalter JM (2010) Carbon sequestration in agroforestry systems. Adv Agron 108:237–307

NRCS (2007) Riparian forest buffer code 391. Natural resources conservation service, conservation practice standard. Available at http://efotg.sc.egov.usda.gov//references/public/NE/NE391.pdf. Accessed Jan 2011

Oelbermann M (2002) Linking carbon inputs to sustainable agriculture in Canadian and Costa Rican agroforestry systems. PhD Dissertation, Dept Land Res Sci, University of Guelph, Canada, 208 p

Oelbermann M, Voroney RP, Kass DCL, Schlönvoigt AM (2006a) Soil carbon and nitrogen dynamics using stable isotopes in 19- and 10-yr old tropical agroforestry systems. Geoderma 130:356–367

Oelbermann M, Voroney RP, Thevathasan NV, Gordon AM, Kass DCL, Schlönvoigt AM (2006b) Soil carbon dynamics and residue stabilization in a Costa Rican and southern Canadian alley cropping system. Agroforest Syst 68:27–36

Pallardy SG, Gibbins DE, Rhoads JL (2003) Biomass production by two-year-old poplar clones on floodplain sites in the lower Midwest, USA. Agroforest Syst 59:21–26

Paudel BR, Udawatta RP, Kremer RJ, Anderson SH (2011) Agroforestry and grass buffer effects on soil quality parameters for grazed pasture and row-crop systems. Applied Soil Ecol 48:132–152

Peichl M, Thevathasan NV, Gordon AM, Huss J, Abohassan RA (2006) Carbon sequestration potentials in temperate tree-based intercropping systems. southern Ontario, Canada. Agroforest Syst 66:243–257

Piedade MTF, Worbes W, Junk WJ (2001) Geological controls of elemental fluxes in communities of higher plants in Amazonian floodplanes. In: McClain ME, Victoria RL, Richey JE (eds.) The biogeochemistry of Amazonian Basin. Oxford University Press, New York, pp 209–234

Qiu Z, Prato T (2001) Physical determinants of economic value of riparian buffers in an agricultural watershed. J Am Water Res Assoc 37:295–303

Reed D (2007) Economic and social benefits of soil carbon management: policy implications and recommendations. In: Kimble JM (ed.) Soil carbon management. CRC Press, New York, pp 13–43

Rhoades CC, Nissen TM, Kettler JS (1998) Soil nitrogen dynamics in alley cropping and no-till systems on ultisols of Georgia Piedmont, USA. Agroforest Syst 39:31–44

Sanchez PA (2000) Linking climate change research with food security and poverty reduction in the tropics. Agric Ecosyst Environ 82:371–383

Sauer TJ, Cambardella CA, Brandle JR (2007) Soil carbon and litter dynamics in a red cedar-scotch pine shelterbelt. Agroforest Syst 71:163–174

Schoeneberger MM (2009) Agroforestry: working trees for sequestering carbon on agricultural lands. Agroforest Syst 75:27–37

Schroeder P (1994) Carbon storage benefits of agroforestry systems. Agroforest Syst 27:89–97
Schultz RC, Isenhart TM, Colletti JP, Simpkins WW, Udawatta RP, Schultz PL (2009) Riparian and upland buffer practices. In: Garrett HE (ed.) North American agroforestry, an integrated science and practice, 2nd edn. American Society of Agronomy, Madison, pp 163–218
Sharrow SH, Ismail S (2004) Carbon and nitrogen storage in agroforests, tree plantations, and pastures in western Oregon, USA. Agroforest Syst 60:123–130
Sharrow SH, Brauer D, Clason TR (2009) Silvopastural practices. In: Garrett HE (ed.) North American agroforestry: An integrated science and practice, 2nd edn. American Society of Agronomy, Madison, pp 105–131
Smith P, Martono D, Cai Z, Gwary D, Janzen H, Kumar P, McCarl B, Olge S, O'Mara F, Rice C, Scholes B, Sirotenko O (2007) Agriculture. In: Metz B, Davidson OR, Bosch PR, Dave R, Meyer LA (eds.) Climate change 2007: mitigation. Cambridge University Press, Cambridge, UK, pp 497–540
Steinbeiss S, Bessler H, Engels C, Temperton VM, Buchmanns N, Roscher C, Kreutziger Y, Baade J, Habekost M, Gleixner G (2008) Plant diversity positively affects short-term soil carbon storage in experimental grasslands. Glob Change Biol 14:2937–2949
Thevathasan NV, Gordon AM (1997) Poplar leaf biomass distribution and nitrogen dynamics in a poplar-barley intercropped system in southern Ontario, Canada. Agroforest Syst 37:79–90
Thevathasan NV, Gordon AM (2004) Ecology of tree intercropping systems in the north temperate region: experiences from southern Ontario, Canada. Agroforest Syst 61:257–268
Tufekcioglu A, Raich JW, Isenhart TM, Schultz RC (1999) Fine root dynamics, coarse root biomass, root distribution, and soils respiration in a multispecies riparian buffer in Central Iowa, USA. Agroforest Syst 44:163–174
Tufekcioglu A, Raich JW, Isenhart TM, Schultz RC (2001) Soil respiration within riparian buffers and adjacent crop fields. Plant Soil 229:117–124
Tufekcioglu A, Raich JW, Isenhart TM, Schultz RC (2003) Biomass, carbon and nitrogen dynamics of multi-species riparian buffers within an agricultural watershed in Iowa, USA. Agroforest Syst 57:187–198
Turnock B (2001) The carbon sequestration potential of prairie and their contribution to a national greenhouse gas mitigation strategy. In: Schroeder W, Kort JK (eds.) Temperate agroforestry: adaptive and mitigation role in a changing physical and socioeconomic climate. Proceedings 7th Biennial conference. on agroforestry in North America and the 6th annual conference plains and prairie forestry association, Saskatchewan, 13–15 Aug 2001, pp 27–33
Udawatta RP, Godsey LD (2010) Agroforestry comes of age: putting science into practice. Agroforest Syst 79:1–4
Udawatta RP, Nygren PO, Garrett HE (2005) Growth of three oak species during establishment in an agroforestry practice for watershed protection. Can J For Res 35:602–609
Udawatta RP, Kremer RJ, Adamson BW, Anderson SH (2008) Variations in soil aggregate stability and enzyme activities in a temperate agroforestry practice. Appl Soil Ecol 39:153–160
Udawatta RP, Kremer RJ, Garrett HE, Anderson SH (2009) Soil enzyme activities and physical properties in a watershed managed under agroforestry and row-crop system. Agric Ecosyst Environ 131:98–104
USDA NRCS (2003) USDA natural resources conservation service. Annual NRI-Land use. Available at http://www.nrcs.usda.gov/technical/NRI. Accessed 28 May 2010
USDA NASS (2008) Crop production. Historical track records. April 2008. USDA national agricultural statistics service, Washington DC. Available at http://usda.mannlib.cornell.edu/usda/nass/htrcp//2000s/2008/htrcp-04-30-2008.pdf. Accessed Feb 2011
USEPA (US Environmental Protection Agency) (2010) Water quality assessment and total maximum daily loads information: National summery of state information. Available at http://iaspub.epa.gov/waters10/attains_nation_cy.control. Accessed 28 May 2010
Watson RT, Noble IR, Bolin B, Ravindranathan NR, Verardo DJ, Dokken DJ (2000) *IPCC special report on landuse, land use change, and forestry*. Available at http://www.rida.no/climate/ipcc/land_use/. Accessed 28 May 2010
Welsh DJ (1991) Riparian forest buffers: function and design for protection and enhancement of water resources. NA-PR-07-91, USDA forest service, Radnor, PA
Young A (1997) Agroforestry for soil management. CAB International, Wallingford, UK, 297 p

Carbon Sequestration in European Agroforestry Systems

M.R. Mosquera-Losada, Dirk Freese, and A. Rigueiro-Rodríguez

Abstract Agroforestry systems (AFS) are recommended for Europe through the European Rural Development Council regulation 1698/2005, in recognition of their role in reducing carbon (C) emissions and promoting C sequestration which would help to fulfil the Kyoto Protocol requirements. These systems have been found to be a good tool to reduce fire risk and C release in southern European countries. The implementation of AFS could also reduce C release to atmosphere because of the value given to non-timber products, thereby reducing chances for clear cutting of trees. Furthermore, the tree component in AFS will add C into the soil through litterfall and root decomposition, which takes place at deeper soil layers than under agronomic crops or pasture. Tree management practices such as regulating tree density and planting arrangement will influence the C sequestered in the system. Compared with the tree components, the understory components of AFS have less impact on the total C sequestration. The higher inputs of residues generated by the trees in AFS than in tree-less systems may cause high soil C sequestration potential, but soil C increase depends on the incorporation and mineralization of C in the soil, which are affected by understory crop management practices.

Keywords Alley cropping • Fire protection • Kyoto protocol • Silvopasture • Soil organic matter (SOM)

M.R. Mosquera-Losada (✉) • A. Rigueiro-Rodríguez
Crop Production Department, Escola Politécnica Superior, Universidad de Santiago de Compostela, Campus de Lugo, 27002 Lugo, Spain
e-mail: mrosa.mosquera.losada@usc.es; antonio.rigueiro@usc.es

D. Freese
Soil Protection and Recultivation, Brandenburg University of Technology, Konrad-Wachsmann-Allee 6, D-03046 Cottbus, Germany
e-mail: freese@tu-cottbus.de

B.M. Kumar and P.K.R. Nair (eds.), *Carbon Sequestration Potential of Agroforestry Systems: Opportunities and Challenges*, Advances in Agroforestry 8,
DOI 10.1007/978-94-007-1630-8_3,

Introduction

During the last century, the climate in Europe has changed more than in other areas of the world (IPCC 2007). Compared to the pre-industrial era, when the mean annual temperature increased by 0.8°C globally, it increased by 1.2°C in Europe. Based on theoretical models, a further increase of 1.0–5.5°C is expected by the end of the twenty-first century (Christensen et al. 2007). The increase in temperature has been most apparent in hilly areas such as the Alps, which tend to have high biodiversity and where temperature increased by 2°C during the twentieth century (EEA 2009a). This is twice the average temperature increase for the northern hemisphere. In addition, the quantity and distribution of precipitation have also changed in Europe during the twentieth century. Although there has been a 20% decrease in rainfall in southern Europe, there has been a 10–40% increase in rainfall in northern Europe. Furthermore, an increase in the frequency of extreme weather events is predicted across the European continent (EEA 2008).

Climate change may lead to an increase in the incidence of wildfire outbreaks, a decrease in biodiversity, and an increase in carbon dioxide (CO_2) emissions. Wildfires are a serious threat to forest ecosystems in Europe (Rigueiro-Rodríguez et al. 2009a), and represent a major source of CO_2 emissions. Any increase in temperature will aggravate the danger of forest fires by increasing the incidence of fire events, the area burnt, and the duration of fire seasons, especially in southern and central Europe (EEA 2008). Moreover, climate change in Europe may modify biodiversity through habitat loss and cause changes in dispersal capacity, phenological characteristics, life cycles, and food sources of native species. Climate change may also provoke the decoupling of predator-prey relationships, new invasions, or the spread of already established invasive alien species (EEA 2009b). It would also lead to a decline in soil organic carbon (C) stocks and an increase in CO_2 emission from soils. Soils may become more susceptible to erosion, especially in the Mediterranean areas where annual soil losses may reach 200 Mg ha^{-1} (Correal et al. 2009). Soil degradation is already intense in parts of the Mediterranean and central Eastern Europe and may contribute to desertification (EEA 2008). Agroforestry systems (AFS) offer solutions to some of these climate change related ecosystem management problems. For example, AFS have proved to be an excellent fire prevention technique in many parts of southern Europe such as France (Etienne 1996; Etienne et al. 1996; Rigolot and Etienne 1996), Greece (Papanastasis et al. 2009), and Spain (Robles et al. 2009; Rigueiro-Rodríguez et al. 2009b). Agroforestry practices are considered good land management tools to enhance biodiversity (Rois-Díaz et al. 2006; Rigueiro-Rodríguez et al. 2011b) and augment C sequestration, compared with tree-less systems worldwide (Nair et al. 2008, 2009).

The Kyoto Protocol establishes that land use, land use change and forestry (LULUCF) activities such as afforestation, reforestation, and deforestation (Article 3.3), and forest land management, cropland management, grazing land management, and revegetation (Article 3.4) can be used to meet the greenhouse gas (GHG) emission reduction goal (UN 1998). Burley et al. (2007) indicated that forest offset projects can be based on two approaches, namely, (a) the absorption of GHG by new vegetation (i.e., sink creation and sink enhancement), and (b) displaced emissions by existing vegetation (i.e., fire risk reduction and avoided deforestation). Emission from timber harvesting, which also negatively affects soil organic matter (SOM), could be reduced by the adoption of agroforestry systems that provide benefits other than timber from forest areas. Austria, Belgium, Denmark, Finland, Ireland, Italy, Luxembourg, the Netherlands, Portugal, and Spain plan to fulfil their assigned C emissions by using the Kyoto mechanisms described in Articles 3.3 and 3.4. However, Spain and Italy are among the EU countries with the greatest focus on increasing atmospheric CO_2 removals by enhancing C sink activities. Therefore, the implementation of AFS in these two countries, aimed at reducing CO_2 in the atmosphere through the two LULUCF activities described by the Kyoto Protocol, should be greater than in other European countries (EEA 2009c).

Considerable efforts in land use change for the reduction of GHG emissions have been carried out in Europe. More than one million hectares of forests were planted between 1994 and 1999 in Europe (Rois-Díaz et al. 2006). According to current targets, it is expected that more than 650,000 ha of agricultural land and about 240,000 ha of non-agricultural land will be afforested in Europe during the period 2007–2013 (EU 2009). This process will involve more than 12,000 landowners. The most recent European Rural Development Report estimates that AFS will cover 60,000 ha of agricultural lands representing 3,000 landowners during the period 2007–2013 (EU 2009) as a result of the council regulation 1698/2005 (EU 2005).

Agroforestry and Carbon Sequestration

The C sequestration potential of AFS is based on live components growing up within the system including the soil, but should also include activities such as forest fire prevention and other multifunctional outputs from the system (Rigueiro-Rodríguez et al. 2009b). The potential of C sequestration in AFS is dependent on the tree component (Nair et al. 2009). Tree presence would increase C sequestration per unit of land due to the C sequestered by the tree itself, the inputs of residues (leaves and branches) it makes on the soil, and the incorporation of roots into the soil. Trees use a greater volume of soil to build up SOM than herbaceous crops, as they are able to explore soils farther from the tree trunk and to a greater depth, assuming small tree density is used (Moreno et al. 2005). The greater soil volume explored by tree roots would enhance belowground organic matter depositions (Howlett et al. 2011). However, understory species may also be positively or negatively affected by the tree presence. The symbiotic or competitive relationship of these components (i.e., tree and understory) depends on specific edapho-climatic conditions (Rigueiro-Rodríguez et al. 2009a; Mosquera-Losada et al. 2010a). Conditions such as adequate water regime, optimal temperatures, and soil nutrient availability would promote tree growth (López-Díaz et al. 2010), but in areas with strong water deficits, usually development of pasture (or other understory species used in the AFS) is reduced due to the presence of trees.

Agroforestry as a land use option has great potential for C sequestration in Europe, as it allows for the sequestration of more C per unit of land, compared with tree-less agronomic systems (Matos et al. 2010a). Agroforestry also results in higher annual economic returns per unit of land through the whole life cycle than in exclusive forestry systems where the revenue is generally only realised at final harvest. These returns could be further increased if appropriate land management practices mainly regulating tree density and distribution are adopted (Sibbald 1996; Fernández-Núñez et al. 2007). The role of AFS in the reduction of C emissions derives from the prevention of forest fires in Mediterranean Europe, as silvopasture agroforestry practices reduce the understory woody biomass (Etienne et al. 1996; Rigueiro-Rodríguez et al. 2009b, 2010). Most AFS have also been shown to reduce soil erosion, and improve nutrient cycling, water availability for crops, soil faunal activities, and soil fertility, while at the same time sustaining high levels of crop production (Grünewald et al. 2007; Quinkenstein et al. 2009; Rigueiro-Rodríguez et al. 2009a).

Some of the most important options to increase C sequestration are those dealing with LULUCF measures. Currently, most European forests are relatively young and they act as a C sink. Growing forests sequester C, but when they reach maturity, the C annually sequestered is reduced. For this reason, forested land conservation that avoids total clear felling should be better at reducing C emissions in the future, not only because of C exported in the harvested trees, but also due to C soil emissions, once the trees are harvested (Nair et al. 2009). However, Dresner et al. (2007) highlighted that if cut timber is worth more than trees still standing in the forest, there is no incentive for farmers to protect the forest. As such, deforestation is likely to occur, regardless of the wider impacts of this such as C emissions. Nonetheless, if the agronomic component of an AFS is valuable for farmers, this would be an additional reason to prevent deforestation and thus reduce CO_2 emissions (Dresner et al. 2007).

Several types of agroforestry practices are currently implemented in Europe. Silvoarable and silvopasture agroforestry practices are the most prevalent in Europe in terms of the area under those practices compared with other agroforestry practices (Eichhorn et al. 2006; Mosquera-Losada et al. 2009). They are mostly carried out in Spain and Portugal, but also in Germany, France, Italy and the UK (Dupraz et al. 2005; Grünewald et al. 2007; Mosquera-Losada et al. 2010a; Quinkenstein et al. 2009).

The tree component of an AFS may be more efficient at CO_2 utilisation from the atmosphere and may have higher C returns to the soil through their litter than herbaceous crops (Gordon et al. 2006). One year after the implementation of an agroforestry system (Böhm et al. (2010), the content of organic C in soil under tree hedgerows was significantly higher as compared to field alleys (Fig. 1) in Germany, due to the higher root development in the hedgerows compared with the field alleys.

In agroforestry systems, C is located in five main pools, namely, aboveground plant biomass (tree and understory), plant roots (tree and understory), litter, microbial, and soil C. These pools interact with each other via different pathways of transformation and translocation, e.g., plants absorbing CO_2 from the atmosphere during

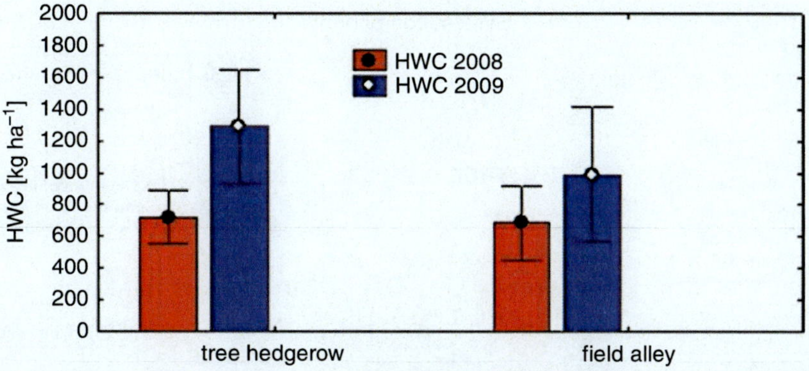

Fig. 1 Hot water extractable organic carbon in the surface (0–30 cm) soil, 1 year (HWC 2008) and 2 years (HWC 2009) after establishing an alley cropping system, in a mining reclamation landscape, Lower Lusatia, north-eastern Germany (Source: Adapted from Böhm et al. 2010)

photosynthesis. Some CO_2 is released back to the atmosphere in the process of plant respiration. Litter falling from plants and dead roots from plant material are decomposed into soil C. Some of the soil C is taken up by microbes and stored, and some becomes mineralised. Soil stores C, but as a result of the mineralisation and root respiration, part of that C is released back to the atmosphere. Furthermore, biodiversity enhancement by AFS facilitate a better nutrient use and therefore increases C sequestration compared with tree-less agronomic systems (Howlett et al. 2011; Rigueiro-Rodríguez et al. 2011b).

Measurement of C sequestration following land use changes from tree-less agriculture to forestland requires the evaluation of the baseline C stocks as well as the nature of the tree component and the modifications the tree causes to the understory and in the soil compartment. If silvopasture agroforestry is carried out, then the animal component and the emission of methane and nitrous oxide gases should also be taken into account (IPCC 2007). The main components and their GHG balance in a silvopasture agroforestry system including grazing animals are presented in Fig. 2.

Tree Component

Land use change through afforestation or reforestation should increase C sequestration per unit of land and the rate of C sequestered by trees within a system will depend on tree species, age, and density (Quinkenstein et al. 2009), besides the edapho-climatic conditions, management, fertilization, and land clearing, among others. Carbon sequestration by an individual tree can be estimated by allometric equations based on the tree diameter that have been recently developed in Spain (Montero et al. 2005) and Europe (Zianis et al. 2005). In their studies, 13 and 24 conifers and 15 and 31 broadleaf trees species were used to estimate the C sequestered in

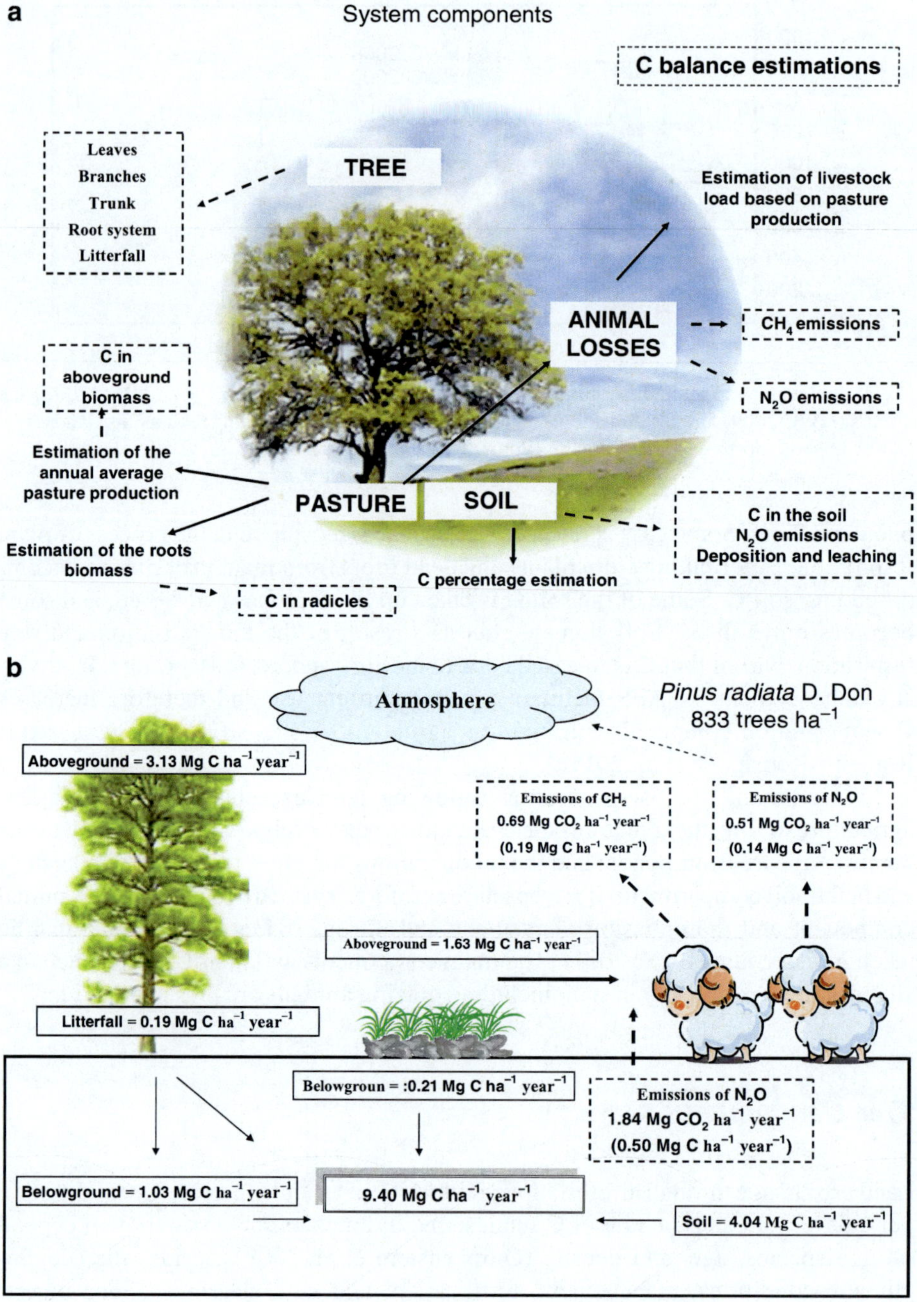

Fig. 2 Carbon pools in a silvopasture system including GHG emissions: (**a**) A schematic diagram showing the different compartments. (**b**) An example of the estimated quantities in each compartments in a 11 years-old *Pinus radiata* D. Don stand in Galicia, NW Spain (Source: Adapted from Fernández-Núñez et al. 2010)

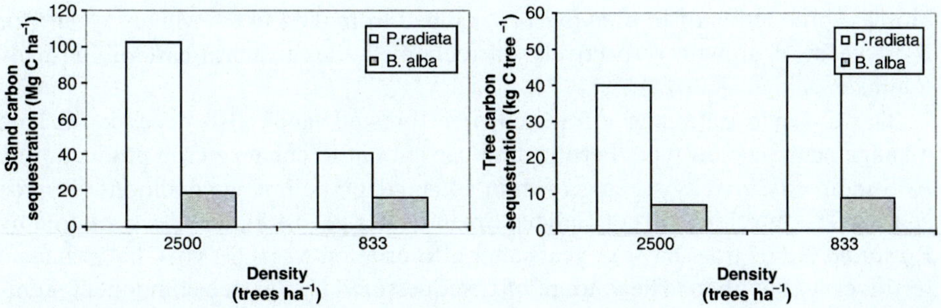

Fig. 3 Carbon stock in the tree stand and individual tree of *Pinus radiata* D. Don and *Betula alba* L. Eleven years after establishment at two densities (833 and 2,500 trees/ha^{-1}) in Galicia, NW Spain

aboveground biomass and in roots, respectively. This was carried out for species from the Mediterranean, mountainous, and Atlantic biogeographic regions of Europe. However, most of the trees used to develop the equations were in mature, dense stands, and therefore, more research is needed to understand how C is sequestered in younger stands (Knopka et al. 2010) and growing at lower densities such as in AFS.

The growth rate of tree species is a significant factor in promoting C sequestration. Annual estimates of C sequestered by tree biomass of *Eucalyptus globulus* Labill, *Pinus pinaster* Ait., *Pinus radiata* D. Don and *Castanea sativa* Mill. in Spain were 5.14, 1.58, 1.11, and 0.52 Mg Cha^{-1}, respectively (Pardos 2010). Differences in growth rates explain why, after 10 years, *P. radiata*, a species with a high growth rate, sequestered eight times more C per tree than *Betula alba* L. at densities of 833 and 2,500 trees ha^{-1} in Spain (Fig. 3). Similarly, species like poplar or eucalyptus were able to sequester C faster than species such as *P. radiata*, *B. alba*, *P. pinaster*. However, these three species grew and sequestered C faster than the other common silvopastoral tree species such as *Pinus sylvestris* L., *Quercus petraea* L., *Quercus robur* L., or *Fagus sylvatica* L. (Pardos 2010). Gordon et al. (2006) highlighted the importance of using fast growing tree species in silvopastoral systems to reduce C emissions in Canada. They estimated that net C sequestration of a poplar-based silvopastoral system was almost three times more than that reached by a monoculture pasture system. However, if trees grow quickly, C sequestered for a given period of time is reduced as trees will be harvested earlier (Fernández-Núñez et al. 2010). The time required for C sequestration to occur is longer for slow growing species than for fast growing species. Therefore, once harvested, the fast growing species emit C into the atmosphere earlier than slow growing species, mainly from SOM mineralization. Moreover, sawn timber production is usually associated with slow growing species, which are retained for longer time than pulp and paper from fast growing species such as *Eucalyptus* spp.

The production of *Robinia pseudoacacia* L. in an alley cropping system has received considerable interest in Germany as an alternative to agricultural crops as well as an additional wood source, while simultaneously acting as a potential C sink to counterbalance greenhouse gases emissions. Average aboveground biomass production of *R. pseudoacacia* ranged from 0.04 to 9.5 Mg ha^{-1} $year^{-1}$ for 1–14 years of growth, respectively on reclaimed sites in north-eastern Germany (Quinkenstein et al. 2011).

The C sequestration of afforested or reforested lands also depends on land management and soil type. Fertilization carried out to enhance crop production in AFS indirectly increases tree growth in some edapho-climatic conditions (Dupraz et al. 2005). In acidic soils of Galicia, Spain (water pH = 4.5), the C sequestered by *P. radiata* (1,667 trees ha^{-1}) 11 years after afforestation was 4.09 Mg C ha^{-1} when no fertilizer was applied. The amount of C sequestered by the tree component significantly increased to 7 Mg C ha^{-1} when sewage sludge was used as fertilizer in the same soil. However, these values were lower than those reported for agricultural lands (initial soil water pH = 6.9), which were afforested at high density (2,500 trees ha^{-1}) (Fig. 3; Fernández-Núñez et al. 2010). Soil fertility improvements usually increase growth rates and symbiosis. However, facilitation between the tree and the understory should be promoted in the early tree ages in order to enhance resource use and increase C sequestration (Mosquera-Losada et al. 2006, 2011b). The use of legumes such as clover (*Trifolium* spp.) in the sown mixture, increased tree growth and was found to promote symbiosis between *P. radiata* and understory (López-Díaz et al. 2010). However, the increase in ryegrass (*Lolium perenne* L.) density during the year of plantation establishment reduced *P. radiata* growth due to competition between the tree and ryegrass for soil resources (Mosquera-Losada et al. 2011b).

Tree density is another factor that affects C sequestration. Fernández-Núñez et al. (2010) reported from Galicia, Spain, that land that had previously been under agriculture when afforested with *P. radiata* at 833 or 2,500 trees ha^{-1} was able to sequester 40.8 and 102.4 Mg C ha^{-1} 11 years after plantation in tree roots and aboveground biomass, respectively, despite the fact that C sequestered per tree was higher at a low density (48 and 40 kg C $tree^{-1}$, respectively: Fig. 3). Similar results were also found for *B. alba* planted at these densities in the same area.

In the Atlantic biogeographic region of Europe, tree stands were established at higher tree densities than in the Mediterranean dehesa area to promote timber production (Serrada et al. 2008). Due to the intraspecific competition in the high density stands, tree roots may not spread far away from the tree trunks compared with low density stands. There have been few published studies where the differences in root system profiles with respect to the distance from the tree for low versus high density stands have been measured. In the Mediterranean environments, Moreno et al. (2005) reported that most fine roots of *Quercus ilex* L. trees were below 80 cm depth, while herbaceous plant roots were mainly located in the top 30 cm soil layer. Drought conditions could have a great effect on tree root distribution within the soil profiles of Mediterranean systems. The same effect could be simulated by competition within the herbaceous layer in more northern European countries if AFS with low tree densities were implemented. If tree roots are located below the herbaceous understory rhizosphere, then competition for soil resources between trees and herbaceous plants is reduced. Implementation of agroforestry could increase the volume of soil explored by roots (the upper part of the soil explored by the herbaceous component and the lower part by the tree component) in low density stands compared with tree-less pastures. The amount of fine roots that are considered to be the main source of organic matter within a soil C pool (Dresner et al. 2007) would also increase.

Tree C sequestration also depends on the species. Evergreen trees retain C in the leaves for longer period of time than deciduous tree species, which cause regular inputs of organic matter into the soil, apart from the roots. Evergreen tree litterfall is usually low until canopy closure. Afterwards, the relatively low understory light levels may cause an accumulation of litter on the forest floor. Density affects the dynamics of the tree litter inputs into the soil. A dense *P. radiata* canopy caused an accumulation of a thick litter layer of several centimetres above the soil a few years after canopy closure, which prevented herbaceous plant establishment and reduced biodiversity and soil C sequestration potential. Litter biomass also depends on tree density, which was higher in high density forests (6.25 Mg ha^{-1} at 2,500 trees ha^{-1}) than in low density stands (4.26 Mg ha^{-1} at 833 trees ha^{-1}) in a *P. radiata*-afforested land 11 years after planting (Fernández-Núñez et al. 2010). The thick litter layer could emit large quantities of C once the forest stand is harvested. However, no accumulation of litterfall on the soil was observed in a silvopastoral system established with *B. alba* at 2,500 trees ha^{-1} or 833 trees ha^{-1} due to the low growth rate of birch as compared with radiata pine. Higher soil temperatures in birch stand, comparable with *P. radiata,* increased birch litter decomposition, promoting soil C sequestration (Howlett et al. 2011).

Understory Component

Compared to the tree and the soil C pools, the amount of C sequestered in the understory component of the AFS is relatively small (Fernández-Núñez et al. 2010). In European AFS, the understory component may be a crop (e.g., a cereal or leguminous crop) in the silvoarable systems or herbaceous or woody plants in the silvopastoral systems. Arable systems have lower C sequestration potential than herbaceous pasture or understory woody plants and involve annual crops that are usually harvested within a year of sowing, and the biomass is exported from the system. Crop management practices such as plowing, liming, and fertilization may cause either soil C increase or losses. Improvement of soil fertility increases the growth of AFS components and therefore soil inputs of C. However, management activities may also result in better aeration, increased pH, and enhanced soil fertility, promoting microbial activity and organic matter mineralisation, in turn, leading to lower SOM levels (Reijneveld et al. 2010).

Perennial grasslands and shrublands may store C within their tissues for a longer period of time than arable crops. The large area of the European Union allocated to grasslands in different biogeographic regions (33% and 25% of the Atlantic and continental biogeographic regions of Europe; EEA 2006) offers a high potential for C sequestration. This potential, however, is dependent on the edapho-climatic conditions and land management practices adopted (Follet et al. 2001; Schanabel et al. 2001). The input of organic matter to grassland soils is very important (Sanderson and Wätzold 2010) and would increase the SOM content. Mature pasturelands, however, show no net annual C uptake when all sources and sinks are considered (Suyker and Verma 2001; Gianelle et al. 2004). Follet et al. (2001) concluded that improved grassland management could enable C sequestration to continue for 25–50 years until a new equilibrium of soil C content is reached. After that, the improved grasslands would no longer serve as C sinks. Even though grasslands may sequester C, grazing by livestock animals may result in CH_4 or N_2O emissions (IPCC 2007). When the animal stocking rate is adjusted to the production of grasslands, the C losses with GHGs are offset by the C sequestered (Fernández-Núñez et al. 2010).

Understory shrubs sequester more C than herbaceous plants. However, the risk of C emissions caused by fires associated with forestlands is increased by these shrubs, making the presence of woody vegetation understory very hazardous in the Mediterranean countries of Europe and in the southern Atlantic biogeographic region of Europe, where summers are too dry and fire risk is high (Rigueiro-Rodríguez et al. 2009b). Prevention of forest fires mitigates C emissions (Burley et al. 2007). Agroforestry practices could be successfully implemented to reduce the emissions of C caused by fires. For instance, shrub grazing by goats in silvopastoral systems reduces the amount of combustible vegetation in the understory and encourages a less flammable herbaceous layer (Rigueiro-Rodríguez et al. 2011a). Understory vegetation transformation from shrubs to a grass is thus promoted by grazing of shrubs by animals as well as by soil nutrient cycling through animal faeces and urine deposition (Rigueiro-Rodríguez et al. 2009b).

Soil Component

The soil represents the most important pool of C storage in terrestrial ecosystems, accounting for about 75% of total stored C (Lal 2005; Dresner et al. 2007). Soil C sequestration depends on edapho-climatic conditions, which may increase or reduce the organic matter inputs (i.e., the quantity of plant residues), incorporation of organic matter into the soil, and organic matter mineralisation (Nieder et al. 2003). Soil properties such as clay content determine the extent of C enrichment in humus. Organic matter inputs usually create a C gradient from the surface to the lower layers of the soil worldwide (Fig. 4; Howlett et al. 2011).

Temperature and humidity are the main drivers of SOM production, incorporation, and mineralisation (Theng et al. 1989). If temperature and humidity are optimal for aboveground biomass production as in the Atlantic climate, the inputs of organic matter into the soil are greater than that in less favourable climatic conditions such as in the Mediterranean climate. For this reason, the higher potential productivity of crops in the Spanish Atlantic region is an important indicator of higher SOM as compared to the Spanish Mediterranean region (i.e., 3 and 0.4 Mg ha^{-1} $year^{-1}$ for

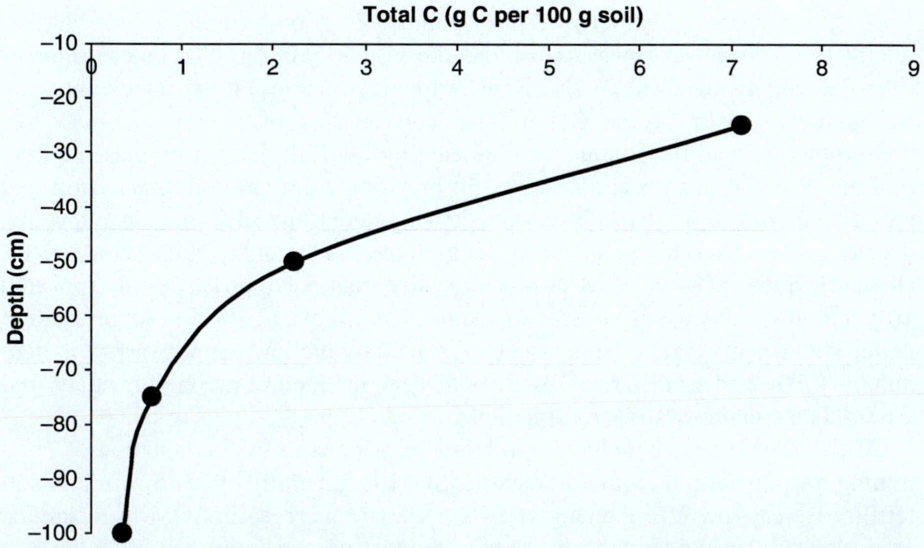

Fig. 4 Total soil carbon concentration by depth in an afforested and non-fertilized soil in Galicia, NW Spain

Atlantic and Mediterranean regions, respectively: Pardos 2010). In the dehesas located in the Mediterranean area, the presence of mature trees and, therefore, the rate of incorporation of their residues are associated with higher SOM levels below the tree than away from the tree in AFS established under trees without canopy closure (Moreno and Obrador 2007). In some cases, the degree of incorporation of plant residues into the soil may be restricted by high humidity and low temperature. In *P. radiata* stands, the closure of tree canopies caused an annual accumulation of about 7 Mg ha^{-1} of litterfall in Galicia (Fernández-Núñez et al. 2010).

Roots are also an important part of the C balance in terrestrial ecosystems because they transfer large amounts of C into the soil. More than half of the C assimilated by the plant is transported belowground via root growth and turnover, root exudates (of organic substances) and litter deposition, and roots may contribute up to 33% to C sequestered in ecosystems (Fernández-Núñez et al. 2010). The dynamics of growth, decay, and root turnover are some of the least understood aspects of below-ground interactions in agroforestry (Nair et al. 1999). There is much information on C sequestration in the topsoil layer of 0–20 cm. However, information on deeper soil layers, where most of the tree roots occur, is lacking in most environments, but some studies have been carried out in the Spanish dehesa agrosilvopastoral system (Moreno and Obrador 2007). Roots of trees and grass or crops have different root length and depth profiles. Tree roots are longer and deeper in soil than grass or crop roots, and in soils under trees, a considerable amount of C is stored below the plow layer (50 cm). This C is also better protected from disturbance, which leads to longer residence time in the soil. Most of the root biomass of annual crops and grasses consists of fine roots (diameter < 2 mm). Fine roots of both trees and crops have a relatively fast turnover (measured in days to weeks), but lignified coarse roots of trees decompose much more slowly once trees are harvested and may contribute substantially to belowground C pools (Vanlauwe et al. 1996).

Carbon inputs to the soil are also affected by litterfall. Higher biomass production per tree and per hectare obtained in previously agricultural lands afforested with *P. radiata* and *B. alba* at high stem density increased soil C more than low tree density 5 years after the establishment of both trees (Fernández-Núñez et al. 2010). However, differences in SOM between density treatments or species disappeared 10 years after afforestation, probably due to the lack of litterfall incorporation under high density stands. This can be explained by the low temperatures and high humidity experiences differentially by both systems. Incorporation of residues into the soil is the first step to increasing SOM.

Alley cropping systems have also come into focus in the reclamation of post-mining areas where the initial content of SOM is generally close to zero and soil fertility is very low (Nii-Annang et al. 2009). The increase in SOM in reclaimed areas depends on the amount of biomass production and return to soil as well as mechanisms for C protection and retention. Due to its high potential for litterfall production and nitrogen fixation, *R. pseudoacacia* improves soil physical, chemical, and biological properties by increasing SOM, thereby converting mine spoils into productive and sustainable soils (Grünewald et al. 2007).

High amounts of litterfall increase fire risk in European Mediterranean areas (Delabraze 1986) and, therefore, the risk of C emissions to the atmosphere. Implementation of silvopasture has been shown to reduce fire risk through the enhancement of litter incorporation into the soil as nitrogen is added with the urine of the animals and C/N relationship is reduced (Etienne et al. 1996; Rigolot and Etienne 1996).

It is well known that soil management activities such as plowing or fertilization may reduce or increase SOM content. Matos et al. (2010a, b) investigated the effect of conversion from silvopasture to arable land and reported lower contents of total organic carbon (TOC) and total nitrogen (TN) in arable soils than silvopasture. The composition and distribution of SOM also differed between these two systems. The light fraction C content declined with depth in silvopasture system, while there were no such depth-related differences in arable system. This can be attributed to tillage in arable systems, which leads to the disturbance of upper soil layers causing an increase in mineralization rates, CO_2 emissions from soils, and the reduction of soil C. Soil management through fertilization also affects soil C storage. Mosquera-Losada et al. (2010b) reported that the addition of sewage sludge (pH around 7) in acidic soils (water pH = 4.5) increased SOM content through the input of organic matter as well as calcium via the sewage sludge (Fig. 5). The SOM content was not modified when mineral nitrogen was added, as incorporation of organic residues through the improvement of soil pH was not promoted. In a *Populus canadensis* Moench silvopastoral system developed on a Galician (Spain) acid soils with pH around 5.5, the SOM content in winter was related to pasture production in the preceding autumn ($r^2 = 0.93$; % SOM = 0.48 autumn production [Mg ha^{-1}] +8.87 $p < 0.05$), which

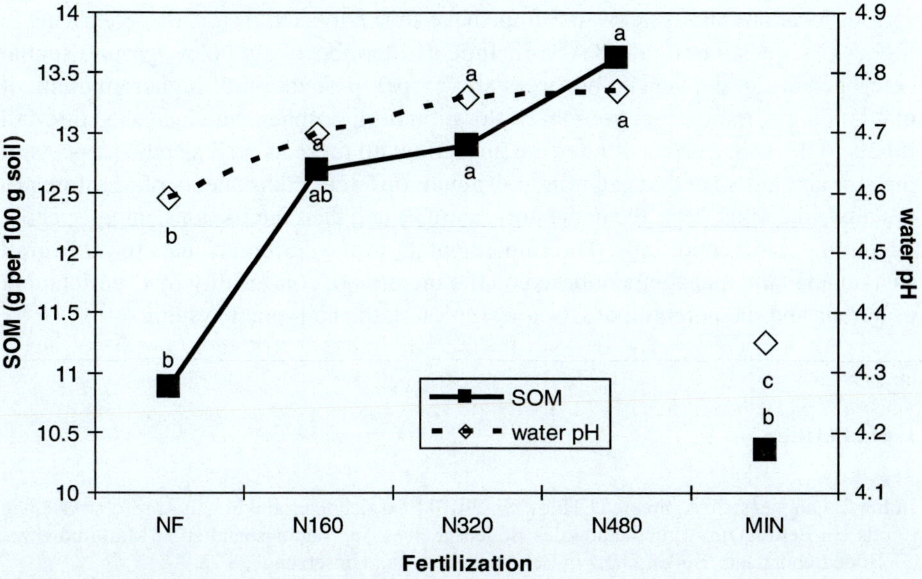

Fig. 5 Soil organic matter (SOM) and pH $_{(water)}$ under different fertilization treatments of *Pinus radiata* D. Don in Galicia. NF: no fertilization; N160, N320 and N480 refer to160, 320, and 480 kg N ha^{-1}, respectively; MIN: inputs of a mineral compound fertilizer 500 kg of 8N: 24P_2O_5: 16 K_2O. The letters on graph indicate significance of differences between treatments at p=0.05 according to Duncan´s multiple range test (Source: López-Díaz et al. 2007; Rigueiro-Rodríguez et al. 2011a)

suggested that organic matter was incorporated but not mineralized (Mosquera-Losada et al. 2011a). Therefore, SOM seems to have increased when herbaceous autumn production was high, causing an increase in organic matter inputs into the soil.

It has been proposed that C stored in the soil could be linked to different soil-size fractions (Lal 2005). However, there have been only very few studies evaluating C storage in different soil-size fractions in treeless versus AFS. Carbon associated with macroaggregates (250–2,000 μm), microaggregates (53–250 μm) and silt clay (<53 μm) can have mean residence time of 1–10, 1–25, and 100–1,000 years, respectively (Parton et al. 1987; Schimel et al. 1994). One study carried out in Galicia, Spain, showed that the broadleaf *B. alba* sequestered more C in the 250–2,000 μm size class as compared to soils under the conifer *P. radiata*. However, pastures had more C than pine silvopasture in soils with finer particle sizes fractions of less than 250 μm (Howlett et al. 2011).

Conclusion

Agroforestry systems have great potential to enhance C sequestration compared with tree-less agronomic systems, and therefore their implementation should be considered as a land use option in Europe. The limited number of studies undertaken so far at various locations and systems in Europe have shown that the factors that contribute to higher C sequestration under AFS include greater above-and below-ground spatial heterogeneity in the vegetation (trees and crops), production of higher amounts of plant biomass, more extensive root exploration of rhizosphere and increased litterfall inputs to the soil. Further studies are needed on all these as well as other aspects of the soil and associated vegetation to evaluate different components of agroforestry systems, including trees, the understory, animals and their interactions, under specific edapho-climatic conditions. The implementation of AFS contributes to an overall sustainable land management based on the increase of soil fertility by C enrichment in humus and the potential of C sequestration in the soil–plant system.

References

Böhm C, Quinkenstein A, Freese D, Hüttl RF (2010) Energieholzproduktion in Agroforstsystemen als ein Beitrag zur Verbesserung des Bodenschutzes. In: Tagungsband zu 6. Marktredwitzer Bodenschutztage "Bodenschutz in Europa – Ziele und Umsetzung", S 78–84

Burley J, Ebeling J, Costa PM (2007) C sequestration as a forestry opportunity in a changing climate. In: Freer-Smith PH, Boradmeadow MSJ, Lynch JM (eds.) Forestry and climate change. CAB International, Wallingford, UK, pp 31–37

Christensen JH, Hewitson B, Busuioc A, Chen A, Gao X, Held I, Jones R, Kolli RK, Kwon WT, Laprise R, Magaña Rueda V, Mearns L, Menéndez CJ, Räisänen J, Rinke A, Sarr A, Whetton P (2007) Regional climate projections. In: Solomon S, Qin D, Manning M, Chen Z, Marquis M, Averyt KB, Tignor M, Miller HL (eds.) Climate change 2007: the physical science basis contribution of working group I to the fourth assessment report of the Intergovernmental Panel on Climate Change. Cambridge University Press, Cambridge, UK, pp 849–940

Correal E, Erena M, Ríos S, Robledo A, Vicente M (2009) Agroforestry systems in southeastern Spain. In: Rigueiro-Rodríguez A, McAdam J, Mosquera-Losada MR (eds.) Agroforestry in Europe: current status and future prospects. Springer, Dordrecht, pp 183–210

Delabraze P (1986) Sylviculture méditeranéenne. In: Lanier L (ed.) Précis de Sylviculture. ENGREF, Nancy, pp 362–376

Dresner S, Ekins P, McGeevor K, Tomei J (2007) Forest and climate change global understandings and possible responses. In: Freer-Smith PH, Boradmeadow MSJ, Lynch JM (eds.) Forestry and climate change. CAB International, Wallingford, UK, pp 38–48

Dupraz CC, Burgess PJ, Gavaland A, Graves AR, Herzog F, Incoll LD, Jacson N, Keesman K, Lawson G, Lecomte I, Mantzanas K, Mayus M, Palma J, Papanastasis V, Paris P, Pilbeam DJ, Reisner Y, van Noordwijk M, Vincent G,van der Werf W (2005) SAFE (Silvoarable Agroforestry for Europe) synthesis report. SAFE project (August 2001–January 2005). Available at http://www1.montpellier.inra.fr/safe/english/results/final-report/SAFE%20Final%20Synthesis%20Report.pdf. Accessed 9 Dec 2010

EEA (2006) Land accounts for Europe 1990–2000. Towards integrated land and ecosystem accounting. Available at http://www.eea.europa.eu/publications/eea_report_2006_11. Accessed 9 Dec 2010

EEA (2008) Impacts of Europe´s changing climate. EEA briefing 2008–03. Available at http://www.eea.europa.eu/publications/briefing_2008_3. Accessed 12 Jan 2011

EEA (2009a) Regional climate change and adaptation. The Alps facing the challenge of changing water resources. Available at http://www.eea.europa.eu/publications/alps-climate-change-and-adaptation-2009. Accessed 12 Jan 2011

EEA (2009b) Progress towards the European 2010 biodiversity target. EEA report no 4/2009. Available at http://www.eea.europa.eu/publications/progress-towards-the-european-2010-biodiversity-target. Accessed 12 Jan 2011.

EEA (2009c) Greenhouse gas emission trends and projections in Europe 2009. Tracking progress towards Kyoto targets. Available at http://www.eea.europa.eu/publications/eea_report_2009_9. Accessed 12 Jan 2011

Eichhorn MP, Paris P, Herzog F, Incoll LD, Liagre F, Mantzanas K, Mayus M, Moreno G, Papanastasis VP, Pilbeam DJ, Pinsanelli A, Dupraz C (2006) Silvoarable systems in Europe - past, present and future prospects. Agroforest Syst 67:29–50

Etienne M (1996) Western silvopastoral systems. INRA, Paris, 276 p

Etienne M, Drezko M, Rigolot E (1996) Browse impact in silvopastoral systems participating in fire prevention in the French Mediterranean region. In: Étienne M (ed.) Western European silvopastoral systems. INRA, Paris, pp 93–102

EU (2005) Council Regulation (EC) No 1698/2005 of 20 September 2005 on support for rural development by the European Agricultural Fund for Rural Development (EAFRD). Available at http://eur-lex.europa.eu/LexUriServ/LexUriServ.do?uri=OJ:L:2005:277:0001:0040:EN:PDF. Accessed 12 Jan 2011

EU (2009) Report on implementation on forestry measures under the rural development regulation 1698/2005 for the period 2007–2013. Directorate-general for agriculture and rural development. Available at http://ec.europa.eu/agriculture/fore/publi/report_exsum_en.pdf. Accessed 12 Jan 2011

Fernández-Núñez E, Mosquera-Losada MR, Rigueiro-Rodríguez A (2007) Economic evaluation of different land use alternatives: forest, grassland and silvopastoral systems. Grassland Sci Eur 12:508–511

Fernández-Núñez E, Rigueiro-Rodríguez A, Mosquera-Losada MR (2010) C allocation dynamics one decade after afforestation with *Pinus radiata* D. Don and Betula alba L. under two stand densities in NW Spain. Ecol Eng 36:876–890

Follet RF, Kimble JM, Lal R (2001) The potential of US grazing lands to sequester carbon and mitigate the greenhouse effect. CRC Lewis Publishers, Boca Raton, 442 p

Gianelle D, Zorer R, Marcolla B, Cescattia A, Vescovo L (2004) Carbon flux measurements in a mountain grassland in the Italian Alps. Grassland Sci Eur 9:157–159

Gordon AM, Thevathasan NV (2006) How much carbon can be stored in Canadian agroecosystems using a silvopastoral approach? In: Mosquera-Losada MR, McAdam J, Rigueiro-Rodríguez A (eds.) Silvopastoralism and sustainable land management. CABI Publishing, Wallingford, UK, pp 210–219

Grünewald H, Brandt BKV, Schneider BU, Bens O, Kendzia G, Hüttl RF (2007) Agroforestry systems for the production of woody biomass for energy transformation purposes. Ecol Eng 29:319–328

Howlett DS, Mosquera-Losada MR, Nair PKR, Nair VD, Rigueiro-Rodríguez A (2011) Soil C storage in silvopastoral systems and a treeless pasture in northwestern Spain. J Environ Qual 40:1–8

IPCC (2007) Climate change 2007 — synthesis report. Contribution of working groups I, II and III to the fourth assessment report of the intergovernmental panel on climate change. Available at: http://www.ipcc.ch/publications_and_data/publications_ipcc_fourth_assessment_report_synthesis_report.htm. Accessed Jan 2011

Knopka B, Patik J, Moravcik M, Lukac M (2010) Biomass partitioning and growth efficiency in four naturally regenerated forest tree species. Basic Appl Ecol 11(3):234–243

Lal R (2005) Forest soils and C sequestration. For Ecol Manag 220:242–258

López-Díaz ML, Mosquera-Losada MR, Rigueiro-Rodríguez A (2007) Lime, sewage sludge and mineral fertilization in a silvopastoral system developed in very acid soils. Agroforest Syst 70:91–101

López-Díaz ML, Rigueiro-Rodríguez A, Mosquera-Losada MR (2010) Influence of pasture botanical composition and fertilization treatments on tree growth. For Ecol Manag 257:1363–1372

Matos ES, Freese D, Mendonca ES, Slazak A, Huttl RF (2010a) C, nitrogen and organic C fractions in topsoil affected by conversion from silvopastoral to different land use systems. Agroforest Syst. doi:10.1007/s10457-010-9314-y

Matos ES, Freese D, Ślązak A, Bachmann U, Veste M, Hüttl RF (2010b) Organic-C and nitrogen stocks and organic-C fractions in soil under mixed pine and oak forest stands of different ages in NE Germany. J Plant Nutr Soil Sci 173:654–661

Montero G, Ruiz-Peinado R, Muñoz M (2005) Producción de biomasa y fijación de CO_2 por los bosques españoles, vol 13, INIA: Serie forestal. INIA, Madrid, 270 p
Moreno G, Obrador JJ (2007) Effects of trees and understory management on soil fertility and nutritional status of holm oaks in Spanish dehesas. Nutr Cycle Agroecosyst 78:253–264
Moreno G, Obrador JJ, Cubera E, Dupraz C (2005) Fine root distribution in dehesas of central-western Spain. Plant Soil 277:153–162
Mosquera-Losada MR, Fernández-Núñez E, Rigueiro-Rodríguez A (2006) Pasture, tree and soil evolution in silvopastoral systems of Atlantic Europe. For Ecol Manag 232:135–145
Mosquera-Losada MR, McAdam J, Romero-Franco R, Santiago-Freijanes JJ, Rigueiro-Rodríguez A (2009) Definitions and components of agroforestry practices in Europe. In: Rigueiro-Rodríguez A, McAdam J, Mosquera-Losada MR (eds.) Agroforestry in Europe. Current status and future prospects. Springer, Dordrecht, pp 3–20
Mosquera-Losada MR, Ferreiro-Domínguez N, Rigueiro-Rodríguez A (2010a) Fertilization in pastoral and *Pinus radiata* D. Don silvopastoral systems developed in forest and agronomic soils of Northwest Spain. Agric Ecosyst Environ 139(4):618–628
Mosquera-Losada MR, Muñoz-Ferreiro N, Rigueiro-Rodríguez A (2010b) Agronomic characterisation of different types of sewage sludge: policy implications. Waste Manage 30(3):492–503
Mosquera-Losada MR, Morán-Zuloaga D, Rigueiro-Rodríguez A (2011a) Effects of lime and sewage sludge on soil, pasture production and tree growth in a six year old *Populus canadensis* Moench silvopastoral system. J Plant Nutr Soil Sci 174:145–153
Mosquera-Losada MR, Cuiña-Cotarelo R, Rigueiro-Rodríguez A (2011b) Effect of understory vegetation management through liming and sewage sludge fertilization on soil fertility and *Pinus radiata* D. Don growth after reforestation. Eur J For Res. doi:10.1007/s10342-011-0489-1
Nair PKR, Buresh RJ, Mugendi DN, Latt CR (1999) Nutrient cycling in tropical agroforestry systems: myths and science. In: Buck LE, Lassoie JP, Fernandes ECM (eds.) Agroforestry in sustainable agricultural systems. CRC Press, Boca Raton, pp 1–31
Nair PKR, Gordon A, Mosquera-Losada MR (2008) Agroforestry. Encyclopaedia Ecol 1:101–110
Nair PKR, Kumar BM, Nair VD (2009) Agroforestry as a strategy for C sequestration. J Plant Nutr Soil Sci 172:10–23
Nieder R, Benbi DK, Isermann K (2003) Soil organic matter dynamics. In: Benbi DK, Nieder R (eds.) Handbook of processes and modelling in the soil-plant system. Haworth Press, Binghamton, pp 346–408
Nii-Annang S, Grünewald H, Freese D, Hüttl R, Dilly O (2009) Microbial activity, organic C accumulation and 13C abundance in soils under alley cropping systems after 9 years of recultivation of quaternary deposits. Biol Fert Soils 45(5):531–538
Papanastasis VP, Mantzanas K, Dini-Papanastasi O, Ispikoudis I (2009) Traditional agroforestry systems and their evolution in Greece. In: Rigueiro-Rodríguez A, McAdam J, Mosquera-Losada MR (eds.) Agroforestry in Europe. Current status and future prospects. Springer, Dordrecht, pp 89–110
Pardos JA (2010) Los ecosistemas forestales y el secuestro de C ante el calentamiento global. INIA, Madrid, 253 p
Parton WJ, Schimel DS, Cole CV, Ojima DS (1987) Analysis of factors controlling soil organic matter levels in Great Plains grasslands. Soil Sci Soc Am J 51:1173–1179
Quinkenstein A, Wöllecke J, Böhm C, Grünewald H, Freese D, Schneider BU, Hüttl RF (2009) Ecological benefits of the alley cropping agroforestry system in sensitive regions of Europe. Environ Sci Policy 12(8):1112–1121
Quinkenstein A, Böhm C, Matos E, Freese D, Hütt RF (2011) Assessing the carbon sequestration in short rotation coppices of *Robinia pseudoacacia* L. on marginal sites in northeast Germany. In: Kumar BM, Nair PKR (eds.) Carbon sequestration potential of agroforestry systems: opportunities and challenges. Springer, Dordrecht, pp 201–216
Reijneveld JA, Kuikman PJ, Oenema O (2010) Changes in soil organic matter content of grassland and maize land in the Netherlands between 1970 and 2009. Grassland Sci Eur 15:3–32
Rigolot E, Etienne M (1996) Litter thickness on tree covered fuel-breaks maintained by grazing. In: Étienne M (ed.) Western European silvopastoral systems. INRA, Paris, pp 111–124

Rigueiro-Rodríguez A, Fernández-Núñez E, González-Hernández P, McAdam JH, Mosquera-Losada MR (2009a) Agroforestry systems in Europe: productive, ecologial and social perspectives. In: Rigueiro-Rodríguez A, McAdam J, Mosquera-Losada MR (eds.) Agroforestry in Europe. Current status and future prospects. Springer, Dordrecht, pp 43–66

Rigueiro-Rodríguez A, López-Díaz ML, Mosquera-Losada MR (2011a) Organic matter and chromium evolution in herbage and soil in a *Pinus radiata* silvopastoral system in North-West Spain after sewage sludge and lime application. Commn Soil Sci Plant 12:1–14

Rigueiro-Rodríguez A, Santiago-Freijanes JJ, Fernández-Núñez E, Mosquera-Losada MR (2009b) Sistemas biológicos de prevención de incendios forestales. Los sistemas agroforestales. In: Vélez R (ed.) La defensa contra incendios forestales. Fundamentos y experiencias, 2nd edn. McGraw-Hill/Intramericana, Madrid, pp 447–451

Rigueiro-Rodríguez A, Rois-Díaz M, Mosquera-Losada MR (2010) Integrating silvopastoralism and biodiversity conservation. Sustainable Agric Rev 5:359–373

Rigueiro-Rodríguez A, Mosquera-Losada MR, Fernández-Núñez E (2011b) Afforestation of agricultural land with *Pinus radiata* D. Don and *Betula alba* L. in NW Spain: effects on soil pH, understorey production and floristic diversity eleven years after establishment. Land Degrad Dev. DOI:10.1002/ldr.1072

Robles AB, Ruíz-Mirazo J, Ramos ME, González-Rebollar JL (2009) Role of livestock grazing in sustainable use, naturalness promotion in naturalization of marginal ecosystems of south-eastern Spain (Andalusia). In: Rigueiro-Rodríguez A, McAdam J, Mosquera-Losada MR (eds.) Agroforestry in Europe. Current status and future prospects. Springer, Dordrecht, pp 211–232

Rois-Díaz M, Mosquera-Losada MR, Rigueiro-Rodríguez A (2006) Biodiversity indicators on silvopastoralism across Europe. Available at http://www.efi.int/files/attachments/publications/tr_21.pdf. Accessed 12 Jan 2011

Sanderson MA, Wätzold F (2010) Balancing trade-offs in ecosystem functions and services in grassland management. Grassland Sci Eur 15:639–648

Schanabel TT, Franzluebbers AJ, Stout WL, Sanderson MA, Stuedemann JA (2001) Pasture management effects on soil carbon sequestration. In: Follet RF, Kimble JM, Lal R (eds.) Potential of US grazing lands to sequester carbon. CRC Lewis Publishers, Boca Raton, pp 291–322

Schimel DS, Braswell BH, Holland EA, Mckeown R, Ojima DS, Painter TH, Parton WJ, Townsend AR (1994) Climatic, edaphic, and biotic controls over storage and turnover of C in soils. Glob Biogeochem Cy 8(3):279–293

Serrada R, Montero G, Reque JA (2008) Compendio de Aselvicultura Aplicada en España. INIA, Madrid, 1178 p

Sibbald A (1996) Silvopastoral systems on temperate sown pastures a personal perspective. In: Etienne M (ed.) Western European silvopastoral systems. INRA, Paris, pp 23–36

Suyker AE, Verma SB (2001) Year-round observations of the net ecosystem exchange of carbon dioxide in a native tallgrass prairie. Glob Change Biol 7:279–289

Theng BKG, Tate KR, Sollins P, Moris N, Madkarni N, Tate R (1989) Constituents of organic matter in temperate and tropical soils. In: Coleman DC, Oades JM, Uehara G (eds.) Dynamics of soil organic matter in tropical ecosystems. College of Tropical Agriculture and Human Resources, University of Hawaii, Honolulu, pp 5–32

UN (1998) Kyoto protocol to the United Nations framework convention on climate change. Available at http://unfccc.int/resource/docs/convkp/kpeng.pdf. Accessed 12 Jan 2011

Vanlauwe B, Nwoke OC, Sanginga N, Merckx R (1996) Impact of the residue quality on the carbon and nitrogen mineralization of leaf and root residues of three agroforestry species. Plant Soil 183:221–231

Zianis D, Muukkonen P, Mäkipää R, Mencuccini M (2005) Biomass and stem volume equations for tree species in Europe. Silva fennica monographs 4. The Finnish society of forest science. Tammer-Paino Oy, Tampere, 63 p

Carbon Sequestration Potential of Agroforestry Systems in Africa

Eike Luedeling, Gudeta Sileshi, Tracy Beedy, and Johannes Dietz

Abstract Agroforestry can raise carbon (C) stocks of agricultural systems, and such increases can potentially be sold as CO_2 emission offsets. We assembled information on the biophysical, technical, economic, and practical potential of agroforestry to sequester C for the West African Sahel, East Africa, and Southern Africa. Agroforestry systems (AFS) such as parklands, live fences, and homegardens had substantial C stocks, but only accumulated 0.2–0.8 Mg C ha^{-1} $year^{-1}$. Rotational woodlots (2.2–5.8 Mg C ha^{-1} $year^{-1}$) and possibly improved fallows in Southern Africa sequestered C relatively faster, but only during the fallow phases. Data on soil C are scarce because most studies only compared soil C under different land uses, which provides limited (and sometimes unreliable) information on sequestration rates. Comparing results from different studies is difficult, because no standard protocols exist. Few studies have evaluated the economic potential of agroforestry to sequester C. However, at prices of \$10 per Mg CO_2-eq or less, the value of stored C in most systems would be less than \$30 ha^{-1} $year^{-1}$, which is a small fraction of annual farm revenue and it needs to cover all transaction measurement reporting and verification costs. Practical constraints to C sequestration (CS) such as land tenure, policy issues, and the opportunity costs incurred by possibly foregoing more profitable land management options have not been fully explored for Africa. For evaluating the challenges and opportunities involved in CS by smallholder farmers, comprehensive studies are needed that explore all C and non-C costs and benefits of agroforestry activities.

E. Luedeling (✉) • J. Dietz
World Agroforestry Centre (ICRAF),
Nairobi, Kenya
e-mail: e.luedeling@cgiar.org; j.dietz@cgiar.org

G. Sileshi • T. Beedy
World Agroforestry Centre (ICRAF), Lilongwe, Malawi
e-mail: Sweldesemayat@cgiar.org; t.beedy@cgiar.org

B.M. Kumar and P.K.R. Nair (eds.), *Carbon Sequestration Potential of Agroforestry Systems: Opportunities and Challenges*, Advances in Agroforestry 8,
DOI 10.1007/978-94-007-1630-8_4,

Keywords Carbon costs • Homegardens • Live fences • Parklands • Rotational woodlots • Smallholder farmers

Introduction: Carbon Sequestration Potential

Most agroforestry systems (AFS) have higher carbon (C) stocks than agricultural monocultures, and expansion of agroforestry practices could raise the C stocks of Africa's terrestrial systems (Albrecht and Kandji 2003). On a global level, Dixon et al. (1993) estimated a sequestration potential by forestry and agroforestry practices of about 1 Pg of C per year, corresponding to about 3.7 Pg CO_2, or roughly one-eighth of annual global emissions. This chapter explores the C sequestration (CS) capacity of African AFS, with particular emphasis on the West African Sahel, East Africa, and Southern Africa. This discussion requires at first a clarification of the term 'carbon sequestration potential', which can be and has been interpreted in different ways.

Referring to soil organic C, Ingram and Fernandes (2001) distinguished between 'potential' CS, which is determined by soil characteristics, 'attainable' CS, which accounts for limiting factors, such as net primary productivity and climate, and 'actual' CS, which is defined by reducing factors, such as removal of crop residue, tillage etc. Along the same lines, Cannell (2003) offered a terminology to differentiate between different assessments of the capacity of land management regimes to sequester C, using the terms 'theoretical potential capacity', 'realistic potential capacity', and 'conservative, achievable capacity'.

In agricultural contexts, studies on the CS potential of agroforestry are often conducted with a view to creating opportunities for smallholder farmers to benefit from international C payment schemes. We therefore use a modification of these two terminologies to guide the content of this chapter. Making decisions about the feasibility of CS activities requires an interdisciplinary approach, exploring the biophysical, technical, economic, and practical potential of land management options to sequester C. Figure 1 outlines these concepts and lists the most important constraints that are considered in quantifying the four types of potential.

1. Studies on the biophysical capacity deal with the general geographic setting of a region and use environmental and/or climatic parameters to estimate the additional amount of C that could be stored in terrestrial systems. This is often based on assessments of C stocks in natural vs. actual vegetation, as well as on the potentially available land area.
2. The technical potential explores the management options that are available, or innovative new options, and their effects on system C stocks. Studies on the technical capacity may include assessments of available technical skills and the availability of necessary inputs.
3. The economic potential to sequester C includes both above steps, but considers potential economic constraints, such as the profitability of a system, as well as estimates of opportunity costs or marginal abatement costs.

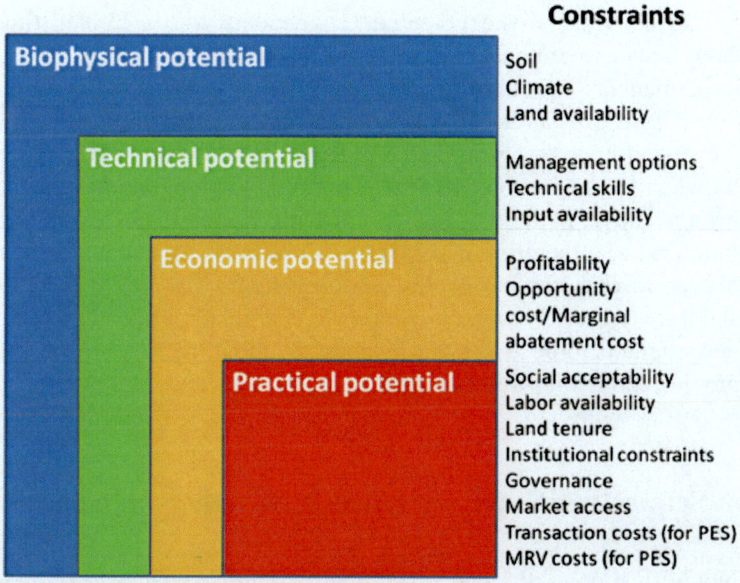

Fig. 1 Schematic illustration of the different types of carbon sequestration potentials, and the constraints that are encountered in their quantification

4. Finally, the practical potential considers additional constraints to system adoption. Examples of these are the social acceptability of the proposed land management option, labor availability, land tenure questions, institutional and governance constraints, as well as market access. For carbon sales and other payment for ecosystem services (PES) schemes, transaction costs, and costs incurred in measurement, reporting, and verification (MRV) can also be important constraints that determine the practical potential.

A thorough assessment of the potential of AFS to sequester C should comprise all four components, but many studies miss one or more of these, limiting the conclusions that can be drawn. For example, a high biophysical capacity to sequester C does not automatically mean that smallholder farmers can benefit from such a scheme, and it is clearly insufficient for guiding development efforts. On the other hand, assessments of the practical capacity are typically very limited in their geographic scope and cannot be used to justify international C payment schemes. A distinction between studies that focus on the different types of potential is therefore imperative for assessing the state of research, and for identifying current knowledge gaps.

In addition to different meanings of the word 'potential', the term 'sequestration' can also be interpreted in a number of ways. For climate change mitigation purposes, the most effective form of sequestration is the incorporation of C in long-lived C pools in the soil, in permanent biomass, or in long-lived wood products. Carbon sequestered into such pools is more or less permanently removed from the atmosphere. In most cases, however, these C pools are limited in their capacity, restricting CS activities – and potential C payments – to a limited time frame. Establishing trees as permanent structures on agricultural fields is an example of relatively permanent sequestration of C from the atmosphere. Other agroforestry systems may have substantial C accumulation rates, but require most of it to be released again after a few years. In improved fallows or rotational woodlots, especially when trees are grown as fuelwood, net C accumulation in the system is low, or even negative, in spite of fast tree growth. Rather than focusing only on C accumulation rates, it therefore makes more sense to examine net C increase rates, averaged over several rotation cycles or, for systems that reach C saturation, to specify the time frame, over which certain C accumulation rates can be sustained. This review attempts to be as specific as possible about the time frames, but not all studies provide enough information on this.

Carbon Sequestration by Agroforestry Systems in Africa

From a biophysical point of view, Africa's agricultural systems clearly have potential for sequestering additional C. Across different eco-zones, Dixon et al. (1994) estimated a C storage potential of agroforestry and integrated land use approaches of between 12 and 228 Mg C ha^{-1} over a 50 year rotation, corresponding to 0.2–4.6 Mg C ha^{-1} $year^{-1}$. They provide two values for Africa, 0.6 Mg C ha^{-1} $year^{-1}$ for establishing agroforestry in the tropical highlands of Congo (presumably Brazzaville), which could be realized at a cost of $69 per Mg C, and a sequestration rate of 1.1 Mg C ha^{-1} $year^{-1}$ for a fuelwood system in the Democratic Republic of the Congo (DRC), at a cost of $4 to 12 per Mg C. Cost estimates here include only costs for establishing and maintaining land management systems, neglecting all other expenses. Especially in the latter system, however, net time-averaged sequestration rates appear to be much lower, because the fuelwood that is produced in the system is burnt after harvest, releasing most of the stored C.

Jarecki and Lal (2003) reviewed various studies on the potential of agroforestry systems to store C, listing a range of 0.25–1.58 Mg C ha^{-1} $year^{-1}$ in the soil and 0.98–6.7 Mg C ha^{-1} $year^{-1}$ in aboveground biomass. Their review does not include explicit estimates for Africa, but mentions a potential of 6.2 Mg C ha^{-1} $year^{-1}$ in aboveground storage for new forests in tropical regions and 0.25–0.50 Mg C ha^{-1} $year^{-1}$ in soil and 2–4 Mg C ha^{-1} $year^{-1}$ aboveground for tree plantations in degraded tropical areas.

More detailed reviews collate information from case studies on the CS potential of AFS. Kuersten and Burschel (1993) provide estimates of the amounts of C sequestered by fuelwood production in AFS of 0.5–2.0 Mg C ha^{-1} $year^{-1}$ for shade trees in coffee (*Coffea* spp.) and cacao (*Theobroma cacao* L.), 2.0–3.6 Mg C ha^{-1} $year^{-1}$ for fuelwood plantations, 0.3–2.0 Mg C ha^{-1} $year^{-1}$ for secondary forests, 0.1 Mg C ha^{-1} $year^{-1}$ for trees in corrals and annual crops, and 1.4 Mg C ha^{-1} $year^{-1}$ for living fences. Nair et al. (2009) estimated potential sequestration rates of 5.9 Mg C ha^{-1} $year^{-1}$ for cacao agroforests of Cameroon, 6.3 Mg C ha^{-1} $year^{-1}$ for

Table 1 Carbon sequestration rates reported for agroforestry systems across the West African Sahel, East Africa, and Southern Africa

Activity	Duration (years)	C sequestration rate (Mg C ha^{-1} year^{-1})	Reference
West African Sahel			
Faidherbia albida plantation in Senegal	50	0.22	Tschakert (2004b)
Optimal agricultural intensification, incl. *Leucaena* prunings in Senegal	50	0.27	Tschakert (2004b)
Restoring degraded grassland to woody grassland in Senegal	20	0.77	Woomer et al. (2004b)
Establishment of new parklands in the Sahel	50	0.4	Data from Takimoto et al. (2008b), Tschakert et al. (2004), Woomer et al. (2004b)
East Africa			
Tree planting to restore highly degraded land	25	0.4–0.8	Batjes (2004a)
Intensification of windrows and tree biomass	20	0.8	Henry et al. (2009)
Conversion of cropland to homegardens	20	0.5–0.6	Henry et al. (2009)
Southern Africa			
Regrowth of woodland on abandoned farms in Mozambique	25	0.7	Walker and Desanker (2004)
Coppiced Miombo woodland in Zambia	16	0.5	Stromgaard (1985)
Coppiced Miombo woodland in Zambia	35	0.9	Chidumayo (1997)
Faidherbia albida plantation in Tanzania	6	1.2	Okorio and Maghembe (1994)
F. albida converted to 50 trees/ha	6	0.22	Okorio and Maghembe (1994)
Rotational woodlots in Tanzania	5	2.6–5.8	Nyadzi et al. (2003)
Rotational woodlots in Tanzania (wood C)	5	2.3–5.1	Kimaro (2009)
Rotational woodlots in Zambia	2	2.15–4.75	Kaonga and Bayliss-Smith (2009)

shaded coffee in Togo and between 0.3 and 1.1 Mg C ha^{-1} year^{-1} for agroforestry in the Sahel. The review papers mentioned above do not provide information on the time-frames, over which the stated sequestration rates can be sustained.

The wide range of estimates in the case studies collected in these reviews may be caused by summarizing studies of different types of potential and by consideration of different C pools. Most reviews do not explicitly state the nature of the studies that are listed, and many are mixtures between different types of potential. We therefore focus on three African regions to provide a more comprehensive overview of existing studies. Biophysical and technical potentials are explored for each region separately. Carbon sequestration rates determined for selected tree-based systems across all three regions are shown in Table 1. Due to the scarcity of studies on the economic and practical potentials, we discuss these for all regions together.

West African Sahel

General Setting

The West African Sahel is the transition zone between the Sahara Desert in the North and the Sudan savanna zone in the South, comprising parts of Senegal, Mauritania, Mali, Burkina Faso, Niger, and Nigeria. It is characterized by mean annual rainfall of between 200 and 600 mm, falling during one summer rainy season, which lasts between 2 and 5 months. Annual rainfall amounts are highly variable between years and on inter-decadal scales, leading to recurrent droughts (Hulme 2001). Livelihood strategies in the Sahel therefore revolve around exploitation of the scarce rainfall, with agricultural systems focused on rainfed production of annual crops, such as maize (*Zea mays* L.), millets, peanut (*Arachis hypogaea* L.), and cowpea (*Vigna unguiculata* (L.) Walp.) or on extensive livestock production in nomadic systems. Agroforestry has a long history in this region, and traditional alongside modern improved systems are in existence. Traditional agroforestry parkland systems dominate the landscape in many parts of the Sahel (Boffa 1999), and novel practices such as live fences and fodder banks are being promoted.

Compared to the global average or even other parts of Africa, C storage potential in Sahelian agroecosystems is relatively low, due to harsh environmental conditions, with high temperatures and low precipitation restricting net primary productivity and thus the supply of C that can be sequestered (Batjes 2001). Hanan et al. (1998) measured an increment in biomass of about 5 Mg ha^{-1} $year^{-1}$ (corresponding to about 2.5 Mg C ha^{-1} $year^{-1}$ at 50% C in biomass) in a Sahelian fallow savanna in Niger. While raising C uptake rates may be possible, many land management options tend to decrease C stocks. In particular when the soil is tilled, soil organic matter is quickly decomposed, lowering C stocks substantially below those of natural systems (Batjes 2001; Tieszen et al. 2004).

Projected climate change in the Sahel may exert additional pressure on system C stocks. While future projections for this region disagree substantially, the majority of projections indicate a drier and hotter climate (Tieszen et al. 2004), which will likely reduce equilibrium C levels, even in the absence of cultivation (Batjes 2001). Lufafa et al. (2008) estimated soil organic carbon (SOC) losses between 21% and 23% in Senegal's Peanut Basin for two climate change scenarios. Adverse impacts on crop yields are also likely (Liu et al. 2004; Tieszen et al. 2004). Woomer et al. (2004a) reported that net losses in C stocks in response to climate change have already occurred in various ecozones of Senegal. On a related note, Gijsbers et al. (1994) and Maranz (2009) reported that existing agroforestry parklands are degrading, which may be attributable to a decline in environmental suitability due to recent climate change (Maranz 2009). Agroforestry and other land management practices may have potential to counteract current trends towards lower C stocks (Batjes 2001; Woomer et al. 2004b). The recent, farmer-driven regeneration or establishment of parklands in parts of Niger (Reij et al. 2009) and the introduction of irrigated AFS along the Senegal River (Venema et al. 1997) are promising steps in this direction.

Biophysical Potential to Sequester Carbon

General estimates of C stocks in the Sahelian ecosystems are difficult, because of the strong dependence of C stocks on environmental conditions. In particular, the soil type is a primary determinant of system C stocks. Batjes (2001) reported large differences in C stock between soil types, even under the same land use. In Senegal, the top meter of a rice (*Oryza sativa* L.) field on a Gleyic Cambisol (US Taxonomy: Tropepts, Inceptisols) may store 34 Mg C ha^{-1}, whereas the same land use of a Dystric Gleysol (Aquepts, Inceptisols) may have 65 Mg C ha^{-1}. Combined with short grassland, soil C stocks in the top meter of a rice field on a Dystric Fluvisol (Entisols) may even reach 301 Mg C ha^{-1}.

It has been argued that not all C in an ecosystem can be considered sequestered, because of widely variable turnover rates among different C pools. Batjes (2001) distinguished seven different soil C pools, with turnover times ranging from 0.1 to 3,000 years. Simpler models distinguish only between stable and labile C pools (Traoré et al. 2008). The turnover time of aboveground C stocks also varies substantially, with annual crops being harvested every year, intensively used trees persisting for up to 10 years, and structural elements of traditional agroforestry and forestry systems remaining in place for many decades. Ideally, ecosystem scale and time-averaged C accounting, in particular when the focus is on climate change mitigation, would consider such differences in C pool stability.

Nevertheless, most studies to date have focused on quantifying total system C stocks, soil C stocks, and/or C stored in aboveground biomass. Figure 2 summarizes results from seven studies, investigating C stocks in a range of natural and agricultural ecosystems of the Sahel. Takimoto et al. (2008b; a in Fig. 2) investigated various agroforestry systems in Ségou, Mali, reporting C stocks (including the top 40 cm of the soil) of 70.8 Mg C ha^{-1} in parklands dominated by *Faidherbia albida* (Delile) A. Chev., which was almost twice as high as when the dominant species was *Vitellaria paradoxa* C.F. Gaertn. Carbon stocks in live fence systems and fodder banks were substantially lower (Takimoto et al. 2008b). Woomer et al. (2004a; b in Fig. 2) determined C stocks in 16 ecosystem types along a transect through Senegal, and reported C stocks from 11 to 112 Mg C ha^{-1}, with lowest values in degraded or cultivated land, followed by pastures, fallow plots, parkland, woodland, and forest. Liu et al. (2004; c in Fig. 2) measured C stocks in different ecosystem types in Senegal, and reported 31.8–52.1 Mg C ha^{-1} for cropland, parklands, and fallows with trees. Tschakert (2004a; d in Fig. 2) estimated total system C stocks of 28 Mg C ha^{-1} in the Old Peanut Basin in Senegal, with 11 Mg C ha^{-1} stored in the top 20 cm of the soil and 6.3 Mg C ha^{-1} in trees. She found that parklands on average contained 9.3 Mg C ha^{-1} more than cultivated land. Finally, Woomer et al. (2004b; e in Fig. 2), investigated C stocks across a range of grasslands and silvopastoral systems along a climate gradient covering the Sahelian transition in Senegal. With increasing aridity, they found that total system C declined from 31.9 Mg C ha^{-1} in shrubland with scattered trees to 19.4 Mg C ha^{-1} in grasslands with scattered shrubs and 12.0 Mg C ha^{-1} in degraded grasslands at the arid end of their transect. It should

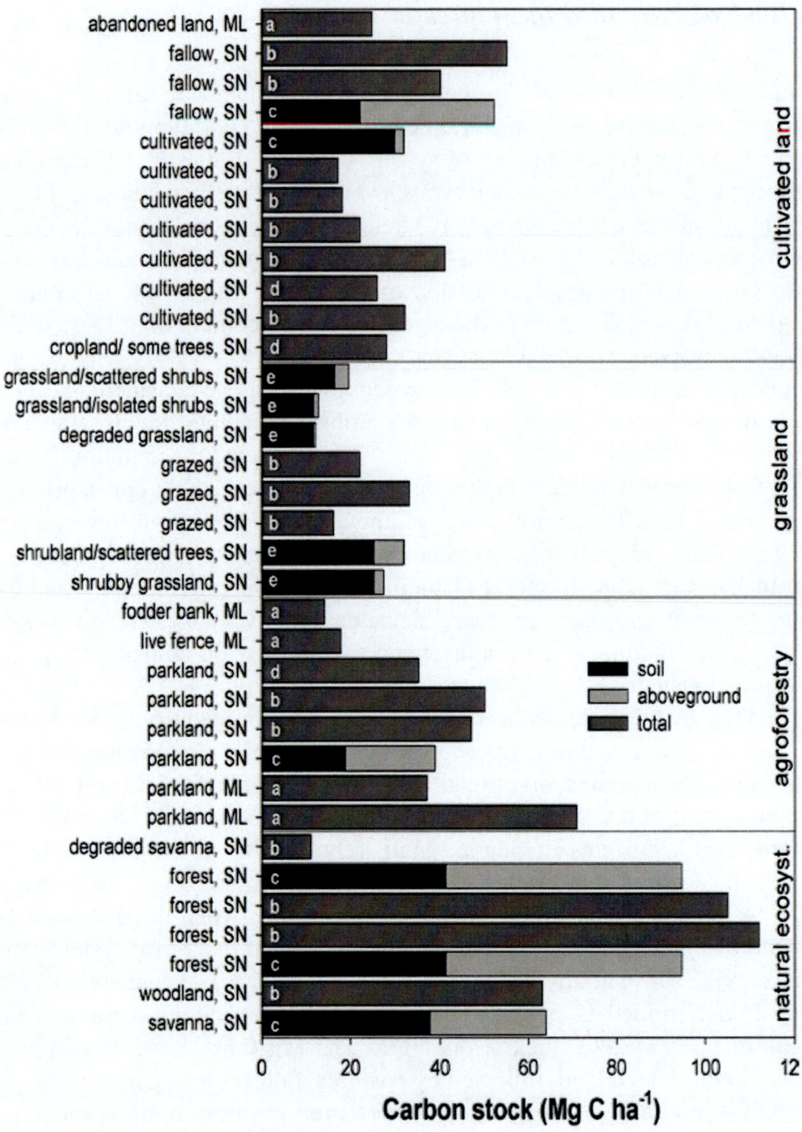

Fig. 2 Carbon stocks in natural and cultivated ecosystems of the Sahel. *ML* Mali, *SN* Senegal. Data sources: a=Takimoto et al. (2008b), b=Woomer et al. (2004a), c=Liu et al. (2004), d=Tschakert (2004a), e=Woomer et al. (2004b)

be noted that in all these studies, C stocks are a result not only of management, but also of site-specific pedological and climatic conditions. Conclusions about the effects of management on C stocks should thus be drawn with caution.

Climates, ecosystem types, and environmental conditions among all these study sites were variable and so were the sampling protocols. For example, the depth to which soil was included was variable and aboveground biomass was determined using different allometric equations, which in some cases were transferred from a different environment due to the lack of site-specific equations. Nevertheless, the wide range of C stock estimates gives an impression of the variation encountered in the field and the difficulty of extrapolating results beyond the immediate sampling sites. It should also be noted that most estimates listed in this section are empirically derived and include the confounding effects of site-specific soil and climate conditions. For deriving the technical potential, different management systems should be compared under similar environmental conditions for a better indication of the effect of individual land use options.

Technical Potential to Sequester Carbon

A wide range of management options can have substantial impacts on C stocks in natural and agricultural ecosystems, when implemented over a sufficiently long time scale. Judging the effectiveness of such a management option for CS, and ultimately the potential of farmers to reap benefits from C payments, requires consideration of both the total effect they may have and the time needed to achieve this effect. Doraiswamy et al. (2007) modeled the effect of various management regimes on soil C stocks in agricultural systems in Mali. In this study, the effect of 25 years of continuous conventional agriculture was a net loss of between 0.5 and 0.7 Mg C ha^{-1}, across four different crops. The best management option in their study, ridge cultivation with incorporation of crop residue and increased fertilization produced net gains between 1.5 and 3.0 Mg C ha^{-1}. Since treatments had to be implemented for 25 years to obtain these results, the amount of C that could theoretically be marketed amounted to less than 0.15 Mg ha^{-1} $year^{-1}$ in all treatments (Doraiswamy et al. 2007).

Tschakert (2004b) used the CENTURY model to evaluate 25 management options on C stocks in the Old Peanut Basin in Senegal. During the first 25 years, net C changes amounted to between −3.2 Mg C ha^{-1} and +10.8 Mg C ha^{-1}. The highest gains were achieved by 'optimal' agricultural intensification (crop rotation, fallow, manure, *Leucaena* prunings, and increased fertilization), followed by plantation of *F. albida* at 250–300 trees per hectare (+5.8 Mg C ha^{-1}). Net C changes thus ranged between −0.13 and +0.43 Mg C ha^{-1} $year^{-1}$. During the second 25 year period of maintaining the same management options, C changes decreased substantially for all management options (−0.74 to +5.30 Mg C ha^{-1}). Over the entire simulation period, annual C gains were thus 0.22 Mg C ha^{-1} $year^{-1}$ for *F. albida* plantations and 0.27 Mg C ha^{-1} $year^{-1}$ for 'optimal agricultural intensification'.

Woomer et al. (2004b) estimated that restoring degraded grasslands in Senegal to woody grasslands over a 20-year time frame may sequester up to 0.77 Mg C ha^{-1} $year^{-1}$. Establishing new parkland agroforestry systems may sequester about 20 Mg C ha^{-1} in addition to C stored in continuous cropland (averaging data from Takimoto et al. 2008b; Tschakert et al. 2004; Woomer et al. 2004b). Assuming 50 years to reach potential C stocks, the annual C stock increment could be fixed at about 0.4 Mg C ha^{-1} $year^{-1}$.

Takimoto et al. (2008a) concluded that further substantial increases in C stocks would not be feasible for existing parkland systems. In all these studies, none of the investigated management options, including agroforestry practices, sequestered more than 0.8 Mg C ha^{-1} $year^{-1}$ (Table 1).

East Africa

General Setting

The East African region extends across Kenya, Uganda, Rwanda, and northern Tanzania. It is comprised ecologically of a narrow coastal strip, arid deserts, semi-arid savannas, and the highlands region, which is densely populated and predominantly used for intensive agriculture. The savanna region is characterized by scarce and irregular rainfall and predominantly used as grazing land. Net primary productivity in this environment has been estimated at 6.2 Mg C ha^{-1} $year^{-1}$ (e.g. Nairobi National Park: Long et al. 1989), but net increases in C stocks are relatively low. Because establishing trees in this environment would require irrigation, agroforestry is not commonly practiced. In contrast, farmers in the East African Highlands practice a wide variety of AFS. Shade trees in coffee plantations, shelter belts (windbreaks) around homesteads and agricultural fields (Stigter et al. 2002), fruit trees, and woodlots on scarce fallow or infertile patches of land are common features of land use systems. Among the most intensively managed AFS in this region are the multi-story Chagga homegardens in northern Tanzania (Fernandes et al. 1985; Hemp 2006).

Biophysical Potential to Sequester Carbon

Because most pure cropping systems have negligible time-averaged C stock changes in aboveground vegetation, tree C stocks can be used to approximate aboveground C gains as a consequence of tree introduction. Studies on biomass in the highly heterogeneous agroforestry-dominated landscapes of the East African highlands are scarce. Glenday (2008) computed a C stock of 19 Mg C ha^{-1} in aboveground biomass for AFS around the Arabuke Sokoke forest on the Kenyan coast, a value that equals the one for woodlands in the same study. Tree planting has also been explored as an option for restoring highly degraded land, where it can sequester 0.4–0.8 Mg C ha^{-1} $year^{-1}$ (Batjes 2004a). For Kenya, Batjes (2004b) integrated CS estimates from various sources to arrive at potential C stock increases between 0 and 0.5 Mg C ha^{-1} $year^{-1}$, for seven agroclimatic zones in the country. For three different scenarios, in which he assumed that improved management practices are introduced on between 10% and 30% of current croplands and on 5–15% of current grasslands. For all of Kenya, he estimated a CS potential of between 5.8 and 9.7 Tg C over 25 years, or 0.23–0.39 Tg C $year^{-1}$.

Technical Potential to Sequester Carbon

Putting the biophysical capacity of C sequestration into a realistic perspective, Henry et al. (2009) estimated current C stocks of 9–11 Mg C ha^{-1} on average for the agroforestry landscapes of Western Kenya. These stocks could be raised by about 16 Mg C ha^{-1} over 20 years or 0.8 Mg C ha^{-1} $year^{-1}$, on average across seven land use types including the introduction and intensification of hedgerows. In their detailed study, Henry et al. (2009) distinguished between several spatially explicit land use types and assessed their potential of tree intensification. Assuming across the board a 20 year time frame for such transitions, they showed that windrows are currently almost at their maximum capacities, while woodlots have the potential to sequester 1.4–3.2 Mg C ha^{-1} $year^{-1}$ and homegardens 0.20–0.25 Mg C ha^{-1} $year^{-1}$, if more trees were introduced. Conversion from food crops to homegardens would result in an aboveground biomass increase of 0.5–0.6 Mg C ha^{-1} $year^{-1}$.

Southern Africa

Agro-Environmental Setting

In Southern Africa, agroforestry research over the past two decades has mainly focused on Malawi, Mozambique, Tanzania, Zambia, and Zimbabwe. Within these countries, efforts have concentrated on the upland plateau zone, which lies between 600 and 1,200 m above sea level. Mean annual rainfall ranges from 500 to 1,200 mm, mainly falling during a single rainy season between December and April, followed by a dry season of 7–8 months duration. Rainfall is greatly variable both within the rainy season and between years, in particular in the drier parts of the region.

The dominant vegetation type is Miombo woodland, the world's largest savanna region covering some 2.7 million km^2 (Campbell et al. 1996; Kanschik and Becker 2001; Lawton 1978). It is comprised of slow growing mainly deciduous trees that form a 15 to 20-m high light-but-closed canopy above a forest floor covered by grasses (Lawton 1978). The traditional land use in this region is slash and burn shifting cultivation, with cropping periods of 3–5 years followed by bush fallow phases of 10–20 years (Nhantumbo et al. 2009). In densely populated areas, shortening fallow periods have led to decreases in soil fertility (Chidumayo 1987; Matthews et al. 1992) and to expansion of farming activities to marginal lands (Abbot and Homewood 1999). Agricultural systems consist mainly of continuous maize-mixed cropping and extensive production of cattle and goats (Chakeredza et al. 2007).

Farmers in Southern Africa use a wide range of AFS, including both traditional and improved practices (Akinnifesi et al. 2008; Campbell et al. 1991; Sinclair 1999). Improved practices that are developed and promoted by researchers and development agencies include various options of fertilizer (Akinnifesi et al. 2008), fruit, fodder (Chakeredza et al. 2007) and fuelwood trees. Traditional agroforestry practices include intensive intercropping in highly diversified, multi-story homegardens, as well as various other systems that integrate trees with food or cash crops.

In some systems, trees are recruited from the natural tree population, and cropping systems resemble the parklands of the West African Sahel. Such systems are common in Malawi, Tanzania, Zambia, and Zimbabwe (Boffa 1999; Campbell et al. 1991). They include the *Faidherbia*/coffee system in Tanzania, and the *Faidherbia*/maize system in riparian settings in Malawi, Zambia, and Zimbabwe (Akinnifesi et al. 2008; Campbell et al. 1991). In other settings, trees are deliberately planted along farm and field boundaries, on soil conservation structures and as terrace risers. Many farmers practice relay fallow intercropping, in which fast growing nitrogen-fixing trees or shrubs (e.g. *Sesbania* spp., *Tephrosia* spp. or *Cajanus cajan* (L.) Millsp. and *Crotalaria* spp.) are planted into a field when annual crops have already been well established (Akinnifesi et al. 2008). Such improved fallows can also take the shape of rotational woodlots (Akinnifesi et al. 2008; Sileshi et al. 2008), in which leguminous trees are grown for about 5 years, then harvested and replaced by food crops (Nyadzi et al. 2003). Another common form of agroforestry is permanent tree-cereal intercropping. Trees in such systems are typically leguminous coppicing species, which are cut regularly. Leaves and twigs are incorporated into the soil to increase soil fertility (Sileshi et al. 2008). The best known manifestation of such a system is the intercropping of *Gliricidia sepium* (Jack.) Kunth. ex Walp. with maize in Malawi and Zambia (Akinnifesi et al. 2008; Sileshi and Mafongoya 2006). In Southern Africa, agroforestry trees provide a range of products and ecosystem services, such as soil fertility, fuelwood, poles, fruits, or shade.

Biophysical Capacity to Sequester Carbon

Carbon stocks of natural and agricultural ecosystems are generally lower than potential stocks, due to a range of human activities, such as C-depleting farming practices (e.g. ridging of soils, burning of crop residues, and inadequate fertilizer use), charcoal production, bush fires (Eriksen 2007) and wood harvesting (Abbot and Homewood 1999; Chidumayo 1987, 1997). In particular in comparison with undisturbed Miombo woodland, C stocks in agricultural systems are low (Walker and Desanker 2004; Williams et al. 2008). Conversion of Miombo woodland to agriculture in Mozambique reduced stem wood C stocks by 19.0 Mg C ha^{-1} and total C stocks by 23% (Williams et al. 2008). In Malawi, such conversion reduced C stocks in the top 150 cm of soil from 82.5 Mg C ha^{-1} to 49.0 Mg C ha^{-1} in fallow land and to 52.2 Mg C ha^{-1} in agricultural soil (Walker and Desanker 2004). Following clearing, Solomon et al. (2000) reported a 56% reduction of soil C content in the cultivated fields in a semiarid area in Tanzania. Reintroducing trees into the landscape can restore some of the lost C. In Mozambique, Williams et al. (2008) showed that on farmland that had been abandoned for more than 20 years, stem C stocks were at 15.7 Mg C ha^{-1} almost as high as in protected woodland (19.0 Mg C ha^{-1}). During 2–25 years of re-growth, wood C stocks accumulated at 0.7 Mg C ha^{-1} $year^{-1}$

in Mozambique (Walker and Desanker 2004). Similarly, mean annual increment was 0.5 Mg C ha^{-1} $year^{-1}$ in 16-year old coppiced Miombo woodland in northern Zambia (Stromgaard 1985) and 0.9 Mg C ha^{-1} $year^{-1}$ over 35 years (Chidumayo 1997). According to Williams et al. (2008) soil C stocks in the top 0.3 m on abandoned land had a narrower range (21–74 Mg C ha^{-1}) than stocks in the Miombo woodland soils (18–140 Mg C ha^{-1}) and with a median C stock of 44.9 Mg C ha^{-1} had reached 78% of median C stocks in Miombo soils (57.9 Mg C ha^{-1}) (Williams et al. 2008). Agroforestry practices are designed to raise system C levels without requiring abandonment of crop production.

Technical Capacity to Sequester Carbon

Although the C sequestration potential of parkland systems in Southern Africa has not yet been studied extensively, these systems are believed to store substantial amounts of C. *Faidherbia albida* at Morogoro, Tanzania accumulated 1.2 Mg C ha^{-1} $year^{-1}$ during 6 years after planting at 6 m spacing (Okorio and Maghembe 1994). If calculated at 50 trees ha^{-1} density, which is more realistic for an agroforestry setting, this C accumulation would amount to an annual rate of 0.22 Mg C ha^{-1}.

Nyadzi et al. (2003) compared the performance of different tree species in rotational woodlots in Tabora and Shinyanga in Tanzania reporting biomass accumulation in the wood between 9.6 and 40.9 Mg biomass ha^{-1} over 5 years, corresponding to mean C accumulation rates between 2.6 and 5.8 Mg C ha^{-1} $year^{-1}$ (assuming 50% C in biomass). In Tanzania, C sequestered in wood ranged from 11.6 Mg C ha^{-1} in *Acacia nilotica* (L) Del. and *A. auriculiformis* A. Cunn. ex Benth. to 25.5 Mg C ha^{-1} in *A. crassicarpa* A. Cunn. ex Benth. after 5 years (Kimaro 2009). The resulting stocks were comparable with wood C (19 Mg C ha^{-1}) reported from protected Miombo forests in Mozambique (Williams et al. 2008). Wood C accumulation rates ranged between 2.3 Mg C ha^{-1} $year^{-1}$ under *A. nilotica* and 5.1 Mg C ha^{-1} $year^{-1}$ under *A. crassicarpa* (Kimaro 2009). These figures are higher than the 1.5–3.5 Mg C ha^{-1} $year^{-1}$ estimated for smallholder AFS in the tropics (Montagnini and Nair 2004), but these rates are of course only reached during the woodlot fallow phases. Soil organic C stocks (within 0–30 cm depth) under 5 year old rotational woodlots (15.8–25.6 Mg C ha^{-1}) in Morogoro were higher than in soils that had been continuously cropped for the same time period (13 Mg ha^{-1}) and fallowed Miombo soils (9–15 Mg C ha^{-1}: Kimaro 2009). However, because initial C stocks were not measured in this study, inferring sequestration rates is not possible.

Kaonga and Bayliss-Smith (2009) compared woodlots of eight different agroforestry species in eastern Zambia, and reported that 4.3–9.5 Mg C ha^{-1} were stored in aboveground biomass after 2 years. Rotational woodlots satisfy household and regional fuelwood demand (Nyadzi et al. 2003) and may thus reduce pressure on adjacent woodland, opening potential opportunities for payments for avoided deforestration and forest degradation. Improved fallows can also store substantial amounts of C in plants and soil (Albrecht and Kandji 2003). In eastern Zambia, Kaonga and Coleman (2008) estimated the annual aboveground plant C input at 2.8 Mg C ha^{-1} $year^{-1}$ for *Tephrosia vogelii* Hook.f., 2.7 Mg C ha^{-1} $year^{-1}$ for *Sesbania sesban* (L.) Merr. and 2.5 Mg C ha^{-1} $year^{-1}$ for *C. cajan*, which was comparable with 2.7 Mg C ha^{-1} $year^{-1}$ recorded for fully fertilized maize. Estimated total SOC stocks under these species were higher at 27.3–31.2 Mg C ha^{-1} (Table 1; Kaonga and Coleman 2008) than under fully fertilized maize (26.2 Mg C ha^{-1}) and unfertilized maize (22.2 Mg C ha^{-1}).

Makumba et al. (2007) compared C sequestration in two fields of *Gliricidia*-maize intercropping conducted for 7 and 10 years with continuous cropping of sole maize. Carbon stocks in the top 200 cm of the soil were about twice as high in the intercropping system as in sole maize. In addition, tree stumps and structural roots stored a total of 17 Mg ha^{-1} of C after 7 years of intercropping (Makumba et al. 2007).

Economic Potential of Carbon Sequestration in Africa

Little work has been done on the economic potential of CS activities. However, rough estimates of the economic benefits that farmers may be able to derive from different kinds of C-sequestering agroforestry practices can be calculated based on a few relatively straightforward factors. The value of C that is sequestered annually depends on the C accumulation rate and the sale price of the C (Fig. 3). Since C prices vary widely among sequestration schemes and future C prices are uncertain, it seems sensible to calculate benefits for a range of values.

While the C value can relatively easily be computed, other economic factors are more site-specific and difficult to estimate. Costs of participating in C markets must be subtracted from the total C value. These costs include expenses for MRV of sequestered C, the cost of registering the C project, and possibly additional transaction costs incurred by C marketing. An economic budget must also include the costs of planting or protecting trees or changing management practices in other ways. Added benefits from higher system C, such as higher crop yield potential due to higher soil organic matter contents, should also be taken into account. Finally, total profit from the C sequestration activity, including the value of all goods that are produced should be compared with profits that could be derived from other activities. Where these opportunity costs, the income foregone by choosing the high C management option, are higher than the profit from the high-CS system, adopting this system may not ultimately benefit farmers.

Most factors are difficult to estimate and require site-specific research, modeling and a range of assumptions. It is, however, relatively easy to estimate C values by multiplying sequestration rates by assumed C prices (while correcting for the fact that C prices are commonly given per Mg CO_2-eq, while sequestration rates are often in Mg C). Figure 4 shows C value estimates for typical AFS in Africa. Carbon accumulation rates are taken from the case studies mentioned above, and contours

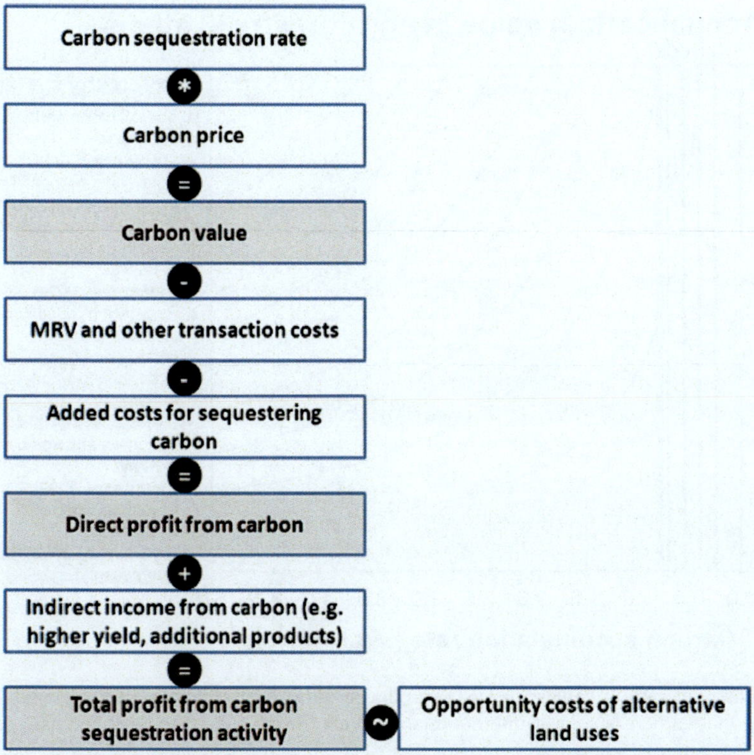

Fig. 3 Factors determining the economic potential of agroforestry systems to sequester carbon

in the figure show C values, as a function of C price (given on the y-axis). For most agroforestry options, C accumulation rates are relatively low, resulting in low C values, in particular at low assumed C prices. Common estimates of C prices in the literature range around 10 USD per Mg CO_2-eq or less, which would translate into C values of less than 30 USD ha^{-1} year^{-1} for typical AFS. The only AFS that were reported to accumulate C at a relatively fast rate were improved fallows and rotational woodlots. At a C price of 10 USD per Mg CO_2-eq, sequestration rates would translate into C values of up to 200 USD ha^{-1} year^{-1}. These values, however, are only produced during the fallow phases of the AFS, after which trees and shrubs are harvested and incorporated into the soil, processed into wood products, or used as fuelwood. Net C accumulation rates, and thus the amount of C that is credibly and permanently (or at least for a long time) sequestered from the atmosphere is thus substantially lower than biomass C buildup suggests. Nyamadzawo et al. (2008) tracked the effects of improved fallow practices on soil C stocks (0–20 cm depth) over two fallow/cropping cycles (2 years improved fallow/3 years cultivation), finding consistently higher soil organic matter contents under rotation than in continuous maize. However, after two full rotations, the improved fallow system had only

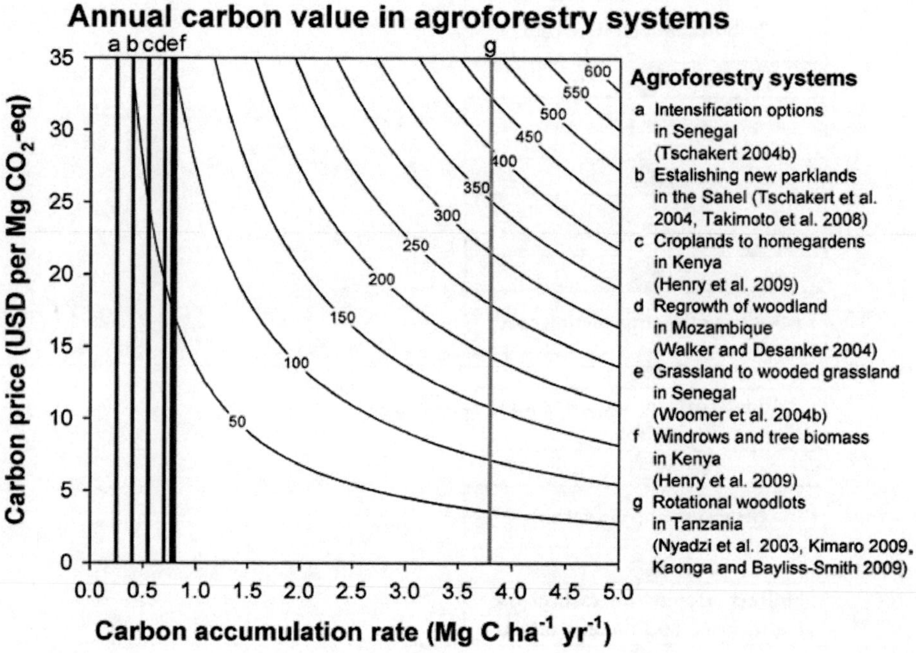

Fig. 4 Potential carbon values (in USD) produced annually by agroforestry systems in Africa, assuming that carbon is sold at international markets. *Black bars* (bars a–f) indicate permanent agroforestry systems, whereas the *grey bar* (bar g) signifies rotation woodlots that are only present for part of the cropping cycle. All bars are *straight lines*, assuming constant CS rates in each AFS, and carbon values for each AFS depend primarily on the carbon price. For example, at a price of 10 USD per Mg CO_2-eq, conversion of croplands to homegardens in Kenya, which can store 0.55 Mg C ha^{-1} $year^{-1}$ (*bar c*), would produce a carbon value of 20 USD $year^{-1}$; at 30 USD per Mg CO_2-eq the value would be 61 USD $year^{-1}$

between 2.2 and 6.6 Mg C ha^{-1} more than the continuous maize system, corresponding to a time-averaged advantage compared to continuous maize of between 0.2 and 0.7 Mg C ha^{-1} $year^{-1}$. Because soils under continuous maize cultivation (the control in this study) likely experienced further depletion of soil C during the study period, the net sequestration rate of improved fallows is even lower. Such considerations will also apply to C dynamics of rotational woodlots. Carbon sequestration rates by such systems depend on processes during the tree and the cultivation phases, as well as the use of the trees. Where trees are processed into long-lived wood products, substantial amounts of C may be sequestered, but when woody biomass is predominantly used as fuelwood, sequestration rates are likely low. Few studies contain sufficient data for calculating time-averaged net CS rates, but we find it unlikely that such rates exceed 1 Mg C ha^{-1} $year^{-1}$, especially for fuelwood systems. It is also worth noting that C finance projects normally only pay for C sequestered *in situ*, and that all C that is removed from the field is considered emitted, even if the wood is preserved elsewhere.

While complete budgets cannot be derived from our back-of-the-envelope calculations, low C values for many AFS indicate that they may not be profitable, unless substantial additional benefits can be produced by the system. Due to transaction and MRV costs, profits from C sales may be quite a bit lower than the net C value produced. Of course, many AFS deliver added benefits, such as yield increases and additional marketable products, and more economic analysis should focus on the importance of C credits in whole farm budgets.

A few studies have explored the economics of CS by African agroforestry in more detail. Henry et al. (2009) found that in East Africa, afforestation is likely among the fastest ways to increase aboveground C stocks, whereas inducing smallholder farmers, with average land holdings of about 1 ha, to plant additional trees without adversely affecting food production is 'a real challenge'. They demonstrate that at the current market price for C and considering average farm sizes in their study area, 140–300 farms (or 170 to >400 ha depending on intensification scenarios) would have to collaborate in C marketing, in order to compensate for the minimum transaction costs incurred by marketing C in Clean Development Mechanism Afforestation/Reforestation (CDM A/R) projects (Henry et al. 2009).

Some site-specific modeling efforts have been undertaken to evaluate the suitability of C sequestration as an income option for farmers in the West African Sahel. Doraiswamy et al. (2007) assumed a carbon price of \$10 per Mg of sequestered C (not CO_2-eq, which is more commonly used), which resulted in annual returns from C sales of between \$0.84 and \$1.46 ha^{-1}. This was between 0.2% and 0.8% of net annual revenue of the modeled farm. In the case of the most economically successful farming option, even a C price 20 times higher than \$10 per Mg C would bring the proportion of farm revenues from C sales to only 4.3%. All these figures assume that no transaction, measurement, reporting, or verification costs are incurred. Tschakert (2004a) also reports low C revenues, amounting to between <\$2 and <\$7 per hectare and year, or between 1% and 4.5% of revenue per hectare (at \$15 per Mg C; once again not CO_2-eq). At a higher C price of \$25 per Mg C, C income would constitute between 1.6% and 7.2% of farm revenue, again without including the costs of C marketing. Assuming a C price of \$42 per Mg C, Takimoto et al. (2008a) calculated that selling C credits would raise the net present value of live fences by \$14 (from \$96 to \$110) and that of fodder banks by \$16.5 (from \$159 to \$175). Carbon revenues would thus amount to between 9 and 13% of net present value. This estimate assumes an accounting method that is favorable to farmers (C revenues drop to 0.2–0.3% of net present value, if the alternative 'tonne-year accounting' is chosen) and that all costs of C marketing are external to landowners.

These figures indicate that payments for CS by agroforestry are unlikely to generate substantial income to smallholder farmers in most cases, unless C payments are combined with payments for other environmental services provided by agroforestry. Carbon prices will also influence the attractiveness of sequestration projects. A macro-economic simulation by Diagana et al. (2007) confirms this impression, finding that the amount of C likely to be sequestered in the Nioro region of Senegal's Peanut Basin ranges between 200 Gg C at a C price of \$0 and 1.3 Tg C at \$200 per Mg C. Future C prices are difficult to predict, but if an efficient global market develops, with abundant participation by small- and large-holder farmers around the world, C prices will likely drop to close to the costs of C sequestration paid by the most efficient sequestration efforts. Carbon is most easily sequestered in ecological zones that are much more productive than the Sahel, e.g. in the humid tropics. It is thus difficult to imagine that at world market prices, C sequestration projects by Sahelian farmers will be competitive. In all studies that we reviewed, C incomes were very low, even when (probably unrealistically) assuming that no costs were incurred by C sales. It is also troubling that, according to the 'additionality' criterion in the Clean Development Mechanism, the most C intensive forms of land use in the region, such as Sahelian parklands, would be excluded because they allow for only little additional C sequestration. From the smallholder perspective, it should also be considered that C sales do not necessarily present a 'win-win' situation, because on Sahelian farms, most resources, including trees, are intensively used. Depending on the opportunity costs of potential income options that are restricted under C sequestration contracts, net benefits compared to a situation without a formal C contract could thus easily turn negative, because farming in the Sahel is often opportunistic and requires farmers to adapt to variable circumstances (Tschakert 2004a).

Practical Potential to Sequester Carbon in Africa

In most situations the practical potential to sequester C is even lower than economic calculations suggest, due to a number of constraints that are often overlooked. Acceptance of new land management options by farmers, for example, has been shown to depend on a variety of factors in addition to the economic bottom line (Ajayi 2007). Even farmers who decide to test a new AFS may choose not to adopt it because of poor tree performance in initial trials, caused by pests, drought, bush fires or other biotic or abiotic factors (Sileshi et al. 2007). Damage of young trees by livestock can also limit adoption rates (Ajayi and Kwesiga 2003).

Commonly encountered constraints to the adoption of tree-based systems also include land tenure, cultural norms, and household power structures in many regions (Chidumayo 2002; German et al. 2009; Mwase et al. 2007; Aquino et al. 2011). For example, land tenure insecurity may result in degradation of open access land and unwillingness of people to plant trees (German et al. 2009; Mwase et al. 2007). Many traditional land use systems include customary land resources, which are exploitable by the entire community through grazing, hunting, settlement areas, crop fields, and graveyards. In Malawi, customary land covers 3.1 million ha and half of the forested area is on customary land, and about two-thirds of the customary land is disturbed (Mwase et al. 2007). The incentive for individuals to plant trees on common land is low, and the distribution of potential benefits from C sales would be complicated. Land tenure has been identified as one of the central impediments to making the CDM work for smallholder farmers (Unruh 2008; Aquino et al. 2011). Moreover, a deficit of information on management options or appropriate inputs may constrain the CS potential. The reasons for this are weak extension capacity, scarcity or lack of appropriate planting material (quality seed, seedling, etc.), and lack of knowledge and skills in tree management in agroforestry.

Even though CS is often mentioned as a means to fight poverty, the poorest households in resource-constrained settings are likely to benefit least from C sales. Studies by Tschakert (2004a, b, 2007) in Senegal indicate that due to the initial investments necessary for implementation of CS activities, poor farmers may not be able to participate in C markets, and if they do, their incomes will be lower than those of rich farmers. Dixon et al. (1994) estimate the costs of establishing productive tree plantings at between $500 and $3,000 per hectare, in various parts of the world. Even at the low end of this range, establishment costs would thus probably be too high for most Sahelian farmers. A pro-poor focus rather than a purely market-based approach is thus needed, if smallholder projects are really to help the poor. Such an approach is more likely to be endorsed by a development agency rather than by a more profit-oriented organization, taking away from the idea of C economics more or less automatically leading to enhanced livelihoods of smallholder farmers (Tschakert 2004b). Unfortunately, such socioeconomic considerations have been absent from most CS studies. In the context of relatively food insecure farmers, for whom risk management is of crucial importance, socioeconomic aspects must be considered when planning and studying CS options and potentials. Tschakert and Tappan (2004) therefore called for a farmer-centered approach to CS that includes not only effects of C storage on climate, but also the impacts of the necessary activities on farmers' livelihoods.

The most striking knowledge gap among all CS studies is the lack of efforts to estimate the transaction costs for implementing C projects, as well as the costs of measurement, reporting and verification of the sequestered C. Institutional and governance constraints, land tenure, and market access are also typically neglected. This is understandable, because most of these factors are difficult to study empirically and many cannot reliably be projected. However, analyzing plausible policy scenarios and including estimates of MRV costs into economic analyses would provide insights that are necessary for making justifiable recommendations about the implementation of C schemes. Moreover, the trade-offs incurred by retaining trees for C sequestration vs. using them as fuel, fodder and fertilizer have not been explored adequately.

Conclusions

For many parts of Africa, some data exist for the biophysical and technical potential of agroecosystems to sequester additional C, but important aspects of CS have not been adequately investigated. Many studies have delivered good estimates of aboveground C accumulation in AFS, but numbers for the soil are scarcer and less reliable. While some studies have compared soil C stocks under agroforestry with stocks under different land use types, experimental studies that have compared soil C levels before and after agroforestry establishment are currently lacking.

Such studies, however, are needed to separate the likely build-up of soil C by agroforestry from the decreases effected by, say, continuous (year-after-year) maize cultivation. Estimates of the economic potential of agroforestry to sequester C are only available for very few locations. The few studies that have investigated this in the relatively unproductive ecosystems of the Sahel indicated limited opportunities for farmers to benefit from C markets.

Complete assessments of the practical potential to implement C schemes for smallholders are currently lacking. Without knowledge about the complete array of potentials, ex-ante assessments of the impact that C schemes would have on African livelihoods are currently data-limited. It is quite clear that large-scale implementation of agroforestry practices in Africa would benefit the global climate, but whether such a move would produce benefits for farmers is unclear. Given that in terms of productivity, the African drylands compare unfavorably with most other ecosystems of the world, the chance to generate substantial incomes from C sales is probably quite slim for farmers in such regions, in particular if C prices are determined by international supply and demand dynamics. Without more comprehensive studies that explore CS potentials at all levels, however, such considerations are highly speculative. Laying the groundwork for involving smallholder farmers in international C schemes will require comprehensive, interdisciplinary, and objective assessments of the potential constraints to adoption, as well as the whole array of impacts that C schemes may have on farmers and farming communities.

References

Abbot JIO, Homewood K (1999) A history of change: causes of Miombo woodland decline in a protected area in Malawi. J Appl Ecol 36:422–433

Ajayi OC (2007) User acceptability of sustainable soil fertility technologies: lessons from farmers' knowledge, attitude and practice in southern Africa. J Sustain Agric 30:21–40

Ajayi OC, Kwesiga F (2003) Implications of local policies and institutions on the adoption of improved fallows in eastern Zambia. Agroforest Syst 59:327–336

Akinnifesi FK, Chirwa PW, Ajayi OC, Sileshi G, Matakala P, Kwesiga FR, Harawa H, Makumba W (2008) Contributions of agroforestry research to livelihood of smallholder farmers in Southern Africa: 1. Taking stock of the adaptation, adoption and impact of fertilizer tree options. Medwell Online Agric J 3:58–75

Albrecht A, Kandji ST (2003) Carbon sequestration in tropical agroforestry systems. Agric Ecosyst Environ 99:15–27

Batjes NH (2001) Options for increasing carbon sequestration in west African soils: an exploratory study with special focus on senegal. Land Degrad Dev 12:131–142

Batjes NH (2004a) Estimation of soil carbon gains upon improved management within croplands and grasslands of Africa. Environ Dev Sustain 6:133–143

Batjes NH (2004b) Soil carbon stocks and projected changes according to land use and management: a case study for Kenya. Soil Use Manag 20:350–356

Boffa JM (1999) Agroforestry parklands in sub-Saharan Africa. FAO Conservation Guide 34. Food and Agriculture Organization of the United Nations, Italy, p. 230

Campbell BM, Clarke JM, Gumbo DJ (1991) Traditional agroforestry practices in Zimbabwe. Agroforest Syst 14:99–111

Campbell B, Frost P, Byron N (1996) Miombo woodlands and their use: overview and key issues. In: Campbell B (ed.) The Miombo in transition: woodlands and welfare in Africa. CIFOR, Bogor, pp 1–10

Cannell MGR (2003) Carbon sequestration and biomass energy offset: theoretical, potential and achievable capacities globally, in Europe and the UK. Biomass Bioenergy 24:97–116

Chakeredza S, Hove L, Akinnifesi FK, Franzel S, Ajayi OC, Sileshi G (2007) Managing fodder trees as a solution to human-livestock food conflicts and their contribution to income generation for smallholder farmers in southern Africa. Nat Resour Forum 31:286–296

Chidumayo EN (1987) A shifting cultivation land-use system under population pressure in Zambia. Agroforest Syst 5:15–25

Chidumayo EN (1997) Miombo ecology and management: an introduction. Intermediate Technology Publications in association with the Stockholm Environment Institute, London, 166 p

Chidumayo EN (2002) Changes in Miombo woodland structure under different land tenure and use systems in central Zambia. J Biogeogr 29:1619–1626

de Aquino AR, Aasrud A, Guimarães L (2011) Can forest carbon finance influence land tenure security in project areas? Preliminary lessons from projects in Niger and Kenya. In: Kumar BM, Nair PKR (eds.) Carbon sequestration potential of agroforestry systems: opportunities and challenges. Springer, Dordrecht, pp 231–246

Diagana B, Antle J, Stoorvogel J, Gray K (2007) Economic potential for soil carbon sequestration in the Nioro region of Senegal's Peanut Basin. Agric Syst 94:26–37

Dixon RK, Winjum JK, Schroeder PE (1993) Conservation and sequestration of carbon: the potential of forest and agroforest management practices. Glob Environ Change Hum Policy Dimens 3:159–173

Dixon RK, Winjum JK, Andrasko KJ, Lee JJ, Schroeder PE (1994) Integrated land-use systems: assessment of promising agroforest and alternative land-use practices to enhance carbon conservation and sequestration. Clim Change 27:71–92

Doraiswamy PC, McCarty GW, Hunt ER Jr, Yost RS, Doumbia M, Franzluebbers AJ (2007) Modeling soil carbon sequestration in agricultural lands of Mali. Agric Syst 94:63–74

Eriksen C (2007) Why do they burn the 'bush'? Fire, rural livelihoods, and conservation in Zambia. Geogr J 173:242–256

Fernandes ECM, Oktingati A, Maghembe J (1985) The Chagga homegardens: a multistoried agroforestry cropping system on Mt. Kilimanjaro (Northern Tanzania). Agroforest Syst 2:73–86

German G, Akinnifesi FK, Edriss AK, Sileshi G, Masangano C, Ajayi OC (2009) Influence of property rights on farmers' willingness to plant indigenous fruit trees in Malawi and Zambia. Afr J Agric Res 4:427–437

Gijsbers HJM, Kessler JJ, Knevel MK (1994) Dynamics and natural regeneration of woody species in farmed parklands in the Sahel region (Province of Passore, Burkina-Faso). Forest Ecol Manag 64:1–12

Glenday J (2008) Carbon storage and emissions offset potential in an African dry forest, the Arabuko-Sokoke Forest, Kenya. Environ Monit Assess 142:85–95

Hanan NP, Kabat P, Dolman AJ, Elbers JA (1998) Photosynthesis and carbon balance of a Sahelian fallow savanna. Glob Change Biol 4:523–538

Hemp A (2006) The banana forests of Kilimanjaro: biodiversity and conservation of the Chagga homegardens. Biodivers Conserv 15:1193–1217

Henry M, Tittonell P, Manlay RJ, Bernoux M, Albrecht A, Vanlauwe B (2009) Biodiversity, carbon stocks and sequestration potential in aboveground biomass in smallholder farming systems of western Kenya. Agric Ecosyst Environ 129:238–252

Hulme M (2001) Climatic perspectives on Sahelian desiccation: 1973-1998. Glob Environ Change Hum Policy Dimens 11:19–29

Ingram JSI, Fernandes ECM (2001) Managing carbon sequestration in soils: concepts and terminology. Agric Ecosyst Environ 87:111–117

Jarecki MK, Lal R (2003) Crop management for soil carbon sequestration. Crit Rev Plant Sci 22:471–502

Kanschik W, Becker B (2001) Dry Miombo – ecology of its major plant species and their potential use as bio-indicators. Plant Ecol 155:139–146

Kaonga ML, Bayliss-Smith TP (2009) Carbon pools in tree biomass and the soil in improved fallows in eastern Zambia. Agroforest Syst 76:37–51

Kaonga ML, Coleman K (2008) Modelling soil organic carbon turnover in improved fallows in eastern Zambia using the RothC-26.3 model. Forest Ecol Manag 256:1160–1166

Kimaro AA (2009) Sequential agroforestry systems for improving fuelwood supply and crop yield in semi-arid Tanzania. PhD thesis, University of Toronto, Toronto, 124 p

Kuersten E, Burschel P (1993) CO_2-mitigation by agroforestry. Water Air Soil Pollut 70:533–544

Lawton RM (1978) Study of dynamic ecology of Zambian vegetation. J Ecol 66:175–198

Liu S, Kairé M, Wood E, Diallo O, Tieszen LL (2004) Impacts of land use and climate change on carbon dynamics in south-central Senegal. J Arid Environ 59:583–604

Long SP, Moya EG, Imbamba SK, Kamnalrut A, Piedade MTF, Scurlock JMO, Shen YK, Hall DO (1989) Primary productivity of natural grass ecosystems of the Tropics - a reappraisal. Plant Soil 115:155–166

Lufafa A, Bolte J, Wright D, Khouma M, Diedhiou I, Dick RP, Kizito F, Dossa E, Noller JS (2008) Regional carbon stocks and dynamics in native woody shrub communities of Senegal's Peanut Basin. Agric Ecosyst Environ 128:1–11

Makumba W, Akinnifesi FK, Janssen B, Oenema O (2007) Long-term impact of a gliricidia-maize intercropping system on carbon sequestration in southern Malawi. Agric Ecosyst Environ 118:237–243

Maranz S (2009) Tree mortality in the African Sahel indicates an anthropogenic ecosystem displaced by climate change. J Biogeogr 36:1181–1193

Matthews RB, Holden ST, Volk J, Lungu S (1992) The potential of alley cropping in improvement of cultivation systems in the high rainfall areas of Zambia. 1. Chitemene and Fundikila. Agroforest Syst 17:219–240

Montagnini F, Nair PKR (2004) Carbon sequestration: an underexploited environmental benefit of agroforestry systems. Agroforest Syst 61–62:281–295

Mwase WF, Bjornstad A, Bokosi JM, Kwapata MB, Stedje B (2007) The role of land tenure in conservation of tree and shrub species diversity in Miombo woodlands of southern Malawi. New For 33:297–307

Nair PKR, Kumar BM, Nair VD (2009) Agroforestry as a strategy for carbon sequestration. J Plant Nutr Soil Sci-Z Pflanzenernahr Bodenkd 172:10–23

Nhantumbo A, Ledin S, Du Preez C (2009) Organic matter recovery in sandy soils under bush fallow in southern Mozambique. Nutr Cycle Agroecosyst 83:153–161

Nyadzi GI, Otsyina RM, Banzi FM, Bakengesa SS, Gama BM, Mbwambo L, Asenga D (2003) Rotational woodlot technology in northwestern Tanzania: tree species and crop performance. Agroforest Syst 59:253–263

Nyamadzawo G, Chikowo R, Nyamugafata P, Nyamangara J, Giller K (2008) Soil organic carbon dynamics of improved fallow-maize rotation systems under conventional and no-tillage in Central Zimbabwe. Nutr Cycle Agroecosyst 81:85–93

Okorio J, Maghembe JA (1994) The growth and yield of *Acacia albida* intercropped with maize (*Zea mays*) and beans (*Phaseolus vulgaris*) at Morogoro, Tanzania. Forest Ecol Manag 64:183–190

Reij C, Tappan GG, Smale M (2009) Agroenvironmental transformation in the Sahel - another kind of "Green Revolution". IFPRI Discussion Paper No. 914:1–52

Sileshi G, Mafongoya PL (2006) Long-term effects of improved legume fallows on soil invertebrate macrofauna and maize yield in eastern Zambia. Agric Ecosyst Environ 115:69–78

Sileshi G, Akinnifesi FK, Ajayi OC, Chakeredza S, Kaonga M, Matakala PW (2007) Contribution of agroforestry to ecosystem services in the Miombo eco-region of eastern and southern African. Afr J Environ Sci Technol 1:68–80

Sileshi G, Schroth G, Rao MR, Girma H (2008) Weeds, diseases, insect pests and tri-trophic interactions in tropical agroforestry. In: Batish DR, Kohli RK, Jose S, Singh HP (eds.) Ecological basis of agroforestry. CRC Press, Boca Raton, pp 73–94

Sinclair FL (1999) A general classification of agroforestry practice. Agroforest Syst 46:161–180
Solomon D, Lehmann J, Zech W (2000) Land use effects on soil organic matter properties of chromic luvisols in semi-arid northern Tanzania: carbon, nitrogen, lignin and carbohydrates. Agric Ecosyst Environ 78:203–213
Stigter CJ, Mohammed AE, Al-amin NKN, Onyewotu LOZ, Oteng'i SBB, Kainkwa RMR (2002) Agroforestry solutions to some African wind problems. J Wind Eng Ind Aerod 90:1101–1114
Stromgaard P (1985) Biomass, growth, and burning of woodland in a shifting cultivation area of South Central Africa. Forest Ecol Manag 12:163–178
Takimoto A, Nair PKR, Alavalapati JRR (2008a) Socioeconomic potential of carbon sequestration through agroforestry in the West African Sahel. Mitig Adapt Strat Glob Change 13:745–761
Takimoto A, Nair PKR, Nair VD (2008b) Carbon stock and sequestration potential of traditional and improved agroforestry systems in the West African Sahel. Agric Ecosyst Environ 125:159–166
Tieszen LL, Tappan GG, Touré A (2004) Sequestration of carbon in soil organic matter in Senegal: an overview. J Arid Environ 59:409–425
Traoré PCS, Bostick WM, Jones JW, Koo J, Goïta K, Bado BV (2008) A simple soil organic-matter model for biomass data assimilation in community-level carbon contracts. Ecol Appl 18:624–636
Tschakert P (2004a) Carbon for farmers: assessing the potential for soil carbon sequestration in the old peanut basin of Senegal. Climate Change 67:273–290
Tschakert P (2004b) The costs of soil carbon sequestration: an economic analysis for small-scale farming systems in Senegal. Agric Syst 81:227–253
Tschakert P (2007) Environmental services and poverty reduction: options for smallholders in the Sahel. Agric Syst 94:75–86
Tschakert P, Tappan G (2004) The social context of carbon sequestration: considerations from a multi-scale environmental history of the Old Peanut Basin of Senegal. J Arid Environ 59:535–564
Tschakert P, Khouma M, Sène M (2004) Biophysical potential for soil carbon sequestration in agricultural systems of the Old Peanut Basin of Senegal. J Arid Environ 59:511–533
Unruh JD (2008) Carbon sequestration in Africa: the land tenure problem. Glob Environ Change Hum Policy Dimens 18:700–707
Venema HD, Schiller EJ, Adamowski K, Thizy JM (1997) A water resources planning response to climate change in the Senegal River Basin. J Environ Manage 49:125–155
Walker SM, Desanker PV (2004) The impact of land use on soil carbon in Miombo Woodlands of Malawi. Forest Ecol Manag 203:345–360
Williams M, Ryan CM, Rees RM, Sambane E, Fernando J, Grace J (2008) Carbon sequestration and biodiversity of re-growing Miombo woodlands in Mozambique. Forest Ecol Manag 254:145–155
Woomer PL, Tieszen LL, Tappan G, Touré A, Sall M (2004a) Land use change and terrestrial carbon stocks in Senegal. J Arid Environ 59:625–642
Woomer PL, Touré A, Sall M (2004b) Carbon stocks in Senegal's Sahel transition zone. J Arid Environ 59:499–510

Soil Carbon Sequestration in Cacao Agroforestry Systems: A Case Study from Bahia, Brazil

E.F. Gama-Rodrigues, A.C. Gama-Rodrigues, and P.K. Ramachandran Nair

Abstract Agroforestry systems (AFS) based on cacao (*Theobroma cacao* L.) may play an important role in capturing carbon (C) aboveground and storing it belowground (soil) through continuous deposition of plant residues. Cacao AFS in Bahia, Brazil, are comprised of cacao planted either with woody species such as *Erythrina* spp. and *Gliricidia* spp. or under tree canopies in natural forest, the latter being known as "cabruca". The large amounts of leaf litter, roots, and woody material from shade species as well as cacao represent a substantial addition of C into these systems, most of which, following decomposition, is stored in the soil. The total C storage in the weathered Oxisols under cacao AFS in Bahia is estimated as 302 Mg ha^{-1} to 1 m depth. Occlusion of C in soil aggregates could be a major mechanism of C protection in these soils. Therefore it is important to know the amount of soil C storage across different soil aggregate classes at different soil depths and identify the extent of the sequestered C that is occluded in the soil aggregates. Furthermore, the deep-rooted nature of cacao and shade trees makes it imperative to look below the surface soil, where most conventional soil studies are focused. Carbon sequestration potential of cacao and other shaded-perennial-crop-based AFS could be a source of income for the farmers of these crops, the majority of whom are smallholders. Understanding the mechanisms of soil C sequestration could lead to proper realization of this potential through better management options.

E.F. Gama-Rodrigues (✉) • A.C. Gama-Rodrigues
Soil Laboratory, Norte Fluminense State University, Campos dos Goytacazes, RJ 28013–602, Brazil
e-mail: emanuela@uenf.br; tonygama@ufl.edu

P.K.R. Nair
Center for Subtropical Agroforestry, School of Forest Resources and Conservation, University of Florida, P.O. Box 110410, Gainesville, FL 32611–0410, USA
e-mail: pknair@ufl.edu

B.M. Kumar and P.K.R. Nair (eds.), *Carbon Sequestration Potential of Agroforestry Systems: Opportunities and Challenges*, Advances in Agroforestry 8,
DOI 10.1007/978-94-007-1630-8_5,

Keywords Cacao cabruca • Ecosystem services • *Erythrina* spp. • Natural forest • Shaded perennial systems • Soil aggregates

Introduction

Research interest in soil organic carbon (SOC) as a potential sink for atmospheric carbon dioxide (CO_2) has increased considerably in recent years (IPCC 2000). It is now well recognized that storage of SOC in soils can be increased directly by increasing C returns to the soil in the form of crop residues, manure, or other organic amendments, and indirectly by management practices that decrease soil organic matter (SOM) decomposition and erosion. Such practices include conservation tillage, converting degraded arable lands to perennial grasslands, afforestation, reforestation, restoration of degraded lands, improved silvicultural techniques to increase growth rates, and the implementation of agroforestry practices in agricultural lands (Lal 2004; Montagnini and Nair 2004; Nair et al. 2009, 2010).

Agroforestry systems (AFS) in which high amounts of organic materials are added to the soil, such as the shaded perennial-crop systems, have special relevance because of their high potential for sequestering carbon (C) in soil (Muñoz and Beer 2001; Montagnini and Nair 2004; Oelbermann and Voroney 2007). The extent to which such organic materials are deposited depends on both species (the crop and the shade tree) and the management systems involved.

Cacao (*Theobroma cacao* L.) is rather unique in this respect because of the relatively high amount of deposition via litterfall, estimated as approximately 7.0 Mg ha^{-1} $year^{-1}$ (Fontes 2006). In its native habitat of central and western Amazonia in Brazil, cacao grows as a forest understory tree. Today it is cultivated over approximately two million ha in many tropical humid lowlands of Central and South America, West Africa, South- and Southeast Asia, and the Caribbean and the Pacific islands (World Cocoa Foundation: www.worldcocoafoundation.org; accessed on 24 January 2011). The world production of cacao beans is approximately 3×10^6 Mg (three million tons), with an average market value of about $5 million (World Cocoa Foundation: www.worldcocoafoundation.org; accessed on 18 January 2011); thus the economic importance of the crop is much more than what its area under cultivation would suggest.

The extent to which the C released from the decomposition of the high amounts of litterfall that is continuously deposited on the soil surface in cacao-based AFS (and other shaded perennial crop systems) is retained (sequestered) in the soil depends to a large degree on the soil properties, especially soil aggregation (Six et al. 2004; Gama-Rodrigues et al. 2010; Nair et al. 2010). The relationship between soil aggregation and the reserve pool of C in soils under cacao, based on a case-study conducted on cacao-AFS in southern Bahia, Brazil, is the focus of this paper. The relevance of such studies for the cacao-AFS and shaded perennial crop systems on a broader scale is also examined.

Carbon Input to Soils Under Cacao Agroforestry Systems

In Brazil, cacao AFS are established over more than 0.7 million ha on soils with differing pedogenesis and land use history. The productive capacity of these soils decreases strongly with each harvest due to the nutrients being removed from the system through cacao beans and other products associated with cacao AFS such as the rubber tree (*Hevea brasiliensis* H.B.K. M.-Arg.), coconut (*Cocos nucifera* L.), and a multitude of other woody and non-woody species. In southern Bahia, which is located in the Atlantic forest region of Brazil, cacao AFS include two main types of planting systems: one known as "cabruca" or cacao planted under a thinned natural forest and shaded by native forest trees, and the other involving complete removal of all naturally occurring trees before planting cacao along with introduced shade trees. The leguminous tree *Erythrina* spp. (hereafter referred to as erythrina) is the most common among such introduced shade trees (cacao + erythrina). Under the cabruca system, vegetation structure and stratification are similar to those of natural forests. In the latter system, the cacao stand density is about 1,100 plants ha^{-1}, which is about two times that under the cabruca system (Müller and Gama-Rodrigues 2007). Cacao is also grown in between or under commercial crops such as banana (*Musa* spp.), cassava (*Manihot esculenta* Crantz), coconut, rubber, peach palm (*Bactris gasipaes* Kunth.), açaí (*Euterpe oleracea* Mart.), coffee (*Coffea canephora* Pierre ex A. Froehner), and a variety of other trees including *Bagassa guianensis* Aubl., *Bertholetia excelsa* H.B.K., *Cordia alliodora* (Ruiz & Pavón) Oken., *Schizolobium amazonicum* Huber ex Ducke, *Swietenia macrophylla* (L.) Jacq., *Tabebuia heptaphylla* (Vell.) Toledo, and *Tectona grandis* L.f. (Müller and Gama-Rodrigues 2007). Other species that are commonly grown with cacao in the cacao-growing region of Brazil include piassava palm (*Attalea funifera* Mart. ex Spreng.), black pepper (*Piper nigrum* L.), clove (*Syzygium aromaticum* (L.) Merr. & Perr.), guarana (*Paullinia cupana* Mart.), vanilla (*Vanilla planifolia* Andr.), cardamom (*Elettaria cardamomum* (L.) Maton), passion fruit (*Passiflora edulis* Sims.), allspice (*Pimenta dioica* (L.) Merr.), and patchouli (*Pogostemon cablin* (Blanco) Benth.). Such crop diversification has been proposed to counterbalance the socioeconomic and ecological risks associated with cacao monocropping (Alvim and Nair 1986).

The effectiveness of cacao AFS for sequestering C depends on factors related to both the environment and the types of planting systems. The large amounts of both above- and belowground biomass accumulation and subsequent turnover of leaf litter, roots, and woody material from the shade species and from cacao provide a continuous stream of organic inputs into the soil (Muñoz and Beer 2001; Müller and Gama-Rodrigues 2007). Fontes (2006) reported that the total C stored in the biomass of the cacao + erythrina and cabruca systems was similar, with a mean of 39.27 Mg ha^{-1} (Table 1). The difference between the two systems was due to aboveground cacao biomass and the lower cacao stand density in the cabruca system. However, the shade trees (55 trees ha^{-1}) in the cabruca system stored 44% more C than the erythrina trees (35 trees ha^{-1}). There was essentially no difference in the

Table 1 Biomass carbon stocks in the systems of cacao + erythrina and cacao cabruca systems in Bahia, Brazil

	Cacao + erythrina		Cacao cabruca	
	Cacao	Erythrina	Cacao	Shade trees
Components	(kg ha^{-1})			
Leaf	1,512	911	756	1,309
Branch	8,589	2,473	4,295	3,552
Trunk	3,150	13,969	1,575	20,067
Fruits	1,020		504	
Seeds	425		210	
Husks	595		294	
Total[a]	15,291	17,353	7,634	24,928
Fine roots	2,816		2,903	
Litter	3,925		3,698	
Total[b]	39,385		39,163	

Source: Fontes (2006)
[a]Total aboveground (leaves + branch + trunk + fruits)
[b]Total biomass (aboveground + fine roots + litter layer)

Table 2 Litterfall C stocks in the systems cacao + erythrina and cacao cabruca systems in Bahia, Brazil

	Cacao + erythrina		Cacao cabruca	
	Cacao	Erythrina	Cacao	Shade tree
Components	(kg C ha^{-1} year^{-1})			
Leaf[a]	1,325	847	531	1,565
Flower[a]	323		368	
Branch	232		312	
Total	2,727		2,776	

Source: Fontes (2006)
[a]Values for cacao + erythrina and cacao + shade tree

fine roots and litter layer across the systems; the average C stored was 2,860 kg ha^{-1} in the fine roots and 3,812 kg ha^{-1} in the litter layer. Regarding the fruit production of cacao, there was very little difference between the amounts of C stored in seeds (210–425 kg ha^{-1}) versus husks (294–595 kg ha^{-1}). There was also no difference between the two systems for total C input in terms of natural litterfall (mean 2.75 Mg ha^{-1} year^{-1}). The litterfall of cacao was the main C input in the cacao + erythrina system (Table 2), with leaf fall accounting for about 80% of the total C input in that system, compared with 56% in the cabruca system (Table 2). The other major source of C input to these systems is fine roots and root hairs (Table 1). Up to 90% of root hairs (Kummerow et al. 1981) and 60% of fine roots (Kummerow et al. 1982; Muñoz and Beer 2001) are found in soil depth of 0–15 cm, and up to 80% of lateral roots (>5–10 mm) are known to occur in the 0–60 cm soil depth (Gama-Rodrigues and Cadima-Zevallos 1991).

The greatest proportion (81%) of total C stored in cacao AFS was found in aboveground biomass and the fine roots accounted for 7.3%, litter layer 9.7%, and fruits 2% of total C storage. The relative quantity of C removed in the harvested beans was low at 0.8%. In many cacao plantations, the husk is left on the ground and recycled, thereby representing another input of organic C into soil. Cacao AFS show the capacity to store C in soil microbial biomass (SMB) in the top soil (0–10 cm). Gama-Rodrigues et al. (2006) reported a mean value of 269 kg C ha^{-1} in SMB, measured according to the fumigation-extraction method. Although this constitutes less than 1% total C in the soil, it is the most important component of SOM, controlling the nature and rate of organic matter transformations. Moreover, it plays a critical role in cycling soil C and accounts for roughly half of soil surface CO_2 efflux through heterotrophic soil respiration (Hanson et al. 2000; Högberg et al. 2001).

Importance of Soil Aggregates in C Soil Storage of Cacao AFS

Organic matter is a major agent in forming and stabilizing soil aggregates (Six et al. 2000; Yamashita et al. 2006). The stabilization processes that protect SOM against biodegradation are therefore of great interest, and in order to enhance the potential of C sequestration in soils, it is important to understand the mechanisms that control the stabilization and release of C (Gregorich et al. 2006; Marschener et al. 2008). Three major factors of SOM stabilization include the physical protection of SOM, chemical protection through interaction with mineral surfaces, and selective preservation of certain recalcitrant organic compounds. The relative contribution of each factor to C protection in soils is not well understood (Six et al. 2002; Sollins et al. 1996). The physical protection of SOM through occlusion within aggregates or small pores is considered an important mechanism to reduce the bioavailability and accessibility of organic matter for soil microorganisms and soil enzymes (Tisdall and Oades 1982; Elliott 1986; Gupta and Germida 1988; Hassink 1992; Goebel et al. 2005). Aggregates protect SOM by forming a physical barrier from microorganisms, microbial enzymes, and their substrates; controlling food web interactions, and influencing microbial turnover (Six et al. 2000).

Aggregates are secondary particles formed through the combination of mineral particles with organic and inorganic substances (Bronick and Lal 2005); they represent a significant pool of soil C (Six et al. 1998, 1999). This is because the inclusion of organic materials within soil aggregates reduces their decomposition rate (Oades 1984; Elliott and Coleman 1988), improves C sequestration, and reduces the rate of increase in CO_2 concentration in the atmosphere. Aggregates are often grouped by their size into macroaggregates (>250 μm) and microaggregates (<250 μm). Different aggregate size groups differ in terms of their levels of labile C; C associated with macroaggregates is more labile and represent the light organic matter, while C associated with microaggregates is more recalcitrant and represents the stable fraction (Tisdall and Oades 1982; Gupta and Germida 1988; Bronick and Lal 2005; Cadish et al. 2006). The degree and duration of the stabilization of SOC within macroaggregates and microaggregates differ (Tisdall and Oades 1982).

Conceptual Explanation for Aggregate Formation in Cacao AFS Soils

Gama-Rodrigues et al. (2010) reported that the cacao AFS in Bahia had a high amount of soil C stock in excess of 300 Mg ha^{-1} across the 0–100 cm soil layer, similar to natural forests (Fig. 1). The distribution of C in different soil depths, especially below 30 cm, was also uniform in all three land use systems studied: the two cacao systems and natural forest (Fig. 2). In order to understand the C sequestration potential of cacao AFS, it is important to evaluate the extent of soil C storage in different soil aggregate classes at different soil depths. The fractionation of slaking-resistant aggregates of different sizes by wet sieving and subsequent use of ultrasonic

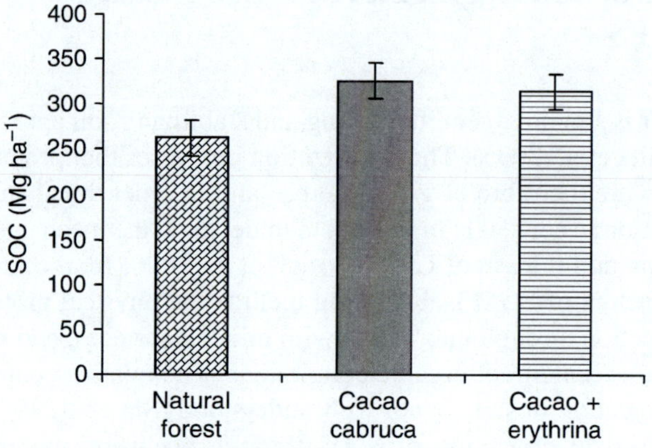

Fig. 1 Soil organic carbon (SOC) storage in the 0–100 cm soil layer in forest, cacao cabruca, and cacao + erythrina land use systems in Bahia, Brazil. There were no statistical differences among the land use systems (Source: Gama-Rodrigues et al. 2010)

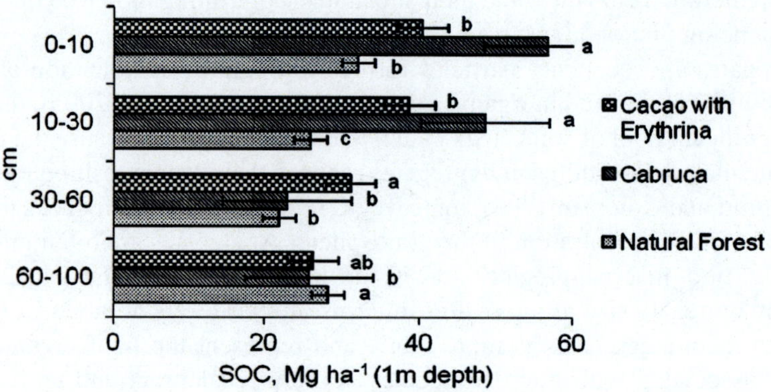

Fig. 2 Soil organic carbon (SOC) storage at different depths in three land use systems in Bahia, Brazil. Values followed by the same letter (s) within each depth are not significantly different according to the Tukey test ($p = 0.05$) (Source: Gama-Rodrigues et al. 2010)

energy and aggregate disruption allow us to determine the amount of fine aggregate occluded C (Gama-Rodrigues et al. 2010).

Various mechanisms have been proposed to account for the binding of soil particles into water stable aggregates, which are the building blocks of soil structure (Tisdall and Oades 1982; Elliott 1986; Oades and Waters 1991; Jastrow 1996). Tisdall and Oades (1982) presented a conceptual model for aggregate hierarchy that described how primary mineral particles are bound together with bacterial, fungal, and plant debris into microaggregates. These microaggregates, in turn, are bound together into macroaggregates by transient binding agents (i.e., microbial- and plant-derived polysaccharides) and temporary binding agents (i.e., roots and fungal hyphae). Three consequences of this aggregate hierarchy are:

1. a gradual breakdown of macroaggregates into microaggregates before they dissociate into primary particles as increasing dispersive energy is applied to soil (Oades and Waters 1991),
2. an increase in C concentration with increasing aggregate size class because large aggregate size classes are composed of small aggregate size classes plus organic binding agents (Elliott 1986), and
3. the presence of higher amounts of younger and more labile organic matter in macroaggregates than in microaggregates (Elliott 1986; Puget et al. 1995; Jastrow et al. 1996).

Oades and Waters (1991) tested the aggregate hierarchy model in different soils by applying a range of treatments to disaggregate soils. They concluded that an Oxisol did not express any hierarchical aggregate structure, probably because oxides, rather than organic materials, were the dominant stabilizing agents. So, we do not expect high C content in the largest aggregate class as compared to microaggregates in Oxisols. Alternatively, Oades (1993) speculated that aggregate hierarchy might exist in soils as a result of a long history of exploration by roots, particularly from grasses; as such, aggregate hierarchy will not apply to very young soils and , perhaps, to soils where inorganic cements are dominant, such as Oxisols.

Zotarelli et al. (2007) did not find differences in C and N content across aggregate-size fractions in Oxisols under crop and pasture systems; based on this, they suggested a conceptual model of macroaggregate turnover in order to determine the stabilization of SOM as fine intra-aggregate particulate organic matter, as proposed by Six et al. (1998, 1999). Six et al. (1999) noted that the reduction in tillage disturbance in no-till systems reduced the rate of macroaggregates turnover and enhanced the formation of highly stable microaggregates within macroaggregates, in which C was stabilized and sequestered in the long term.

Focusing on cacao AFS Oxisol, Gama-Rodrigues et al. (2010) found that macroaggregates comprised the most abundant fraction within the soil, with a high amount of C in the 0–100 cm layer, followed by microaggregates and silt-and-clay fractions. In addition, they found that 70% of C in the cacao AFS was fine C occluded within macroaggregates, which is a physically protected form of C. According to these authors, the constant addition of organic materials via litterfall and the presence of leguminous plant roots (Haynes and Beare 1997), coupled with the absence of tillage

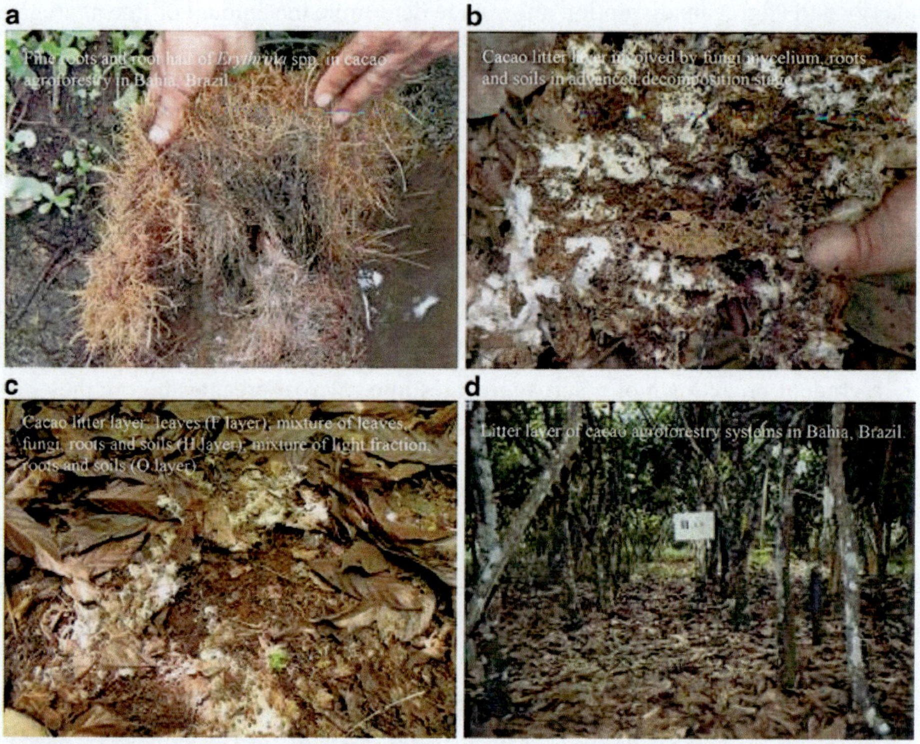

Fig. 3 Photographs showing the roots, root hairs and fungi mycelium in the litter layer of cacao agrofoestry system in Bahia, Brazil. (**a**) Fine roots and root hairs of *Erythrina* spp. in cacao agroforestry in Bahia, Brazil, (**b**) Cacao litter layer involving fungal mycelium, roots, and soils in advanced stages of decomposition, (**c**) Cacao litter layer: leaves (F layer); mixture of leaves, fungi, roots and soils (H layer); mixture of light fraction, roots and soils (O layer), (**d**) Litter layer of cacao agroforestry systems in Bahia, Brazil

and use of machinery in no-till systems such as cacao AFS, help maintain the binding effect and increases the number of water-stable macroaggregates. Although SOC inside the macroaggregates is more subject to disturbance than that in microaggregates and silt and clay fractions, the extent of such disturbances is low in cacao systems; therefore, C contained in this fraction can be expected to become more stabilized over time. Thus, cacao AFS seem to play an important role in environmental protection by mitigating GHG emission through the storage of high amounts of well-protected organic C in the soils.

Root growth increases aggregation through different mechanisms; roots, root hairs, and fungal hyphae form an extensive network (Fig. 3a–c), which physically enmesh fine particles of soil into aggregates even after their death. Also, roots and fungal hyphae exude polysaccharides, which can act as binding agents, thereby helping aggregation. In addition, roots can supply large quantities of organic material to soils (Tisdall and Oades 1982; Haynes and Beare 1997). Thus, soils under crops with the greatest root mass often show the greatest aggregate formation (Oades 1993; Haynes and Beare 1997).

According to Six et al. (1998), when fresh residue containing a high percentage of easily available C is applied to the soil, microbial activity increases and available C is assimilated in the form of extracellular polysaccharides, which leads to aggregate formation (Fig. 3d). These newly applied residues function as nucleation sites for the growth of fungi and other soil microbes, resulting in the binding of residue and soil particles into macroaggregates. The formation and stabilization of macroaggregates play an important role in the protection and subsequent accumulation of SOC (Liao et al. 2006). In land use systems with no tillage, a slow macroaggregate turnover allows time for the formation of fine occluded particulate organic matter that gradually becomes encrusted with clay particles and microbial products to form microaggregates containing young crop-derived C within macroaggregates (Six et al. 1998, 1999, 2000). The occluded C that is physically protected within soil aggregates represents a relatively more stable pool of C, though it is not strongly associated with soil particles (Christensen 1992; John et al. 2005). The constant replacement of organic material in land use systems, such as cacao AFS, maintains the binding effect and increases the number of water-stable macroaggregates (Gama-Rodrigues et al. 2010). Furthermore, previous studies have reported that the high concentration of fine roots at the surface soil up to a depth of 15 cm (Muñoz and Beer 2001) together with lignified coarse roots in subsurface soils up to 100 cm (Gama-Rodrigues and Cadima-Zevallos 1991) contribute substantially to belowground C stocks in cacao AFS. The presence of leguminous plant roots also promotes soil aggregation (Haynes and Beare 1997). Additionally, the cacao AFS is a no-till system that facilitates the maintenance of a high SOM level and macroaggregate formation due to the continuous input of organic material into the soil via litterfall (Isaac et al. 2005; Müller and Gama-Rodrigues 2007) and sloughed-off roots (Gama-Rodrigues and Cadima-Zevallos 1991; Kummerow et al. 1982).

The location of organic matter within the aggregate is a key factor for the stabilization and storage of SOM (Six et al. 2004). The protective effect of clay on SOM involves the interaction of SOM with the surface of the clay particles (including cation bridges, hydrogen bonds, electrostatic, and van der Waals interactions) and the occlusion of organic material in the matrix of soil aggregates (Hassink and Whitmore 1997). Whether the C pool is protected from silt and clay depends on the silt and clay proportions in soils; 1:1 clay-mineral-dominated soils, such as Oxisols, have a low level of silt-and-clay-protected C pool (Six et al. 2002). Therefore, clay content alone is not necessarily an appropriate indication of C protection in these soils. Gama-Rodrigues et al. (2010) suggested that occlusion of C in soil aggregates can be a major mechanism of C protection in Oxisols under cacao AFS. According to these authors, a strong correlation between organic C and aggregate size suggests that very high levels of organic matter could lead to a change in the dominant binding agents of these aggregates from oxides to organic molecules in these soils. The Oxisols under such systems may not present a classic hierarchical model for the formation of aggregates. More research is needed to compare soils under cacao AFS and other crop systems in order to reach a more definitive conclusion. Nevertheless, we can affirm that the development of SOM conservation practices in agricultural systems such as no tillage practices, the incorporation of trees, and the continuous input of organic matter are important to improve soil aggregation and, consequently, the accumulation of soil C in the highly weathered soils of the humid tropics.

Implications for Other Cacao-Growing Regions and Shaded-Perennial Agroforestry Systems

The above case study shows that the biomass of cacao and shade trees (aboveground and belowground) represents a significant source of input to soil C. The C captured in the biomass and deposited on the soil surface is indirectly sequestered as SOC following its decomposition. Furthermore, the soil has sink pools of C with microbial biomass and soil aggregates within them; but this pool is seldom considered in estimates of C sequestration. As reported by Nair et al. (2010), the available estimates of the C sequestration potential of AFS are derived by combining data on the aboveground, time-averaged C stocks and the soil C values, and the estimates are generally not rigorous.

Unfortunately, there are no standardized procedures and protocols for measuring and reporting C sequestration estimates in AFS; the cacao AFS are no exception. Some of the available reports on C sequestration in cacao AFS presented in Table 3 account for only the C stock in the aboveground biomass and roots (Duguma et al. 2001; Cotta et al. 2008; Hertel et al. 2009); some others include the C stock only in the soil surface (Fassbender et al. 1991; Isaac et al. 2005; Oelbermann et al. 2006; Barreto et al. 2010). Only a few studies have reported soil C stock to 1 m depth (e.g., Smiley and Kroschel 2008) and consider the large sink pool of C as soil aggregate (Gama-Rodrigues et al. 2010). Because of these variations in study procedures and lack of a uniform set of parameters that are being reported, even the limited amounts of data on C sequestration in cacao- (and other shaded-perennial-crop)-AFS from different countries cannot, unfortunately, be compared and contrasted. Therefore, the actual magnitude of C sequestration in soils under cacao AFS remains uncertain. Future research efforts should focus on a standardized protocol to study C sequestration in soils under cacao AFS: some of these issues include standardizing the sampling procedures including soil sampling to about 1-m depth, determining the soil organic C by the dry combustion method, reporting soil bulk density in the calculation of C stock, and quantifying the C in the fraction-size classes and inside the aggregates.

In addition to the environmental benefits of GHG mitigation and biodiversity conservation, the C sequestered in cacao AF systems could also provide an added income stream for cacao farmers around the world. Small family-farms are at the heart of cacao industry, with five million to six million smallholder farmers providing more than 85% of the world's crop. Typically, each cacao farmer owns less than 2 ha of land and may grow approximately 1,000 cacao trees (Shapiro and Rosenquist 2004). Using a simulation model designed to simulate the value of terrace and agroforestry investments in the highland tropics of Peruvian Andes, Antle et al. (2007) showed that participation in C contracts could increase adoption of terraces and

Table 3 Summary of literature values on soil carbon stock under cacao agroforestry systems

Country/region	Age (year)	Soil type[a]	Depth (cm)	Shade tree	Biomass (kg ha^{-1} $year^{-1}$)	C stock biomass (Mg ha^{-1})	C stock soil (Mg ha^{-1})	Reference
West Africa	n.d.[b]	Ferric Acrisol/ Haplic Ferralsol (Oxisols/ Ultisols)	n.d.	Timber and fruit trees	10,000	n.d.	n.d.	Gockowski and Sonwa (2010)
Indonesia	15	n.d.	0–100	*Gliricidia sepium* (Jack.) Kunth. ex Walp	n.d.	31,450	160	Smiley and Kroschel (2008)
Costa Rica	24	Eutric Cambisol (Inceptisols)	0–40	*Erythrina poeppigiana* (Walp.) Cook	n.d.	40	162	Oelbermann et al. (2006)
Indonesia	10	Cambisol (Inceptisols)	0–20	*Erythrina subumbrans* (Hassk.) Merr., *G. sepium, Syzygium aromaticum* (L.) Merr. & Perr.), *Nephelium lappaceum* L.	4,500[c]	n.d.	44	Hertel et al. (2009)
West Africa	20	n.d.	0–15	Timber and fruit trees	10,400 [d]	n.d.	18.2	Isaac et al. (2005)
Brazil	30	Oxisols	0–50	*Erythrina glauca* Willd	n.d.	n.d.	93.79	Barreto et al. (2010)

[a]The approximate soil orders according to US Soil Taxonomy are indicated in parentheses
[b]No data
[c]Fine root input
[d]Litterfall input

agroforestry practices in northern Peru, with the rate of adoption depending on the C accumulation rate and key factors affecting terrace productivity. Takimoto et al. (2008) conducted a cost-benefit analysis of C sequestration in two improved agroforestry systems (live fence and fodder bank) and the traditional parkland agroforestry systems in semiarid Mali and concluded that C credit sale was likely to contribute to economic development of the subsistence farmers in the region. It is likely therefore that the data presented in this case study, supplemented with additional studies of this nature in multiple locations, could go a long way in providing economic benefit to these farmers, when and if an effective carbon-trading system becomes operational.

Conclusions

Continuous deposition of plant litter, fine roots, and root hairs are the principal inputs for C content in cacao AFS soils. Soil microbial biomass, which although accumulated a mean of 269 kg ha^{-1} on the soil surface in our study, is an important long-lived pool of C and has substantial influence on processes related to C dynamics in the soil. Cacao AFS are comparable to natural forests with respect to the accumulation of high amounts of SOC. Seventy percent of C in the cacao AFS was fine C occluded within aggregates, which is a physically protected form of C. The vast majority of SOC was present in macroaggregate fractions throughout soil layers up to 1 m in depth. The development of SOM conservation practices in agricultural systems such as no tillage practices, the incorporation of trees, and the continuous input of organic matter are important to improve soil aggregation and, consequently, the accumulation of soil C in the highly weathered soils of the humid tropics which represent the major soil type of the cacao-growing regions of the world. The lack of standardized procedures for studying and reporting soil C sequestration in different agroforestry systems is currently a serious drawback in realizing this unexplored and unappreciated environmental benefit. Rigorous and coordinated research efforts with a holistic view of the interrelationships among various factors that contribute to the complexities of soil C sequestration will be needed to change this unsatisfactory situation.

References

Alvim PT, Nair PKR (1986) Combination of cocoa with other plantation crops: an agroforestry system in Southeast Bahia, Brazil. Agroforest Syst 4:3–15

Antle JM, Stoorvogel JJ, Valdivia RO (2007) Assessing the economic impacts of agricultural carbon sequestration: Terraces and agroforestry in the Peruvian Andes. Agric Ecosyst Environ 122:435–445

Barreto PAB, Gama-Rodrigues EF, Gama-Rodrigues AC, Fontes AG, Polidoro JC, Moço MKS, Machado RCR, Baligar VC (2010) Distribution of oxidizable organic C fractions in soils under cacao agroforestry systems in Southern Bahia. Brazil Agroforest Syst. doi:10.1007/s10457-010-9300-4

Bronick CJ, Lal R (2005) Soil structure and management: a review. Geoderma 124:3–22

Cadish G, Mutuo P, Mercado A, Hairiah K, Nyamugafata P, Boye A, Albrecht A (2006) Organic matter management in tropical agroforestry systems: soil quality, soil C storage and soil atmosphere gas exchange. In: Gama-Rodrigues AC, Barros NF, Gama-Rodrigues EF, Freitas MSM, Viana AP, Jasmin JM, Marciano CR, Carneiro JGA (eds.) Sistemas agroflorestais: bases científicas para o desenvolvimento sustentável. UENF, Campos dos Goytacazes, pp 275–290

Christensen BT (1992) Physical fractionation of soil and organic matter in primary particle size and density separates. Adv Soil Sci 20:1–90

Cotta MK, Jacovine AG, Paiva HN, Soares CPB, Filho ACV, Valverde SR (2008) Biomass quantification and emission reduction certificates for rubber-cocoa intercropping. Rev Árvore 32:969–978

Duguma B, Gockowski J, Bakala J (2001) Smallholder cacao (*Theobroma cacao* Linn.) cultivation in agroforestry systems of West and Central Africa: challenges and opportunities. Agroforest Syst 51:177–188

Elliott ET (1986) Aggregate structure and carbon, nitrogen and phosphorus in native and cultivated soils. Soil Sci Soc Am J 50:627–633

Elliott ET, Coleman DC (1988) Let the soil work for us. Ecol Bull 39:23–32

Fassbender HW, Beer J, Heuveldop J, Imbach A, Enriquez G, Bonnemann A (1991) Ten year balances of organic matter and nutrients in agroforestry systems at CATIE, Costa Rica. For Ecol Manag 4:173–183

Fontes AG (2006) Nutrient cycling in cacao agroforestry systems in south of Bahia, Brazil. Ph.D. dissertation, North Fluminense State University, RJ, Brazil, 71 p

Gama-Rodrigues AC, Cadima-Zevallos A (1991) Efectos de fertilización sobre sistema radicular de cacao en suelos de "tabuleiros" del sur de Bahia, Brasil. Turrialba 41:135–141

Gama-Rodrigues EF, Moço MKS, Gama-Rodrigues AC, Machado RC (2006) Atributos biológicos em solos sob sistemas agroflorestais de cacau: um estudo de caso. In: Gama-Rodrigues AC, Barros NF, Gama-Rodrigues EF, Freitas MSM, Viana AP, Jasmin JM, Marciano CR, Carneiro JGA (eds.) Sistemas agroflorestais: bases científicas para o desenvolvimento sustentável. UENF, Campos dos Goytacazes, pp 243–256

Gama-Rodrigues EF, Nair PKR, Nair VD, Gama-Rodrigues AC, Baligar V, Machado RCR (2010) Carbon storage in soil size fractions under two cacao agroforestry systems in Bahia, Brazil. Environ Manag 45:274–283

Gockowski J, Sonwa D (2010) Cocoa intensification scenarios and their predicted impact on CO_2 emission, biodiversity conservation, and rural livelihoods in the Guinea rain forest of west Africa. Environ Manag. doi:10.1007/s00267-010-9602-3

Goebel M-O, Bachmann J, Woche SK, Fischer WR (2005) Soil wettability, aggregate stability, and the decomposition of soil organic matter. Geoderma 128:80–93

Gregorich EG, Beare MH, McKim UF, Skjemstad JO (2006) Chemical and biological characteristics of physically uncomplexed organic matter. Soil Sci Soc Am J 70:975–985

Gupta VVSR, Germida JJ (1988) Distribution of microbial biomass and its activity in different soil aggregate size classes as affected by cultivation. Soil Biol Biochem 20:777–786

Hanson PJ, Edwards NT, Garten CT, Andrews JA (2000) Separating root and microbial contributions to soil respiration: a review of methods and observations. Biogeochemistry 48:115–146

Hassink J (1992) Effects of soil texture and structure on carbon and nitrogen mineralization in grassland soils. Biol Fertil Soils 14:126–134

Hassink J, Whitmore AP (1997) A model of the physical protection of organic matter in soils. Soil Sci Soc Am J 61:131–139

Haynes RJ, Beare MH (1997) Influence of six crop species on aggregate stability and some labile organic matter fractions. Soil Biol Biochem 29:1647–1653

Hertel D, Harteveld A, Leuschner C (2009) Conversion of a tropical forest into agroforest alters the fine root-related carbon flux to the soil. Soil Biol Biochem 41:481–490

Högberg P, Nordgren A, Buchmann N, Taylor AFS, Ekblad A, Högberg MN, Nyberg G, Ottosson-Löfvenius M, Read DJ (2001) Large-scale forest girdling shows that current photosynthesis drives soil respiration. Nature 411:789–792

IPCC (2000) Land use, land use change, and forestry. A special report of the IPCC. Cambridge University Press, Cambridge, UK, 375 p

Isaac ME, Gordon AM, Thevathasan N, Oppong SK, Quashie-Sam J (2005) Temporal changes in soil carbon and nitrogen in West African multistrata agroforestry systems: a chronosequence of pools and fluxes. Agroforest Syst 65:23–31

Jastrow JD (1996) Soil aggregate formation and the accrual of particulate and mineral-associated organic matter. Soil Biol Biochem 28:656–676

Jastrow JD, Boutton TW, Miller RM (1996) Carbon dynamics of aggregate-associated organic matter estimated by carbon-13 natural abundance. Soil Sci Soc Am J 60:801–807

John B, Yamashita T, Ludwig B, Flessa H (2005) Storage of organic carbon in aggregate and density fractions of silty soils under different types of land use. Geoderma 128:63–79

Kummerow J, Kummerow M, Alvim PT (1981) Root biomass in mature cacao (*Theobroma cacao* L.) plantation. Rev Theobroma 11:77–85

Kummerow J, Kummerow M, Silva WS (1982) Fine-root growth dynamics in cacao (*Theobroma cacao*). Plant Soil 65:193–201

Lal R (2004) Soil sequestration in dryland ecosystems. Environ Manag 33:528–544

Liao JD, Boutton TW, Jastrow JD (2006) Storage and dynamics of carbon and nitrogen in soil physical fractions following woody plant invasion of grassland. Soil Biol Biochem 38:3184–3196

Marschener B, Brodowski S, Dreves A, Gleixner G, Gude A, Grootes PM, Hamer U, Heim A, Jandl G, Ji R, Kaiser K, Kalbitz K, Kramer C, Leinweber P, Rethemeyer J, Schäffer A, Scmidt MWI, Schwark L, Wiesenberg GLB (2008) How relevant is recalcitrance for the stabilization of organic matter in soils? J Plant Nutr Soil Sci 171:91–110

Montagnini F, Nair PKR (2004) Carbon sequestration: an under-exploited environmental benefit of agroforestry systems. Agroforest Syst 61:281–295

Müller MW, Gama-Rodrigues AC (2007) Sistemas agroflorestais com cacaueiro. In: Valle RR (ed.) Ciência, tecnologia e manejo do cacaueiro. CEPLAC, Ilhéus, pp 246–271

Muñoz F, Beer J (2001) Fine root dynamics of shaded cacao plantations in Costa Rica. Agroforest Syst 51:119–130

Nair PKR, Kumar BM, Nair VD (2009) Agroforestry as a strategy for carbon sequestration. J Plant Nutr Soil Sci 172:10–23

Nair PKR, Nair VD, Kumar BM, Showalter JM (2010) Carbon sequestration in agroforestry systems. Adv Agron 108:237–307

Oades JM (1984) Soil organic matter and structural stability: mechanisms and implications for management. Plant Soil 76:319–337

Oades JM (1993) The role of biology in the formation, stabilization and degradation of soil structure. Geoderma 56:377–400

Oades JM, Waters AG (1991) Aggregate hierarchy in soils. Aust J Soil Res 29:815–828

Oelbermann M, Voroney RP (2007) Carbon and nitrogen in a temperate agroforestry system: using stable isotopes as a tool to understand soil dynamics. Ecol Eng 29:342–349

Oelbermann M, Voroney RP, Naresh V, Thevathasan NV, Gordon AM, Kass DCL, Schlönvoigt AM (2006) Carbon dynamics and residue stabilization in a Costa Rican and southern Canadian alley cropping system. Agroforest Syst 68:27–36

Puget P, Chenu C, Balesdent J (1995) Total and young organic-matter distributions in aggregates of silty cultivated soils. Eur J Soil Sci 46:449–459

Shapiro H-Y, Rosenquist EM (2004) Public/private partnerships in agroforestry: the example of working together to improve cocoa sustainability. Agroforest Syst 61:453–462

Six J, Elliot ET, Paustian K, Doran JW (1998) Aggregation and soil organic matter accumulation in cultivated and native grassland soils. Soil Sci Soc Am J 63:1350–1358

Six J, Elliot ET, Paustian K (1999) Aggregate and soil organic matter dynamics under conventional and no tillage systems. Soil Sci Soc Am J 63:1350–1358

Six J, Elliot ET, Paustian K (2000) Soil macroaggregate turnover and microaggregate formation: a mechanism for C sequestration under no-tillage agriculture. Soil Biol Biochem 32:2099–2103

Six J, Conant RT, Paul EA, Paustian K (2002) Stabilization mechanisms of soil organic matter: implications for C-saturation of soils. Plant Soil 241:155–176

Six J, Bossuyt H, Degryze S, Denef K (2004) A history of research on the link between (micro) aggregates, soil biota, and soil organic matter dynamics. Soil Tillage Res 79:7–31

Smiley GL, Kroschel J (2008) Temporal change in carbon stocks of cocoa–gliricidia agroforests in Central Sulawesi, Indonesia. Agroforest Syst 73:219–231

Sollins P, Homann P, Caldwell BA (1996) Stabilization and destabilization of soil organic matter: mechanisms and controls. Geoderma 74:65–105

Takimoto A, Nair PKR, Nair VD (2008) Carbon stock and sequestration potential of traditional and improved agroforestry systems in the West African Sahel. Agric Ecosyst Environ 125:159–166

Tisdall JM, Oades JM (1982) Organic matter and water-stable aggregates in soils. J Soil Sci 33:141–163

Yamashita T, Flessa H, John B, Helfrich M, Ludwig B (2006) Organic matter in density fractions of water-stable aggregates in silty soils: effect of land use. Soil Biol Biochem 38:3222–3234

Zotarelli L, Alves BJR, Urquiaga S, Boddey RM, Six J (2007) Impact of tillage and crop rotation on light fraction and intra-aggregate soil organic matter in two Oxisols. Soil Sci Soc Am J 69:482–491

Carbon Sequestration Potential of Silvopastoral and Other Land Use Systems in the Chilean Patagonia

Francis Dube, Naresh V. Thevathasan, Erick Zagal, Andrew M. Gordon, Neal B. Stolpe, and Miguel Espinosa

Abstract This study was undertaken to quantify the carbon (C) sequestration potentials in three predominant ecosystems on the volcanic soils in Patagonia, Chile. The systems were: *Pinus ponderosa* Dougl. ex P. Laws. – based silvopastoral systems arranged in strips (silvopasture), 18-year-old managed exotic stands (plantation), and natural prairie (prairie), in Patagonia, Chile. Most of the data used in the construction of C models were derived from experimental plots, where litterfall, decomposition, soil respiration, and soil C were measured. The values for greenhouse gas (GHG) emissions by cattle and fertilizer application were obtained from the literature. In the silvopasture and the plantation, total above- and belowground tree C stock accounted for 69% and 64% of the total system C, respectively. Total above- and belowground C pools were 224, 199, and 177 Mg C ha^{-1}, with the aboveground: belowground C pool ratios of 1:10, 1:5, and 1:177, respectively, for silvopasture, plantation, and prairie. Soil respiration decreased in the order prairie >silvopasture >plantation. The C leached beyond the root zone (in leachate collected at 80 cm soil depth) decreased in the order plantation >prairie >silvopasture.

F. Dube (✉) • M. Espinosa
Department of Silviculture, Faculty of Forest Sciences, University of Concepción, Victoria 631, Casilla 160-C, Concepción, Chile
e-mail: fdube@udec.cl; mespinos@udec.cl

N.V. Thevathasan • A.M. Gordon
School of Environmental Sciences, University of Guelph, Guelph, ON N1G 2W1, Canada
e-mail: nthevath@uoguelph.ca; agordon@uoguelph.ca

E. Zagal • N.B. Stolpe
Department of Soils and Natural Resources, Faculty of Agronomy, University of Concepción, Vicente Méndez 595, Casilla 537, Chillan, Chile
e-mail: ezagal@udec.cl; nstolpe@udec.cl

B.M. Kumar and P.K.R. Nair (eds.), *Carbon Sequestration Potential of Agroforestry Systems: Opportunities and Challenges*, Advances in Agroforestry 8,
DOI 10.1007/978-94-007-1630-8_6,

Estimated system net C flux was +1.8, +2.5, and −2.3 Mg C ha^{-1} $year^{-1}$ for the silvopasture, plantation, and prairie, respectively. Based on this study it is estimated that establishing silvopastoral systems with cattle over a land area of approximately 481 km^2 or 0.33% of the Chilean Patagonia territory would be adequate to offset all C losses from cattle-based livestock systems.

Keywords Andisols • C pools and fluxes • Greenhouse gases • *Pinus ponderosa*

Introduction

Between 2000 and 2010, the atmospheric concentration of carbon dioxide (CO_2) has increased from 369 to 388 ppm, a 5.1% increase over the last 10 years, let alone 280 ppm in 1850 (Tans 2010). Land use changes and fossil fuel combustion are two important anthropogenic factors that have contributed to this increase. The influence of land management on the carbon (C) content in soils and biomass is well documented worldwide (Ross et al. 2002; Huygens et al. 2005; Dube et al. 2009). Land use changes not only affect C sources and sinks, but also impact methane (CH_4) and nitrous oxide (N_2O) emissions.

The Climate Change 2007 Synthesis Report (IPCC 2007) proposes key mitigation practices in the agricultural sector. Among them, the use of proven crop and grazing land management to increase soil C storage, restoration of degraded lands, improved livestock and manure management to reduce CH_4 and N_2O emissions are a few practices related to this study. Agroforestry systems (AFS) rank high for all of these strategies. Well designed and managed AFS can be effective CO_2 sinks, especially with the use of perennial crops and fast growing tree species (Nair et al. 2010). Recent studies performed in temperate regions have shown that AFS have greater C sequestration potential than monocropping systems, prairies, or forest plantations, and should be considered as real C sinks (Montagnini and Nair 2004; Sharrow and Ismail 2004; Gordon and Thevathasan 2005; Peichl et al. 2006; Bambrick et al. 2010). This chapter will focus on C sequestration potentials in three distinct ecosystems in the Chilean Patagonia region (Fig. 1a).

Between 1920 and 1940, large areas of the Chilean Patagonia were burned down and converted into pastures for cattle, leaving the slopes exposed to an inexorable erosion and degradation of soils formed by volcanic ash deposits. Overgrazing has also contributed to forest destruction.[1] The deforested areas, especially those located on steeper slopes suffer extreme erosive processes, which complicate in some cases the reestablishment of native forest. The recent ratification of the Kyoto Protocol by Chile coupled with C sequestration potentials of Patagonian AFS opens the possibility for the progressive adoption of silvopastoral systems, well adapted to the climatic conditions and economic reality of Patagonia.

[1] Silva (2004).

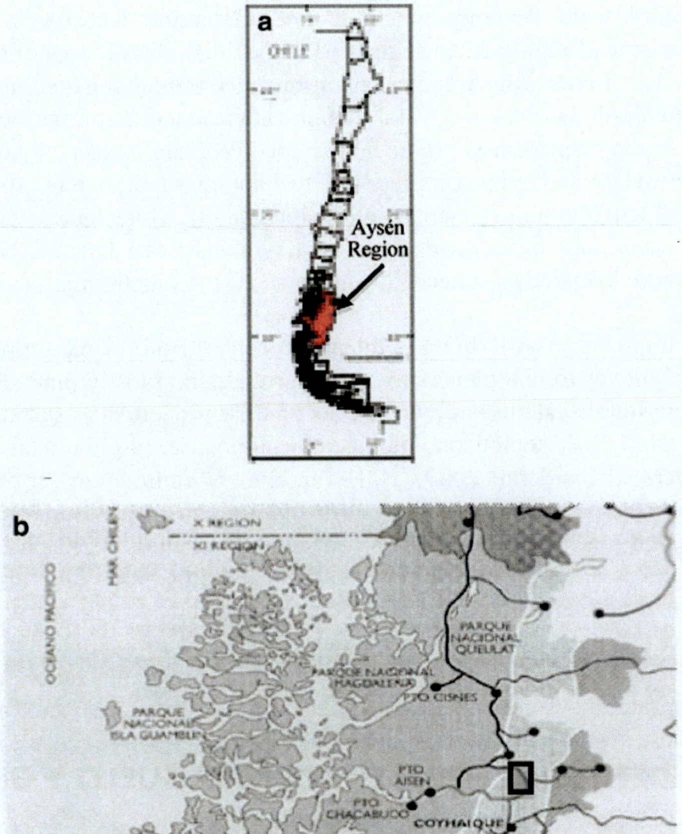

Fig. 1 Map of Chile (**a**) and location of Patagonia (in black) and the Aysén Region (in *red*) in Patagonia (Source: INE 2007b); (**b**) location of the Mano Negra Sector (*inset*) in the Intermediate Temperate ecological zone, Aysén Region, Chilean Patagonia where data were collected in 2007–2009 (Source: Teuber and Ganderats 2009)

Profitability from farming, ranching, and plantation forestry in Chile has decreased recently, mainly because of market globalization (Teuber and Ganderats 2009). The harsh prevailing weather conditions, geographical isolation, high costs of production, and low technological development make it difficult for the implementation and development of new production sectors. Profitability can only be improved through innovation and the incorporation of technologies that increase the efficiency of traditional activities, one of them being the integration of forest practices and ranching on the same unit of land, resulting in a symbiosis that benefits both sectors. However, more knowledge is needed to understand the functioning of the resulting systems.

The Instituto Forestal-INFOR (Chilean Forestry Institute) has implemented a series of incentives to landowners to adopt agroforestry, mostly pine (*Pinus* spp.)-based silvopastoral systems and windbreaks on their properties as sustainable practices that satisfy their socioeconomic needs while conserving the natural resource base (Teuber and Ganderats 2009). However, there is limited number of studies on Chilean AFS and a general lack of scientific research on C sequestration aspects. In addition, since pine plantations already occupy more than 30,000 ha in Chilean Patagonia and natural prairies over one million ha, their C pools and fluxes must be examined as soon as possible. Therefore, the current study was undertaken, perhaps as the first of this nature, to evaluate and model C sequestration potentials in a natural prairie, a managed *Pinus ponderosa* Dougl. ex P. Laws. plantation, and a pine-based silvopastoral system on Andisols in the Chilean Patagonia, and to determine which of these systems has the best potential for long term C sequestration.

Materials and Methods

Site Description and Experimental Design

The research site in the San Gabriel Agroforestry Unit within the Mano Negra Sector (Figs. 1b and 2) was established in 2002 by INFOR 30 km north of the city of Coyhaique in the Aysén Region, on a western exposed slope at 730 m altitude, 45°25′ S, and 72°00′ W. The study location is in the Intermediate Agroecological Zone of the Aysén Region of Chilean Patagonia. The annual precipitation varies from 1,000 to 1,500 mm. However, only 15% of the precipitation occurs between December and February, coinciding with the warmest and windiest period. Mean temperatures fluctuate between 12°C and 14°C in summer and 2°C and 3°C in winter (Dube et al. 2009). During summer, strong westerly winds cause seasonal water deficits and wind erosion, which may diminish soil organic matter. The mineral soil horizons have Andic soil properties that include low bulk density (<0.9 g cm^{-3}), high P fixation values (65–89%), and high water content at 1,500 kPa tension relative to the measured clay content. The soil has been classified as medial, amorphic, mesic Typic Hapludands (Stolpe et al. 2010).

Fig. 2 Aerial photograph of the pine-based silvopastoral system arranged in trip (Silvopasture), surrounded by the managed natural prairie (Prairie) and the 18-year old Ponderosa pine plantation (Plantation), in the Mano Negra Sector, Aysén Region of the Chilean Patagonia (Source: Dube 2010)

Three land uses, hereafter referred to as treatments were studied: (1) managed natural pasture (Fig. 3a) with traditional cattle grazing (prairie), (2) 18-year-old thinned and pruned *P. ponderosa* (Fig. 3b) exotic stands (plantation), (3) pine-based silvopastoral systems arranged in strips (silvopasture), where the width of pasture alley was 21 m (Fig. 3c). The entire study area was initially covered with native forest, dominated by *Nothofagus pumilio* (Poepp. et Endl.) Krasser.

In 1991, *P. ponderosa* plantations were established over pasture, at 2×2.5 m spacing giving a stand density of 2,000 trees ha^{-1}. By 2003, the density had naturally declined to 1,514 trees ha^{-1}, with a mean tree height of 6.7 m, diameter at breast height (DBH) 11.4 cm, basal area 15.3 m^2 ha^{-1}, and crown cover 90%. Part of the plantation (5 ha) was thinned in 2003 to 800 trees ha^{-1} (homogeneous spacing) while another section was thinned down to 400 trees ha^{-1} and converted into a silvopastoral system arranged in strips (5 ha). Although the overall tree density in silvopasture was 400 trees ha^{-1}, actual density within the tree strip (6 m wide) was approximately 1,444 trees ha^{-1} with an average spacing of 2.3×3 m, because that portion of tree stand in the silvopastoral system was not thinned. The prairie and

Fig. 3 Permanent plots established in the (**a**) managed natural prairie (prairie) near Cerro Rosado volcano, fall 2008, (**b**) thinned and pruned ponderosa pine plantation (plantation), and (**c**) silvopastoral system with Black Angus cattle grazing between strips of ponderosa pine (silvopasture) on volcanic soil (mesic Typic Hapludands) (Source: Dube 2010)

Fig. 3 (continued)

silvopasture had a stocking density of 0.5 cows ha^{-1}, and consisted of a mixture of perennial grasses (*Dactylis glomerata* L., *Holcus lanatus* L., *Poa pratensis* L.), leguminous pasture (*Trifolium pratense* L., *T. repens* L.), and other herbaceous species (*Acaena magellanica* (Lam.) Vahl., *A. pinnatifida* Ruiz et Pav., *Hypochoeris radicata* L., *Taraxacum officinale* Weber). According to a recent inventory, perennial grasses, leguminous pasture, weeds, and dead material represented 36%, 30%, 19%, and 15%, respectively, of total dry matter in the prairie, whereas in the silvopasture, the corresponding percentages were 29, 40, 8, and 23.

Treatments were established in October 2007 in a completely randomized design with three replicates (or sampling plots). Given that the Agroforestry Unit was initially established as a demonstration site, no experimental design was considered at that time. The randomly distributed sampling plots were replicated within every treatment of the demonstration site itself but far from each other (at least 50 m). In this study, spatial interspersion of replications together with the use of a systematic design was used to alleviate possible pseudo-replication problems (Stamps and Linit 1999; K. Saez, 2007 and 2010, personal communication). Each plot measured 15×27 m and was located at 5 m from the border of the treatment. In the silvopasture, each plot included three strip rows of pines (6 m wide) and a half strip of pasture on either side of the tree rows (10.5 m in length along the tree rows). In both the plantation and prairie, the plots had only pine and pasture, respectively. All the measurements took place during 2 years from November 2007.

Tree and Pasture Biomass and Carbon Content

An inventory of the plantation and silvopasture was performed in 2007–2009 in which DBH and height of all trees were measured using a diameter tape and clinometers, respectively. Three trees each in the plantation (average height: 8.1 ± 0.4 m; mean DBH: 23.2 ± 1.4 cm) and silvopasture (average height: 8.3 ± 0.4 m; mean DBH: 25.1 ± 1.9 cm) were randomly selected for destructive sampling in 2009 using the average tree method (Teller 1988). Once the trees were felled, the fresh biomass of trunks, branches, twigs, needles, and cones were determined using a 45 kg dynamometer (Salter Brecknell Electro Samson Scale, Raco Industries, Cincinnati, OH, USA). Three sub-samples from each tree component (including superficial coarse roots extracted with a shovel and saw at 0–30 cm depth) were taken to determine the moisture and C concentrations. Cross sectional disks of tree stems were obtained at the initial crown height, breast, and stump heights. The sub-samples were weighed, oven dried at 65°C for five days and weighed again, and the mean dry weight of the distinct subsamples was then extrapolated to the entire stand (Peichl et al. 2006). The C concentration of all the sub-samples (trunks, branches, twigs, roots, leaves) was determined by grinding the samples using a Cyclotec 1093 Sample Mill (Tecator, Sweden), and analysis for total C using a Fisons EA1108 CHNS-O Elemental Analyzer (Fisons Instrument, CA, USA), following the dynamic flash combustion technique (Fisons Instrument 1990). Coarse root biomass was estimated using a subterranean biomass function that relates root biomass to stem DBH (Dube et al. 2009):

$$\text{Root biomass} = -13{,}2750 + e^{(2{,}4148+0{,}0743*\text{DBH})},$$

where e is the base of the natural logarithm (2.71828) and DBH is the diameter at breast height (cm). The annual production of fine root (<5 mm diameter) biomass was estimated as a percentage of litterfall (Abohassan 2004). The dry weight of dead pine branches left after pruning was estimated using nine 25 m^2 subplots per treatment. Decomposing trunks and stumps of *N. pumilio* left on the site were cut in pieces and recollected from eighteen 25 m^2 plots. Carbon contents of organic materials were determined by dry combustion using a Fisons EA1108 CHNS-O Elemental Analyzer (Fisons Instrument, CA, USA).

The total standing aboveground pasture biomass was harvested from three randomly placed quadrats (0.5 m^2 each) per sampling plot (nine per treatment) to determine the aboveground net primary productivity. The grazing material was harvested three times a year during the growing season over a 2-year period. Since it was not possible to measure belowground net primary productivity, it was estimated using a known algorithm (Gibson 2009):

$$\text{BNPP} = \text{BGP} * (\text{Live BGP} / \text{BGP}) * \text{turnover},$$

where BNPP is the belowground net primary productivity (g m^{-2} $year^{-1}$), BGP is the belowground productivity (g m^{-2} $year^{-1}$), Live BGP/BGP = 0.6, and turnover = 90% $year^{-1}$ (Stolpe et al. 2010). In addition, BGP = 0.79 * (AGBIO) − 33.3 *

(MAT + 10) + 1,289, where AGBIO is the peak aboveground live biomass (g m^{-2}) during the growing season, and MAT is the mean annual temperature (°C) of observed belowground biomass. The strength of this algorithm is given by $R^2 = 0.54$ and $p = 0.01$.

Litterfall and Decomposition of Organic Substrates

Circular conic 1 m^2 traps, 60 cm tall with 1 mm mesh screen, were used in the silvopasture and plantation to collect the litterfall (Berg and Laskowski 2006). The amount of litterfall was sampled monthly over 2 years, and weekly during the rainy seasons. In order to quantify the decomposition of litterfall, mixed grass root biomass, and cattle faeces, polyester bags measuring 20 × 20 cm with 0.5 mm mesh were filled with the respective substrates (Berg and Laskowski 2006) and placed on the Oe horizon (needles; faeces) or buried at 15 cm depth (root biomass). The pine needles were sampled every 6 months for 2 years, whereas the sampling of the grass roots and cattle faeces was performed every 3 months for 1 year. The annual C contribution to the soil in the silvopasture and prairie by cattle was estimated using the quantity of faeces produced per animal per day (Yang et al. 2003; Byrne et al. 2007), the fecal C concentration, and the animal stocking rates.

Soil Carbon and Nitrogen

In March 2009, soil samples were taken at 0–5, 5–20, and 20–40 cm depths with a split core soil sampler to determine total, organic, and inorganic C content as well as total N in each treatment (Dube et al. 2009). Given the genesis of volcanic soils, the inorganic C content was almost non-existent and it was concluded that total C was the same as organic C.

Carbon Content in Leached Soil Solution and Soil Respiration

The C concentrations in the leached soil solution under the pasture and pine roots were measured using tension lysimeters permanently installed at a slight angle to a soil depth of 80 cm. Sampling was done on a monthly basis and weekly during the rainy seasons. Dissolved C in rainwater and snowfall was determined six times a year during the months of greatest precipitation from three samplers located at random. Total soluble organic C was analyzed by combustion at 675°C using a TOC-V CPN Total Organic Carbon Analyzer (Shimadzu Corp., Kyoto, Japan). Carbon leaching represents the sum of different inputs, one of them being dissolved C in rain and snowfalls throughout the year. Knowing the C concentration in atmospheric depositions and using C concentration data of leached soil solution (Dube 2010), the contribution from the system itself to the total leached C could then be determined.

Additionally, total soil respiration was quantified with the *in situ* technique of CO_2 absorption by soda lime in a closed chamber (Edwards 1982). Circular chambers were installed at 5 cm depth in the soil, assuring that the area was free of "live" organic matter. The measurements were done at weekly intervals in the summer and at monthly intervals during the rest of the year. The quantity of CO_2 produced in 24 h was calculated for the chamber area and converted to hectares.

Air and Soil Temperature and Soil Moisture

The soil moisture (0–20 cm depth), soil surface temperature (0–5 cm depth), and air temperature above the soil (+5 cm) were measured every 2 h over a 24 month period using Decagon Devices EM-5B Data Loggers, EC-20 soil moisture sensors and ECT soil temperature sensors, respectively (Decagon Devices Inc., Pullman, WA, USA).

Statistical Analyses

All treatments were analyzed with the General Linear Model procedure of SAS v.9.0 (SAS Institute Inc. 2003) for completely randomized design to test the effect of treatments (Peichl et al. 2006). All data were examined for homogeneity of variance and normality. Analysis of variance was conducted using the ANOVA procedure. Student's *t* test for independent populations was used to check for significant differences between the treatment means (comparison by pairs) ($p < 0.05$).

C Flows from Cattle, Decomposition of Woody Debris, and Emissions from Fertilizers

Most of the data were derived from the experimental plots. However, the values for cattle respiration, methane emissions from enteric fermentation and nitrous oxide from dung and urine patches, annual mass loss for decomposition of dead branches of *P. ponderosa* and boles/stumps of *N. pumilio*, annual leaching rate, and annual emissions of N_2O from N fertilizer application were derived from the literature.

Models of C Pools and Fluxes

All the data of C pools and fluxes within the *P. ponderosa*-based silvopastoral system arranged in strip, the 18-year old pine plantation, and the managed natural prairie determined in this study were compiled and complemented by data from the literature in order to build the models of carbon pools and fluxes. The assumptions used in the construction of C models and detailed calculations are given in Appendix.

Results

Carbon Pools

The C pools and fluxes within the *P. ponderosa* based silvopastore, the 18 year old pine plantation, and the managed natural prairie are illustrated in Fig. 4a–c, respectively. Total carbon storage was 224, 199, and 177 Mg C ha^{-1} in the silvopasture, plantation, and prairie, respectively. These C pools do not include *N. pumilio* (lenga) coarse woody debris (CWD) and pine dead branches since they undergo a slow but constant decomposition process over the years and are therefore temporary pools. In addition, given the high annual grass root turnover (90%) in all ecosystems

a

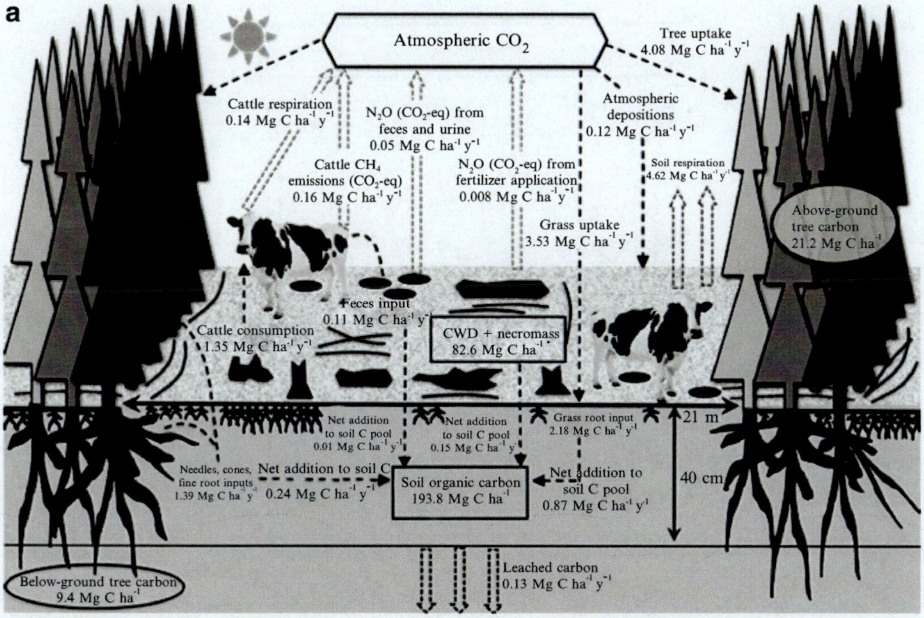

Fig. 4 Models of C pools and fluxes for (**a**) a ponderosa pine-based silvopastoral system arranged in strip (silvopasture), (**b**) an 18-year old *Pinus ponderosa* plantation (plantation), and (**c**) a managed natural prairie (prairie) in the Chilean Patagonia. All C pools appear in boxes and C fluxes are indicated by *arrows*. *The CWD + necromass values indicated in the boxes are informative only; coarse woody debris and dead branches do not represent real C pools where storage occurs as they undergo a slow but constant decomposition process over the years. For more information on calculations, see Appendix

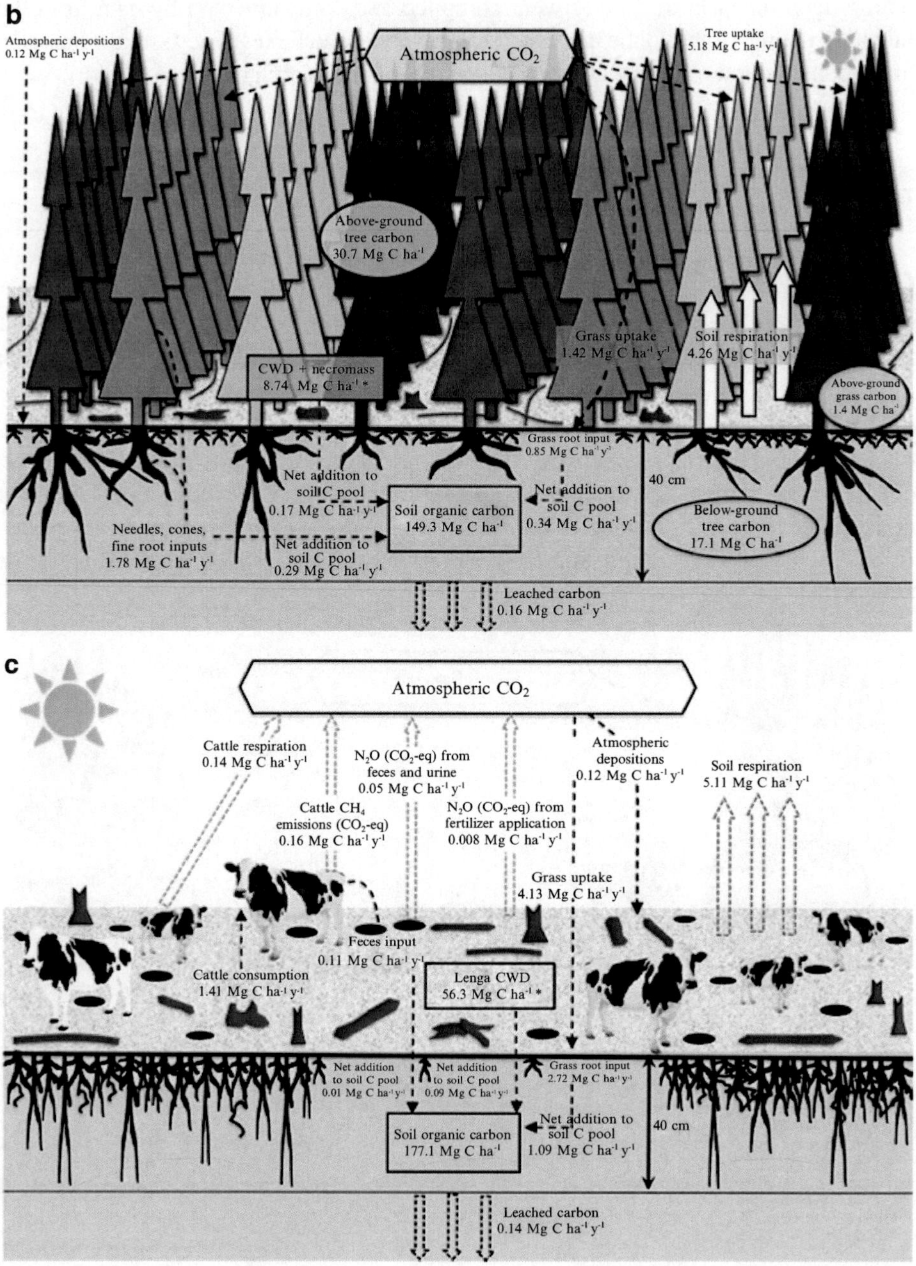

Fig. 4 (continued)

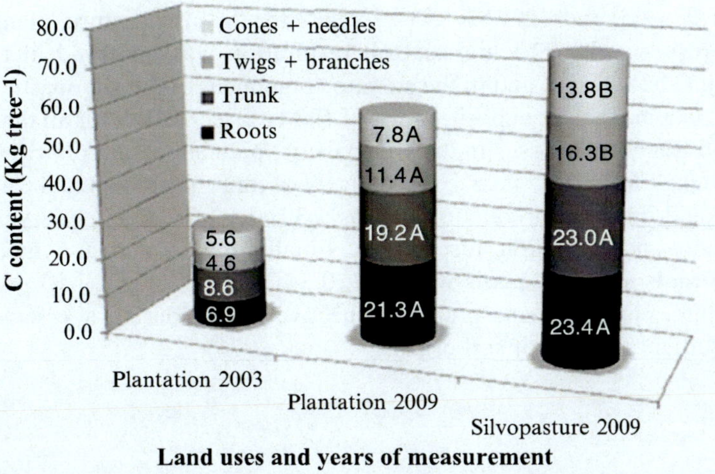

Fig. 5 Distribution of C content (kg) per tree compartment before thinning the pine plantation in 2003 and in the resulting Plantation and Silvopasture 6 years after thinning. Values with the same higher case letter within a tree component and between land uses in 2009 are not significantly different (Student's t test, $^{**}P < 0.01$)

(Stolpe et al. 2010) in spring, summer and fall, C in fine root biomass of grasses was considered as an annual flux where C is added to the soil C pool and not as sequestered C in grass roots (Gordon and Thevathasan 2005). Only perennial tree roots with a diameter larger than 5 mm were considered as the belowground C pool (Abohassan 2004). The aboveground to belowground C pool ratio was approximately 1:10, 1:5, and 1:177 for the silvopasture, plantation, and prairie, respectively.

Eighteen years after the establishment of the pine plantation (i.e. in 2009), the total C stored in the silvopasture and plantation was 27% and 12% greater, respectively, than that in the prairie. However, 6 years after the conversion (2003–2009) of the plantation into the silvopastoral system, the total C storage in the silvopasture increased by 13%. As depicted in Fig. 5, individual trees in the silvopasture sequestered almost 30% more C in the total tree biomass, compared with trees in the plantation. On per tree basis, the C content was higher in every tree component in the silvopasture as compared with the plantation, but significant differences ($p < 0.01$) were found only for cones + needles (70% higher) and twigs + branches (40% higher).

Carbon Fluxes

Net C flux in tested ecosystems over a 2 year measurement period were based on the following quantifications: net assimilation by trees and grass, decomposition of woody detritus, soil respiration, C leaching and atmospheric depositions, animal consumption, faeces input, and fertilizer applications, and was found to be as +1.8, +2.5, and −2.3 Mg C ha^{-1} $year^{-1}$ for the silvopasture, plantation and prairie, respectively. The highest soil respiration was observed in the prairie (5.11 Mg C ha^{-1} $year^{-1}$) and in the pasture component of the silvopasture. Annual C input via atmospheric deposition was 0.12 Mg C ha^{-1} $year^{-1}$ for all three ecosystems, and leaching C losses in the silvopasture, plantation, and prairie were 0.13, 0.16, and 0.14 Mg C ha^{-1} $year^{-1}$, respectively. With respect to annual C input from litterfall and fine root production, it was 1.39 and 1.78 Mg C ha^{-1} $year^{-1}$ in the silvopasture and plantation, respectively. Finally, the net annual C incorporation by grass roots to soil C pools was 0.87, 0.34, and 1.09 Mg C ha^{-1} $year^{-1}$ in the silvopasture, plantation and prairie, respectively, considering a 40% addition to the recalcitrant fraction of soil organic C (Falk 1976).

Discussion

Carbon Pools

The aboveground to belowground C pool ratio depicts the preponderance of soil organic C (SOC) pools belowground. The absence of perennial woody species is responsible for the large ratio observed in the prairie (1:177; Fig. 4c). When comparing pools in the silvopasture and plantation systems, even though the tree density in silvopasture was only 50% of that in the plantation, the aboveground to belowground C pool ratio was higher in the silvopasture. The presence of highly active aerial and subterranean C cycles in the silvopastoral system (Sharrow and Ismail 2004) could have contributed to the higher ratio in the silvopasture in spite of lower tree density. The ratio obtained in this study for silvopasture was the same as that reported by Peichl et al. (2006) for a spruce (*Picea abies* (L.) Karst.)-barley (*Hordeum vulgare* L.) intercropping system in southern Ontario, Canada. In their study, above and belowground C pools were about three times smaller than those obtained in our study, which was to be expected given the lower tree density (111 trees ha^{-1}) and non-volcanic soils in which their system was established. Volcanic soils with allophanes tend to capture larger amounts of C than non-volcanic ones (Buol et al. 1997).

After its conversion, the silvopasture has taken only a third of the time compared to the plantation, since its establishment, to reach similar C gains in their above and belowground pools, perhaps due to positive interactions between cattle, tree, and pasture components. The large C storage potential in the silvopasture can also be explained by higher tree growth (Fig. 5). Additionally, the proportions of total tree C pools and SOC pools in relation to the total C sequestered in the plantation were similar to those reported by Dube et al. (2009) for the plantation in a previous study.

If no animal grazing was allowed in all systems, the total above and belowground grass C pools would represent 6%, 2%, and 10% of total C sequestered in silvopasture, plantation, and prairie, respectively. The relative contribution of belowground root biomass to these values ranged from 80% to 90%. This is in agreement with the findings reported by Raich and Tufekcioglu (2000), where large proportions of the photosynthates produced by prairies were allocated to belowground biomass (roots). Furthermore, it has also been shown that grass growing in harsh environments, such as in Patagonia, tend to develop larger root systems where energy reserves can be stored (Gibson 2009).

The total C storage potentials indicated in this research, 177–224 Mg C ha^{-1}, exceed those reported in other recent studies (Dixon et al. 1994; Sharrow and Ismail 2004; Gordon and Thevathasan 2005; Peichl et al. 2006), demonstrating the potential of temperate agroforestry systems in C sequestration. The high C storage observed in this study could be attributed to the high C sequestration capacity of volcanic soils and their large contribution to the total system C pools. In southern Canada, Peichl et al. (2006) found total system C pools of 97, 75, and 69 Mg C ha^{-1} (0–20 cm soil depth) for hybrid poplar (*Populus deltoides* Bartr. ex Marsh × *Populus nigra* L.) and spruce intercropping and for barley sole cropping systems, respectively. Gordon and Thevathasan (2005) estimated that the C stored in all pools of a poplar-based silvopastoral system with sheep at 62 Mg C ha^{-1} compared to 44 Mg C ha^{-1} (0–5 cm soil depth) in a monoculture pasture system. In Oregon, Sharrow and Ismail (2004) found that in Douglas fir (*Pseudotsuga menziesii* (Mirb.) Franco)/ryegrass (*Lolium perenne* L.)/clover (*Trifolium subterraneum* L.) silvopastoral system, monoculture plantation, and pasture systems of the same species sequestered 109, 101, and 103 Mg C ha^{-1}, respectively (0–45 cm soil depth). Dixon et al. (1994) estimated that agroforestry systems in temperate regions could capture between 15 and 198 Mg C ha^{-1}.

Carbon Fluxes

The C fluxes in this study were found to be similar to those reported by Peichl et al. (2006), who found net C fluxes of 13.2, 1.1, and −2.9 Mg C ha^{-1} $year^{-1}$ for alley cropping systems with poplar and spruce and for a barley monocropping system, respectively. In this study and in the study mentioned above, conifer-based agroforestry systems have demonstrated net positive C fluxes. It was 1.8 Mg C ha^{-1} $year^{-1}$ (present study) as against 1.1 Mg C ha^{-1} $year^{-1}$ (Peichl et al. 2006) – 60% higher in the Patagonian silvopasture. Part of the difference observed could perhaps be explained by the better growth performance of pine in relation to spruce in their respective environments, but several other factors such as the type of crop must also be taken into consideration before a direct comparison can be made. For instance, Gordon and Thevathasan (2005) estimated the net annual C sequestration potential in a silvopastoral system with sheep grazing under poplars to be 2.7 Mg C ha^{-1} $year^{-1}$ compared to 0.99 Mg C ha^{-1} $year^{-1}$ in a monoculture pasture system. However, their study did not take into account C losses through leaching, N_2O emissions from fertilizer application and long-term soil respiration data as we did in this study, which may explain why they obtained a larger flux for the AFS and a positive flux for the pasture.

Table 1 Mean annual soil moisture (0–20 cm depth), superficial air temperature above the soil (+5 cm) and soil surface temperature (0–5 cm depth) measured over a two-year period between November 2007 and 2009

Treatment	Soil moisture 0–20 cm (% VWC)	Air temperature +5 cm (°C)	Soil temperature 0–5 cm (°C)
Prairie	8.7	6.7	6.5
Plantation	8.0	6.8	6.6
Silvopasture (within tree strip)	6.7	6.6	6.4
Silvopasture (2m from tree strip)	13.6	8.0	6.9

VWC volumetric water content

In the silvopasture and plantation, the net C fluxes were positive, which indicate that these systems are true C sinks. A negative net C flux in the prairie indicates that this system is a net C source to the atmosphere. Six years after the conversion of the plantation into the silvopasture, the C flux of plantation was only 39% higher than that of the silvopasture, in spite of higher tree density and the absence of GHG emission from animals and fertilizer application. Carbon inputs from trees and grass were 7.6 and 6.6 Mg C ha^{-1} year^{-1} within the silvopasture and plantation, respectively. These C inputs were higher than C outputs, but in the prairie, C outputs exceeded C inputs. In the winter months, soil respiration was offset by tree photosynthesis in the silvopasture and plantation, although at a lower rate, but was not so in the prairie, resulting in net CO_2 emissions during the winter season.

Soil Respiration

The presence of trees in the silvopasture did help to offset the high soil respiration in the pasture portion of the system, given their higher C assimilation and lower contribution to soil respiration than grass, as reflected by the values obtained within the tree strip (Dube 2010). Raich and Tufekcioglu (2000) also reported respiration rates 20% higher in grasslands than forests growing on the same soil type and under similar environmental conditions. Soil temperature and moisture are largely responsible for differences in soil respiration rates, and the moisture and temperature variations that were observed among ecosystems and at distinct locations within the silvopasture (Table 1) may help to understand the differences recorded.

Other factors such as root activity and density, the presence of mycorrhizae (Kimmins 2004), and the availability of C substrates for microbial biomass (Dube et al. 2009) may also have influenced soil respiration rates. In addition, it should be noted that the Aysén Region of the Chilean Patagonia was under indigenous forest in the past and had large quantities of coarse woody debris (CWD) on the ground, covering approximately one third of the total study area based on a CWD inventory data (Dube 2010). The presence of significant amounts of decaying woody material left on the soil could have contributed to elevated soil respiration in the three studied systems, which exceeded grass C assimilation in the Prairie. Thus, it seems that the prairies of the Patagonia region extending over more than one million ha are acting as a C source.

Atmospheric Depositions

Although the annual C input via atmospheric deposition value appears to be insignificant in the presented models (0.12 Mg C ha^{-1} $year^{-1}$; Fig. 4a–c), it is important to assess the atmospheric annual C inputs to the overall C budgets for the test site (Chilean Patagonia region), and verify if recent volcanic activity caused additional C depositions to the soil. Having said this, it should be mentioned that only dissolved C via rainfall and snow will remove atmospheric CO_2 and any C addition as a result of volcanic activities will not have any effects on atmospheric CO_2 removal. Monthly measurements of dissolved C in rain and snowfall after the eruption of the Chaitén volcano in May 2008 did not show evidence of additional C input from ashes (Dube 2010), although it is located at 400 km north of the research site, probably because of the prevailing winds blowing eastward.

Leached Carbon

With respect to total annual C losses as leachates, its contribution to the C budgets also seems very small as compared with other studies. In Guelph, Canada, Peichl et al. (2006) encountered leachate outputs of 1.8, 1.5, and 1.8 Mg C ha^{-1} $year^{-1}$ in alley cropping systems with poplar and spruce and a barley mono-cropping system, respectively. The values reported by these authors were 7–15 times larger than those from our models, although annual precipitation and leaching rate in Patagonia were only 45% higher, the trees were older and their density was higher. The larger leaching C losses in the Guelph studies could be attributed to the calcareous soil caused by the $CaCO_3$ bedrock. Additionally, Undurraga et al. (2009), working with annual crops on a volcanic soil in Chile, reported dissolved organic C contents of 67.8–151.7 mg l^{-1}, similar to 40–56 mg l^{-1} obtained in this study. Volcanic soils contain a special type of clay known as allophanes, which have the capacity to retain larger amounts of organic C, resulting in lower C concentrations in leachates (Woignier et al. 2007). The Patagonian ecosystems would be more efficient in reducing C losses to ground waters than ecosystems on non-volcanic soils, and thereby reduce the amount of soluble C source in aquatic ecosystems, which is required for denitrification by microbes. As a result, the lower content of dissolved C may aid in the reduction of N_2O emissions from aquatic ecosystems.

The contribution from the system to the total leached C, excluding dissolved C from atmospheric depositions, was 77.3%, 82.1%, and 79.6% in the silvopasture, plantation, and prairie, respectively (Dube 2010). The silvopasture is the most efficient in terms of dissolved C retention and the plantation tends to release higher amounts of dissolved C to the environment. Changes in soil pH could affect soil fungi populations among ecosystems, which in turn may cause variations in dissolved C contents in leachates (W. Blum, 2010, personal communication), but additional work is needed to test this hypothesis.

Litterfall and C Input Through Decomposition

These annual C input from litterfall and fine root production values were obtained from data collected over a 24 month sampling period, which permitted to take into consideration the effect of periodical weather patterns and forest management activities. Measurement of litterfall, fine roots, and cattle faeces inputs helped to quantify the annual C input to soil C pool through decomposition processes. As time constraints impeded the realization of long term decomposition experiments, the annual mass losses observed *in situ* were complemented by data from the literature. Considering a maximum mass loss of 84.5% after a period of 6 years (Berg and Laskowski 2006), the amount of C incorporated into the stable soil C pools will be 0.04 Mg C ha^{-1} year^{-1} in the silvopasture and 0.05 Mg C ha^{-1} year^{-1} in the plantation. Notwithstanding, these values do not consider accumulated litterfall from the previous years that are gradually decomposing and also being added to the soil C pool. Therefore, the sum of net annual additions to stable soil C pools were 0.24 and 0.29 Mg C ha^{-1} year^{-1} for the silvopasture and plantation, respectively.

The larger belowground biomass yield usually results in greater C addition to soil C pool, as reflected in the prairie and silvopasture. Gordon and Thevathasan (2005) reported net additions of 0.99 Mg C ha^{-1} year^{-1} in a poplar-based silvopastoral system and a ryegrass prairie in southern Canada, in agreement with the values indicated for the silvopasture and prairie. The differences that were observed could be explained by the larger grass root input in the prairie and lower input in the silvopasture, since approximately 25% of the spatial area is occupied by trees, whereas only 16% of the area was occupied by ryegrass in the poplar silvopastoral system.

Greenhouse Gases Mitigation Potential

Stocking Rates

Using net C sequestration values from three tested ecosystems and global warming potential (GWP) values for methane and nitrous oxide, it is possible to determine how many cows per hectare could potentially be grazed in the prairie and the silvopasture without resulting in net C emissions to the atmosphere. The net C sequestration values were 3.80 and 1.09 Mg C ha^{-1} year^{-1} for the silvopasture and prairie, respectively (Fig. 4a, c). Based on these net C sequestration values, the hypothetical number of cattle that can be "C-neutrally grazed" was five cows per hectare in the silvopasture and only two in the prairie (see Appendix). This is ten and four times more than the actual stocking rate of 0.5 cows per hectare in the silvopasture and prairie, respectively (Sotomayor et al. 2009), due to the incorporation of trees into the system. The actual stocking rate could be increased provided pasture production can support the new stocking density. In a previous study done in 2006, the prairie had a stocking rate of two cows per hectare (Dube et al. 2009). It must be noted however that these maximal stocking densities are derived based on results from this study, where total C tree uptake (assimilated in woody components and returned to soil via litterfall and fine root turnover) represents the mean above and belowground C sequestration rate over 18 years during the 1991–2003 and 2003–2009 growth periods. The actual stocking rate of cattle would probably be lower during pine establishment and senescence. Also, as the trees mature, their crowns become larger and will create more shadow on the border of the strip, which may affect pasture growth at the edges of the alleys and thereby can reduce the stocking density due to lower pasture production. Higher stocking rates can also be maintained with grazing of goats, sheep, or horses, given their lower individual GHG emissions (Yang et al. 2003).

The stocking rate affects the C budget because of CH_4 emissions by ruminants. Digestible C losses of 5% occur due to the CH_4 emissions from enteric fermentation, which contributes between 16% and 23% of global CH_4 emissions (Soussana et al. 2004). Well managed prairies, using better quality grasses that increase the digestive efficiency will reduce CH_4 emissions because the food remains less time in the rumen, producing less CH_4 (DeRamus et al. 2003). A silvopastoral system with low input sustainable practices, which minimize vegetative and soil disturbances, promote the presence of perennial vegetation, recover or recycle emissions, and will contribute to the preservation of C and N pools during decades or centuries (Lal 2005).

Land Area Under Silvopastoral Management

The results from this investigation and published reports permit the estimation of the total land area required under silvopastoral management so that cattle raising could become C neutral in the Chilean Patagonia. There are approximately 260,000 cows in Patagonia, out of which 199,000 are found in the regions of Aysén (O. Teuber, 2010, personal communication) and 61,000 in Magallanes (INE 2007a). Using the same net C sequestration data, and considering that more than three million ha are either abandoned or degraded land resulting from severe forest fires in the last century, only 48,127 ha under silvopastoral systems with cattle would be needed in the Chilean Patagonia to offset all C losses from cattle-based livestock systems and become C neutral, out of which 36,752 and 11,375 ha are in the Aysén and Magallanes Regions, respectively. Since the Aysén and Magallanes Regions (INE 2007b) cover an area of 108,494.4 and 38,400.8 km^2, respectively, the total area needed would be only 0.33%. However, the land area required using a natural prairie approach would be a total of 167,783 ha, with 128,125 ha in Aysen and 39,658 ha in Magallanes. That is 3.5 times more land area required as monoculture pasture systems, and represents 1.14% of the Chilean Patagonia. However, these estimated areas only consider silvopastoral systems with cows; smaller areas would be needed if sheep only were grazed, but larger areas would be necessary if both cattle and sheep were included. In addition, it is worth mentioning that CH_4, CO_2, and N_2O emissions contribute to 46%, 39%, and 15%, respectively, to the GWP from the areas mentioned above. On the other hand, the use of fast growing tree species well adapted to Patagonian conditions, such as *Populus trichocarpa* Torr. & A. Gray (A. Sotomayor, 2009, personal communiaction) could result in better short term C storage or greater C storage than pines at rotation age. In addition, poplar leaves may annually release more N to the soil than pine needles (Thevathasan and Gordon 1997) because of the different substrate quality and could enhance pasture growth. Based on this study, and knowing the numerous benefits of agroforestry for soil conservation (Gordon et al. 2009), it may be a policy option to be considered to introduce tree-based pasture systems in Patagonia replacing the monoculture pasture systems (prairie) that are currently acting as a C source.

Conclusions

In the Chilean Patagonia, the adoption of silvopasture appears to be a sustainable land use management practice that preserves and increases soil C pools, contributes to reduce atmospheric CO_2, and permit to offset GHG emissions from animal grazing and fertilizer applications, and thus could convert the entire region into effective C sinks rather than C sources, which they are now. Our results indicate that individual trees in the silvopasture are using the site resources more efficiently and have sequestered almost 30% more C in total above- and belowground biomass when compared with trees in the plantation. Sustainable increase in tree density could enhance C sequestration and have an added benefit in terms of biomass production for bioenergy.

As the thinning operation resulted in higher C sequestration rates in the silvopasture, any new establishment of silvopastoral systems in the region may follow the recommendations from this study in terms of C sequestration. We can expect significant gains in SOC in the future resulting from remaining pine stumps and coarse-root decomposition. The aboveground: belowground C pools ratio show the preponderance of C pools belowground and the key role played by volcanic soils in the capture of large amounts of C. Besides the higher stem density in the plantation, the synergistic effect resulting from the combination of trees and pasture led to more C being sequestered in the silvopasture soil.

The C fluxes suggest that the plantation can annually sequester only 40% more C than the silvopasture, in spite of twice as much tree density. In the silvopasture and plantation systems, the net C fluxes were positive, which indicate that these systems are true C "sinks". A negative net C flux in the prairie system indicates that this system is a net C "source" to the atmosphere. Based on this study, the actual cattle stocking rate could be increased to five cows per hectare in the silvopasture and only two in the prairie in order to be "C-neutrally grazed", provided pasture production can support the new stocking density. On the other hand, only 481 km^2 (0.33% of total area) under silvopasture would be needed to offset all C losses from cattle-based livestock systems in the Chilean Patagonia and become C neutral. Given that large deforested areas are currently subject to soil erosion coupled with poor and inferior quality pasture production, the adoption of silvopasture over large tracts of grazing lands should not be a problem in Patagonia nor a threat to other types of land uses. As the prairie is acting as a C source, pine-based silvopastoral systems could contribute enormously towards Chilean strategies to mitigate climate change.

Ackowledgements We sincerely wish to acknowledge and express our sincere thanks to Mr. Victor "Titi" Mata for hosting the research sites in his property near Coyhaique, to Silvia Marchetti and Luis Alvarez for assistance in the field, to Forestal Mininco S.A. - Aysén Project for logistics, to the Center of Biotechnolgy, Laboratory of Soil Science and Natural Resources, and Laboratory of Environmental Organic Chemistry of the University of Concepción, and Celulosa Arauco S.A. - Arauco Pulpmill for processing the samples and numerous laboratory analysis. Special thanks are also given to the Trapananda Center (University Austral of Chile), and Center for Investigations of Ecosystems in the Patagonia (CIEP) for logistics and support. In addition, we sincerely thank INFOR for sharing valuable information and data and authorizing the establishment of the plots in the San Gabriel Agroforestry Unit. The excellent review recommendations from two anonymous reviewers are also appreciated. This study was funded by a grant (Project No. 207.142.025-1.0) from the Direction of Investigations of the University of Concepción. Additional financial support was also obtained from CONICYT of the Chilean Government through a doctoral scholarship to F.D. and indirectly from SAI Global Inc.

Appendix

The following assumptions and calculations were used in order to build the models of carbon pools and fluxes within the pine-based silvopastoral system arranged in strip, the 18-year old *Pinus ponderosa* plantation, and the managed natural prairie:

1. The silvopasture and plantation have a tree density of 400 and 800 stems ha^{-1}, respectively. Pine strips in the silvopasture occupy 22% of the area available for pasture and have never been thinned. All trees were pruned to heights of 2.8 and 4 m in 2006 and 2009, respectively.
2. Using inventory data obtained since the establishment of the pine plantations, the mean above and belowground C sequestration rates were calculated for the 1991–2003 period at the initial tree density, and then for the 2003–2009 periods after thinning to a density of 800 trees ha^{-1} in the Plantation and 400 trees ha^{-1} in the silvopasture.
3. Aboveground tree C pools include trunks, branches, twigs, needles and cones. Belowground C pools include thick roots superior to 5 mm diameter.
4. It was assumed that cattle will consume most of the aboveground pasture biomass produced during the year in the silvopasture and prairie (part of it returning to the system as faeces, methane, nitrous oxide, and respiration) and that only belowground biomass C will be added to the soil C pool. The aboveground

grass biomass present in the plantation remains in the system as no grazing occurs therein. The C content of aboveground net primary productivity (ANPP) was determined after manually harvesting grazing material three times a year over a 2-year period. The C content of belowground net primary productivity and net annual C sequestration by pasture alone could then be calculated, considering a 40% addition to the recalcitrant soil C pool (Falk 1976).

5. Knowing the stocking rates and the amount of faeces produced annually and C content, the cattle respiration (kg CO_2 ha^{-1} $year^{-1}$), methane emissions from enteric fermentation, and nitrous oxide emissions from dung and urine patches, as well as their CO_2-equivalents (IPCC 2001) were calculated using data published by Flessa et al. (2002), Yang et al. (2003), and Byrne et al. (2007). Carbon dioxide, methane, and nitrous oxide emissions from a single animal are estimated to be 996, 56, and 1.29 kg $head^{-1}$ $year^{-1}$, respectively, and depend on the amount and kind of feed that is consumed. The reference weight per head unit is 500 kg. CO_2-equivalents were calculated using the Global Warming Potentials (GWP), which determine the relative contribution of a given gas to the greenhouse effect. The GWP values represent how many times more deleterious than CO_2 in a 100 year period are CH_4 (21) and N_2O (310) in terms of global warming. In addition, the number of cows per hectare in order to attain C neutrality was calculated as follows, using the net C sequestration values of 3.80 and 1.09 Mg C ha^{-1} $year^{-1}$ for the silvopasture and prairie. For the silvopasture, 3.80 = [(56*#cows/1,000*21)/3.67) + (996*#cows/1,000/3.67) + (1.29*#cows/1,000*310)/3.67)]. For the Prairie, 1.09 = [(56*#cows/1,000*21)/3.67) + (996*#cows/1,000/3.67) + (1.29*#cows/1,000*310)/3.67)].
6. The annual mass loss values of cattle faeces obtained after a 12 month litterbag decomposition experiment made possible the calculation of net C additions to soil C pools, considering that 2.2 and 2.5 years are required to get a maximum decomposition in the silvopasture and prairie, respectively. The net addition to soil C pool in each treatment represents therefore the sum of annual C incorporations over these periods. Hirata et al. (2009) reported similar results, where cattle dung reached an average decomposition of 79.1% after 2.2 years.
7. Annual litterfall and needle decomposition in the plantation and the silvopasture were obtained from field measurements over a 2-year period to illustrate the importance of annual C inputs and net additions to soil C pools. It was assumed that annual fine root C turnover in pines is 30% of litterfall (Abohassan 2004).
8. Since time constraints did not permit to undertake a long term experiment for the decomposition of the ponderosa pine needles, a maximum mass loss of 84.5% for Scots pine needles in Scandinavia was assumed to be representative of the situation, considering the similar climatic conditions encountered and values of initial N and lignin contents found in green litter (Berg and Laskowski 2006; Dube 2010). Theoretically, the C contribution to soil from litterfall and fine root turnover for the last 18 years was 25 and 32 Mg C ha^{-1} in silvopasture and plantation, respectively. However, an average of 14.2% in the silvopasture and 14.5% in the plantation of the C added annually via litterfall and fine roots

was released back into the atmosphere through microbial decomposition. Based on the results of the decomposition experiment, approximately 6 years (84.5%/14.2% $year^{-1}$ in the silvopasture and 84.5%/14.5% $year^{-1}$ in the plantation) would be required to obtain maximum needle decomposition. This represents 0.20 Mg C ha^{-1} $year^{-1}$ (25 Mg C ha^{-1}/18 years * 14.2% $year^{-1}$) in the silvopasture and 0.26 Mg C ha^{-1} $year^{-1}$ (32 Mg C ha^{-1}/ 18 years * 14.5% $year^{-1}$) in the plantation that are lost due to decomposition. Considering a maximum mass loss after a period of 6 years, the amount of C incorporated into the stable soil C pools will be 0.04 Mg C ha^{-1} $year^{-1}$ (25 Mg C ha^{-1}/18 years * 15.5%/6 years) in the silvopasture and 0.05 (32 Mg C ha^{-1}/18 years * 15.5%/6 years) in the plantation. However, these values represent what is lost and gained from the annual litterfall, and do not consider accumulated litterfalls from the previous years that are gradually decomposing and also being added to the soil C pool. Taking this process into account, the sum of annual losses as decomposition during the 6 year period in the silvopasture and plantation reached 1.2 and 1.6 Mg C ha^{-1} $year^{-1}$, respectively, whereas the net annual additions to stable soil C pools were 0.24 and 0.29 Mg C ha^{-1} $year^{-1}$ for the silvopasture and plantation, respectively.

9. With respect to decomposition of necromass and coarse woody debris (CWD), the annual mass loss was determined using published *k* values ($year^{-1}$) for decomposition of dead branches of *Pinus ponderosa* (Hart et al. 1992; Yin 1999; Hall et al. 2006) and boles/stumps of *N. pumilio* (Frangi et al. 1997). Knowing the dry weight of dead branches after 2 years of decomposition and CWD, and assuming a 95% loss of initial weights, it was possible to calculate their mass losses and net addition to soil C pools. It should be noted that decomposition of the duff needle layer is not considered here as it has already been accounted for in the calculation of annual litterfall decomposition.
10. Soil C sequestration for the upper 0–40 cm layer was determined using weighted averages of C contents at three measured depths and a bulk density of 0.9 g cm^{-3} (Dube et al. 2009). In the silvopastoral system, an average value was calculated from the C contents obtained within the tree strips and at 2.5 m intervals on either side of strips (up to 10.5 m, corresponding to the middle of the 21 m wide pasture strip).
11. Soil respiration values refer to total respiration, including tree root, mycorrhizae and microbial respiration, and annual decomposition losses of needles, fine roots, cattle faeces, necromass, and coarse woody debris. Annual soil respiration for the three ecosystems was calculated from the monthly respiration rates presented in this study. For the months that soil respiration was not measured, estimates were done as follows: A regression between soil respiration and air temperature (+5 cm) was adjusted for every treatment ($R^2 = 0.94$), using the values obtained in the field. Knowing the mean monthly superficial air temperatures, these equations were then used to estimate monthly soil respiration and check the values calculated initially, the differences being less than 5%. Within the silvopasture, it was assumed that soil respiration in the tree strip accounts for 22% from the spatial area, while respiration from 1 and 7.5 m from the tree

strip accounts for 78%. Since respiration chambers were installed within the pine strips, at 1 m and at 7.5 m from the strips, tree roots growing into the grass band could be taken into consideration in the calculations.

12. In order to determine the annual amount of leaching C, it was assumed that 24% of the annual rainfall leaches to the ground water (Gisi 1997; Peichl et al. 2006). Annual rainfall at the research site is 1,206 mm out of which 290 mm ha^{-1} year^{-1} is lost as leaching. The mean annual total C concentrations of leached soil solution from the land uses were then used to estimate the annual leached C losses in conjunction with total annual leaching losses. As above, C leaching within the tree strip was assumed to account for 22% of the spatial area, whereas leaching from 1 and 7.5 m from the tree strip account for 78% of the area.
13. The annual atmospheric C deposition to the systems was determined as follows: knowing that the annual rainfall is 1,206 mm year^{-1}, the volume occupied by this amount over 1 ha was 1.2×10^4 m^3. Since the density of water is 1 g cm^{-3}, 1% C of 1 l leaching soil solution is equivalent to 10 g C. Therefore, 0.12 Mg C ha^{-1} year^{-1} represents the amount of atmospheric deposition.
14. Since approximately 1.25% of N fertilizer applied to the soil is lost in the form of N_2O emissions (IPCC 1997), and knowing the amount of N fertilizer applied to the pasture every 3 years, annual emissions of N_2O and CO_2-equivalent were estimated.
15. Carbon storage in ecosystems pools was calculated using the following equation:

$$C_{pools} = C_{agt} + C_{bgt} + C_{agg} + C_{soil},$$

where C_{pools} = total carbon stored in ecosystem pools, C_{agt} = aboveground tree carbon, C_{bgt} = belowground tree carbon, C_{agg} = aboveground grass carbon in the Ponderosa pine plantation and C_{soil} = soil organic carbon pool.

16. Positive or negative carbon flux into or out of the ecosystems was calculated using the following equation:

$$C_{flux} = C_{TrU} + C_{GrU} + C_{AtD} + C_{FecS} + C_{CwdS} - C_{SRes} - C_{Lch} - C_{Fert} - C_{AnC}$$

where C_{flux} = net carbon flux in the ecosystem, C_{TrU} = carbon input via total tree uptake, C_{GrU} = carbon input via total grass uptake, C_{AtD} = carbon input through atmospheric depositions (rain and snow), C_{FecS} = net addition to soil carbon pool via faeces input, C_{CwdS} = net addition to soil carbon pool via coarse woody debris and necromass decomposition, C_{SRes} = carbon output via total soil respiration, C_{Lch} = carbon leachate output from the soil solution, C_{Fert} = volatile carbon-equivalent output from fertilizer application, and C_{AnC} = carbon output through pasture consumption by animals (divided between cattle fattening, faeces production and GHG emissions). Therefore, losses as animal respiration, CH_4 emissions from enteric fermentation and N_2O from faeces have already been accounted for in cattle consumption.

References

Abohassan RA (2004) Carbon dynamics in a temperate agroforestry system in southern Ontario, Canada. M.Sc. thesis, Department of Environmental Biology, University of Guelph, Guelph, 122 p

Bambrick AD, Whalen JK, Bradley RL, Cogliastro A, Gordon AM, Olivier A, Thevathasan NV (2010) Spatial heterogeneity of soil organic carbon in tree-based intercropping systems in Quebec and Ontario, Canada. Agroforest Syst 79:343–353

Berg B, Laskowski R (2006) Litter decomposition: a guide to carbon and nutrients turnover. Adv Ecol Res, vol 38. Elsevier Academic Press, San Diego, USA, 428 p.

Buol SW, Hole FD, McCracken RJ, Southard RJ (1997) Soil genesis and classification, 4th edn. Iowa State University Press, Ames, 527 p

Byrne KA, Kiely G, Leahy P (2007) Carbon sequestration determined using farm scale carbon balance and eddy covariance. Agric Ecosyst Environ 121:357–364

DeRamus HA, Clement TC, Giampola DD, Dickison PC (2003) Methane emissions of beef cattle on forages: efficiency of grazing management systems. J Environ Qual 32:269–277

Dixon RK, Wimjum JK, Lee JJ, Schroeder PE (1994) Integrated systems: assessing of promising agroforest and alternative land use practices to enhance carbon conservation and sequestration. Clim Change 30:1–23

Dube F (2010) Carbon dynamics and other interactions in *Pinus ponderosa* (Dougl. ex Laws) and natural prairie agroecosystems established on degraded volcanic soils in the Aysén Region, Chilean Patagonia. Ph.D. thesis, Faculty of Forest Sciences, University of Concepción, Concepción, Chile, 140 p

Dube F, Zagal E, Stolpe N, Espinosa M (2009) The influence of land use change on the organic carbon distribution and microbial respiration in a volcanic soil of the Chilean Patagonia. For Ecol Manag 257:1695–1704

Edwards NT (1982) The use of soda-lime for measuring respiration rates in terrestrial systems. Pedobiologia 23:321–330

Falk JH (1976) Energetics of a suburban lawn ecosystem. Ecology 57:141–150

Fisons Instrument (1990) Instruction manual for the EA 1108 elemental analyzer. Rev. A2-90. American Society of Agronomy, San Carlos, 44 p

Flessa H, Ruser R, Dörsch P, Kamp T, Jimenez MA, Munch JC, Beese F (2002) Integrated evaluation of greenhouse gas emissions (CO_2, CH_4, N_2O) from two farming systems in southern Germany. Agric Ecosyst Environ 91:175–189

Frangi JL, Richter LL, Barrera MD, Aloggia M (1997) Decomposition of Nothofagus fallen woody debris in forests of Tierra del Fuego, Argentina. Can J For Res 27:1095–1102

Gibson DJ (2009) Grasses and grasslands ecology. Oxford University Press, New York, 313 p

Gisi U (1997) Bodenökologie (2. Auflage). Georg Thieme Verlag, Stuttgart, 351 p

Gordon AM, Thevathasan NV (2005) How much carbon can be stored in Canadian agroecosystems using a silvopastoral approach? In: Mosquera-Losada MR, Rigueiro-Rodriguez A, McAdam J (eds) Silvopastoralism and sustainable land management. CABI Publishing, Wallingford, UK, pp 210–218

Gordon AM, Thevathasan NV, Nair PKR (2009) An agroecological foundation for temperate agroforestry. In: Garrett HE (ed) North American agroforestry: an integrated science and practice, 2nd edn. American Society of Agronomy, Madison, pp 25–44

Hall SA, Burke IC, Hobbs NT (2006) Litter and dead wood dynamics in Ponderosa pine forests along a 160-year chronosequence. Ecol Appl 16(6):2344–2355

Hart SC, Firestone MK, Paul EA (1992) Decomposition and nutrient dynamics of ponderosa pine needles in a Mediterranean-type climate. Can J For Res 22:306–314

Hirata M, Hasegawa N, Nomura M, Ito H, Nogami K, Sonoda T (2009) Deposition and decomposition of cattle dung in forest grazing in Kyushu, Japan. Ecol Res 24:119–125

Huygens D, Boeckx P, Van Cleemput O, Oyarzun C, Godoy R (2005) Aggregate and soil organic carbon dynamics in South Chilean Andisols. Biogeosciences 2:159–174

INE (2007a) VII Censo agropecuario y forestal. Existencia de ganado en las explotaciones agropecuarias y forestales por especie. Santiago, Chile (in Spanish). http://www.censoagropecuario.cl/noticias/09/07042009.html. Accessed 20 Aug 2010

INE (2007b) División Político-Administrativa y Censal, 2007. Santiago, Chile (in Spanish). Available at http://www.ine.cl/canales/chile_estadistico/territorio/division_politico_administrativa/pdf/DPA_COMPLETA.pdf. Accessed 18 Oct 2010

IPCC (1997) Revised 1996 IPCC guidelines for national greenhouse gas inventories. IPCC/OECD/IEA, IPCC, Geneva

IPCC (2001) Climate change 2001: the scientific basis. In: Houghton JT, Ding Y, Griggs DJ, Noguer M, van der Linden PJ, Dai X, Mashell K, Johnson CA (eds) Contribution of working group I to third assessment report of the intergovernmental panel on climate change. Cambridge University Press, Cambridge, UK, 881 p

IPCC (2007) Climate change 2007: synthesis report. Available at http://www.ipcc.ch/pdf/assessment-report/ar4/syr/ar4_syr.pdf. Accessed 6 Oct 2010

Kimmins JP (2004) Forest ecology: a foundation for sustainable forest management and environmental ethics in forestry, 3rd edn. Prentice Hall, Upper Saddle River, 611 pp

Lal R (2005) Forest soils and carbon sequestration. For Ecol Manag 220:242–258

Montagnini F, Nair PKR (2004) Carbon sequestration: an underexploited environmental benefit of agroforestry systems. Agroforest Syst 61:281–295

Nair PKR, Nair VD, Kumar BM, Showalter JM (2010) Carbon sequestration in agroforestry systems. Adv Agron 108:237–307

Peichl M, Thevathasan NV, Gordon AM, Huss J, Abohassan R (2006) Carbon sequestration potentials in temperate tree-based intercropping systems, southern Ontario, Canada. Agroforest Syst 66:243–257

Raich JW, Tufekcioglu A (2000) Vegetation and soil respiration: correlations and controls. Biogeochemistry 48:71–90

Ross DJ, Tate KR, Scott NA, Wilde RH, Rodda NJ, Townsend JA (2002) Afforestation of pastures with *Pinus radiata* influences soil carbon and nitrogen pools and mineralization and microbial properties. Aust J Soil Res 40:1303–1318

SAS Institute Inc (2003) SAS user's guide: statistics. SAS Institute Inc, Cary

Sharrow SH, Ismail S (2004) Carbon and nitrogen storage in agroforests, tree plantations, and pastures in western Oregon, USA. Agroforest Syst 60:123–130

Silva F (2004) Final report on the ecological status of the Aysén Region. SAG – Agricultural Ministry, Government of Chile, Santiago, 143 p (in Spanish)

Sotomayor A, Teuber O, Moya I (2009) Resultados y experiencia sobre manejo de sistemas silvopastorales en la Region de Aysén. In: Teuber O (ed) Sistemas agroforestales para la región de Aysén. INIA-INFOR, Coyhaique, 211 p (in Spanish)

Soussana JP, Loiseau P, Vuichard N, Ceschia E, Balesdent J, Chevallier T, Arrouays D (2004) Carbon cycling and sequestration opportunities in temperate grassland. Soil Use Manage 20:219–230

Stamps WT, Linit MJ (1999) The problem of experimental design in temperate agroforestry. Agroforest Syst 44:187–196

Stolpe NB, Dube F, Zagal E (2010) Calibration of CO2FIX to native forest, pine plantation, and pasture on a volcanic soil of the Chilean Patagonia. Acta Agric Scand B Soil Plant Sci 60(3):235–244

Tans P (2010) Trends in atmospheric carbon dioxide – Mauna Loa. NOAA/ESRL. Available at http://www.esrl.noaa.gov/gmd/ccgg/trends/. Accessed 6 Oct 2010

Teller A (1988) Biomass, productivity and wood waste evaluation in a spruce (*Picea abies*) forest (Strinchmps 983). Commonw For Rev 7(2):129–148

Teuber O, Ganderats S (2009) Características geográficas y edafoclimáticas de la región de Aysén. In: Teuber O (ed) Sistemas agroforestales para la región de Aysén: Cortinas cortaviento y silvopastoreo. INIA-INFOR, Coyhaique, 211 p

Thevathasan NV, Gordon AM (1997) Poplar leaf biomass distribution and nitrogen dynamics in a poplar-barley intercropped system in southern Ontario, Canada. Agroforest Syst 37:79–90

Undurraga P, Zagal E, Sepúlveda G, Valderrama N (2009) Dissolved organic carbon and nitrogen in Andisol for six crop rotations with different soil management intensity. Chil J Agric Res 69(3):445–454

Woignier T, Pochet G, Doumenc H, Dieudonné P, Duffours L (2007) Allophane: a natural gel in volcanic soils with interesting environmental properties. J Sol-Gel Sci Technol 41:25–30

Yang SS, Liu CM, Liu YL (2003) Estimation of methane and nitrous oxide emission from animal production sector in Taiwan during 1990–2000. Chemosphere 52:1381–1388

Yin X (1999) The decay of forest woody debris: numerical modeling and implications based on some 300 data cases from North America. Oecologia 121:81–98

Carbon Pools in Tree Biomass and Soils Under Rotational Woodlot Systems in Eastern Tanzania

A.A. Kimaro, M.E. Isaac, and S.A.O. Chamshama

Abstract Landscape approaches to carbon (C) accounting in agriculture, forest, and other land uses are being promoted as a win-win option for integrating climate change mitigation with sustainable rural development. However, limited data on the C sequestration potential of agroforestry systems in the semiarid tropics imply that subsistent farmers may not fully benefit from this opportunity. This chapter quantifies C stocks in biomass and soils in semiarid Morogoro, Tanzania to assess the potential of rotational woodlot systems to sequester C in the soil and offset carbon dioxide (CO_2) emissions. Carbon levels in native vegetation fallows and forests were used as a reference to evaluate the efficacy of this system to minimize forest degradation and balance CO_2 emissions. After a 5 year rotation, wood yield (23–51 Mg C ha^{-1}) was sufficient to meet household demand for fuelwood. Carbon stocks in the highly productive fallows of *Acacia crassicarpa* A. Cunn. ex Benth., *Acacia leptocarpa* A. Cunn. ex Benth., and *Acacia mangium* Willd. (18–26 Mg ha^{-1}) were similar to those in the Miombo forest reserves. Based on C accumulation rates, it would take 4–9 years for these fallows to recover C lost through forest clearance for agricultural expansion, compared to two or three decades for re-growing miombo

A.A. Kimaro (✉)
Department of Soil Science, University of Saskatchewan, 51 Campus Drive, Saskatoon, SK, S7N 5A8, Canada
e-mail: anthony.kimaro@usask.ca

M.E. Isaac
Department of Physical and Environmental Sciences, University of Toronto, Scarborough, 1265 Military Trail, Toronto, ON M1C 1A4, Canada
e-mail: marney.isaac@utoronto.ca

S.A.O. Chamshama
Faculty of Forestry and Nature Conservation, Department of Forest Biology, Sokoine University of Agriculture, P.O. Box 3010, Morogoro, Tanzania
e-mail: chamstz@yahoo.com

B.M. Kumar and P.K.R. Nair (eds.), *Carbon Sequestration Potential of Agroforestry Systems: Opportunities and Challenges*, Advances in Agroforestry 8,
DOI 10.1007/978-94-007-1630-8_7, © Springer Science+Business Media B.V. 2011

woodlands. Tree fallows also enriched the soil organic C (16–26 Mg ha^{-1}), in some cases (e.g., *A. mangium*) close to the reported value for miombo forest soils (28 Mg C ha^{-1}). Overall, this study demonstrates the significant contributions of rotational woodlot systems to reduce forest degradation and offset CO_2 emissions through on-farm wood supply. However, policies and programs that consider comprehensive approaches to avoid deforestation are needed to take full advantage of this system for climate change mitigation and adaptation.

Keywords Avoided deforestation • Carbon sequestration • Tree fallows • Woodfuel

Introduction

Deforestation and other land use changes in developing countries are responsible for ~74% of greenhouse gas (GHG) emissions (Funder 2009). And specifically for Tanzania, deforestation accounts for ~87% of total GHG emissions (Makundi 2001). In addition, limited economic opportunities for forest dependent people and small scale farmers in the developing countries encourage forest degradation (i.e., loss of forest biomass without noticeable changes in the forest cover). Both deforestation and degradation of forests raise concerns about accelerated GHG emissions in the tropics and call for measures to meet the needs of local communities in an environment friendly and sustainable manner (Mountinho et al. 2005; Swallow et al. 2007). This issue has prompted debates in the post-Kyoto climate change agreements to include mechanisms for providing financial incentives to reduce deforestation and degradation of tropical forests through an international carbon (C) market (Mountinho et al. 2005; Samek et al. 2011). One such mechanism being negotiated is the United Nation's collaborative programme on Reducing Emissions from Deforestation and Forest Degradation in Developing Countries (UN-REDD Programme), which allows countries to claim compensation for avoided deforestation associated with implementing forest conservation programs (Mountinho et al. 2005). However, accounting exclusively for deforestation and forest degradation in developing nations may not effectively reduce carbon dioxide (CO_2) emissions as agricultural management is also a major source of GHG emissions (van Noordwijk et al. 2008). Thus, comprehensive C trading schemes that promote emission reductions in the Agriculture, Forestry and Other Land Uses (AFOLU) sectors are being developed to achieve system or area based approaches for climate change mitigation and adaptation (Negra and Ashton 2009). It is anticipated that the presently negotiated REDD and AFOLU initiatives will fund sustainable rural development and enhance natural resource conservation by devolving revenues from C credits to the local communities (Funder 2009).

Tanzania, one of pilot countries in the UN-REDD programme, is developing programs and policies to decrease deforestation rates, which are estimated at 91,000 ha year^{-1} (FAO 2007). Participatory forest management programs initiated in the mid-1990s are promoted as a strategy to involve local communities to benefit from the REDD scheme (Zahabu 2008; URT 2009). While these programs have been successful over the years, it should be noted that current woodfuel extraction rates from unprotected forests in public lands in the country (47% of total forested land) are not sustainable (Luoga et al. 2002; Mwampamba 2007). A similar trend is also noted in other sub-Saharan African countries. It is estimated that ongoing forest degradation in eastern Africa, as a result of shifting cultivation and unsustainable extraction of wood for firewood and charcoal, could lead to more CO_2 emissions than all previously recorded deforestation (Skutsch et al. 2008).

Agroforestry has shown a high potential to reduce soil and forest degradation through rapid replenishment of soil fertility and provision of off-forest tree resources, especially fuelwood. Through implementation of agroforestry practices, agricultural expansion can be minimized and harvesting pressures on native forests reduced (Kimaro et al. 2007; Jama et al. 2008), simultaneously sequestering atmospheric CO_2 (Isaac et al. 2005; Verchot et al. 2007). Agroforestry thus contribute to climate change mitigation in developing countries and diversifies income resources, particularly when mechanisms for accounting and compensating for C sequestered in agroforestry become widely available to the small-scale farmers. Moreover, on-farm wood production via agroforestry can also reduce the risk of *leakage* by providing alternative fuelwood sources. Under the clean development mechanisms (CDM), leakage may occur when net CO_2 emissions is noted outside the CDM-C project and is attributable to the implementation of the project (van Noordwijk et al. 2008). In the context of forest dependent communities in developing countries, this phenomenon could be a result of inadequate provision of alternative sources of forest products, especially woodfuel.

Early studies promoting agroforestry as a C sequestration strategy focused on C rich multistrata agroforestry systems (AFS) in the humid tropical forest margins (Palm et al. 2004). There is, however, scarcity of information on the C sequestration potential of dry lands (Negra and Ashton 2009), and in particular about AFS in the semiarid Africa (e.g., Takimoto et al. 2008; Kaonga and Bayliss-Smith 2009). Estimating C sequestration potential of AFS in the dry lands is crucial for C accounting purposes. Due to poor vegetation cover and inherently low soil C levels, these areas have low C stocks (Lal 2003). However, they seem to possess an enormous potential to sequester C when converted to agroforestry land use. Moreover, the extent to which planted tree fallow systems, such as rotational woodlots, reduce harvesting pressure of the native forests in semiarid zones and thereby offset CO_2 emissions has been minimally investigated.

The rotational woodlot system consists of three interrelated management phases: a tree-crop intercropping phase aimed at generating intermediate crop products while establishing the woodlot (establishment phase); a tree alone phase to buildup wood biomass and provide secondary benefits including dry season fodder, beekeeping, and soil nutrient replenishment (tree fallow phase); and a final post fallow phase characterized by wood harvesting and sequential cropping (Kimaro et al. 2007). Unlike other tree fallows established at high plant density (10,000 stems per ha; Kaonga and Bayliss-Smith 2009) to optimize foliage biomass for soil nutrient replenishment, tree density in rotational woodlots is comparatively low (625–1,111 stems per ha). As such, this system has shown promise to supply fuelwood and increase crop production in semiarid areas and may considerably reduce degradation and over utilization of dry forests (Ramadhani et al. 2002; Nyadzi et al. 2003; Kimaro et al. 2007). This chapter quantifies C stocks in biomass and soils in semi-arid Morogoro, Tanzania in order to assess the potential of rotational woodlot systems to sequester C in the soil and offset CO_2 emissions through on-farm wood supply. Carbon levels in native vegetation fallows and forests (Miombo woodland) were used as references to evaluate the efficacy of this system to minimize forest degradation and offset CO_2 emissions. Finally, policy implications for C management through AFS are discussed to give recommendations for enhancing the contribution of these systems to climate change mitigation and adaptation in the country.

Methods

Study Site

This research was carried out at Mkundi village (6° 40′ S, 37° 39′ E), Morogoro, Tanzania located in a semiarid zone with elevation of about 475 m above sea level, with a mean annual precipitation and temperature of 800 mm and 24°C, respectively. The soils are classified as Regosol (FAO Classification System) or Entisol (USDA system) and have inherently low fertility for crop production. The natural vegetation at the study site is degraded miombo woodland dominated by scattered tree species of *Sclerocarya birrea* (A. Rich.) Hochst., *Dalbergia melanoxylon* (Guill. and Perr.), *Balanites aegyaptiaca* (L.) Del., *Dichrostachys cinerea* (L.) Wight and Arn., *Acacia* spp., and *Albizia* spp.

Experimental Design and Management

The rotational woodlot experiment was established in March 1999 in a randomized complete block design with three replications. The experiment evaluated wood supply, soil fertility replenishment, and crop productivity after 5 year fallows of *Acacia auriculiformis* A. Cunn. ex Benth., *Acacia crassicarpa* A. Cunn. ex Benth., *Acacia julifera* Beth., *Acacia leptocarpa* A. Cunn. ex Benth., *Acacia mangium* Willd., *Acacia nilotica* (L.) Del., *Acacia polyacantha* Willd., *Gliricidia sepium* (Jacq.) Kunth. ex Walp., and *Leucaena diversifolia* (Schldl.). Natural fallow and continuous maize (*Zea mays* L.) cropping systems were included as controls. Details of the experimental establishment and management as well as assessment of soil fertility and wood and crop yields can be found in Kimaro et al. (2007, 2008). This paper focuses on C stocks in biomass and soils to estimate C sequestration potential of the

Table 1 Wood basic density of tree species in planted tree fallows and natural forests in Tanzania and Malawi

Tree species	Basic density (kg m^{-3})	Source
Acacia crassicarpa	584	Luhende et al. (2006)
Acacia mangium	570	Ali et al. (1997)
Acacia polyacantha	705	Wickens et al. (1995)
Gliricidia sepium	470	Ngulube (1994)
Acacia nilotica	700	Malimbwi et al. (1994)
Acacia auriculiformis	617	Ali et al. (1997)
Acacia leptocarpa	693	Luhende et al. (2006)
Acacia julifera	627	Luhende et al. (2006)
Leucaeca diversifolia	450	Orwa et al. (2009)

tested tree fallows. Soil composite samples were collected at 0–15 cm depth from five randomly selected points in each plot. The soil samples were air dried, sieved through a 2 mm sieve and ground to a fine powder for organic C determination by Walkley and Black method (Anderson and Ingram 1993).

Although some controversy exists on accurate percent C conversions, we estimated wood C at 50% of oven dry weight of stem and branches of sampled trees. Resultant values were extrapolated to a per hectare basis. Percent soil organic C was converted to Mg ha^{-1} based on bulk density and mass of the soil within the top 15 cm depth (Kimaro et al. 2007). Annual wood demand and area of forest cleared annually to supply wood for fuelwood (6,960 m^3 year^{-1} and 417 ha year^{-1}) and charcoal (27, 896 m^3 year^{-1} and 1,671 ha year^{-1}) at Kitulangalo area, Morogoro (Luoga et al. 2000, 2002) and for tobacco curing (4,551 Mg year^{-1} and 8,675 ha) in Tabora district (Ramadhani et al. 2002) were estimated based on published information. These two sites were chosen because experimental station and farmer's field trials involving the rotational woodlot system conducted for over a decade have shown high productivity and adoption potential (Ramadhani et al. 2002; Nyadzi et al. 2003; Kimaro et al. 2007, 2008). Total area required to produce wood to meet this demand was estimated based on biomass yield and wood basic density of the tested tree species (Table 1). The area was then scaled down to a household level to assess land availability for the rotational woodlot system. This approach employed secondary data on the population size (4,640 people) at Kitulangalo area (Luoga et al. 2000) and percentage of small scale tobacco farmers (60% of the population) in Tabora (Ramadhani et al. 2002). Basic density of tree species was used to convert biomass to volume since this was the unit used to report productivity and extraction rates of wood in the native forests (Luoga et al. 2002).

Statistical Analysis

A mixed model procedure in statistical analysis system (SAS Institute 2000) was used to run a one way analysis of variance (ANOVA) after confirming normality and constant variance of residuals for tree biomass yield, wood C, and soil organic C using graphical analysis. Tree species was designated as a fixed effects variable while block and block x species interaction were designated as random effects variables in the model. Following the ANOVA, significant treatment means were ranked according to the least significance difference (LSD) at $\alpha = 5\%$.

Results and Discussion

Wood Biomass

After a 5 year fallow period, wood yield (23.2–51.0 Mg ha^{-1}) differed significantly among tree species (Table 2). High productivity of the Australian acacias, especially *A. crassicarpa* in semiarid Tanzania has been attributed to the combined effects of a high water table at Tabora (Kwesiga et al. 2003) and tolerance to low soil fertility (efficient nutrient acquisition and utilization; Doran et al. 1997). These species form symbiotic associations with mycorrhizae and rhizobia that enhance their access to immobile soil nutrients including P and atmospheric N_2 (Kwesiga et al. 2003), contributing to efficient use of nutrients for biomass production (Kimaro et al. 2007).

Wood is mainly used as an energy source (fuelwood and charcoal) for domestic and small scale industrial operations such as tobacco curing, smoking fish, and brick burning. Often these activities can lead to substantial loss of forest cover (Luoga et al. 2002; Zahabu 2008). For example, approximately 70% of deforestation in Tanzania is related to woodfuel extraction, either through direct removal of wood (43%)

Table 2 Biomass and carbon accumulation in wood of tree species under 5 year-old rotational woodlot systems in semiarid Morogoro, Tanzania

	Biomass[a]		Carbon	
Species	(Mg ha^{-1})	(Mg ha^{-1} year^{-1})	(Mg ha^{-1})	(Mg ha^{-1} year^{-1})
Acacia auriculiformis	23.2e[b]	4.64e	11.6e	2.32e
Acacia crassicarpa	51.0a	10.2a	25.5a	5.10a
Acacia julifera	30.8cd	6.16cd	15.4cd	3.08cd
Acacia leptocarpa	38.3b	7.66b	19.2b	3.83b
Acacia mangium	37.7b	7.54b	18.9b	3.77b
Acacia nilotica	23.2e	4.64e	11.6e	2.32e
Acacia polyacantha	36.0cb	7.20cb	18.0cb	3.60cb
Gliricidia sepium	29.1ed	5.82ed	14.5ed	2.91ed
Leucaena diversifolia	33.7cbd	6.74cbd	16.8cbd	3.37cbd
LSD[c]	6.52	1.30	3.25	0.65
Pr>F	<0.0001	<0.0001	<0.0001	<0.0001

[a]Kimaro et al. (2007, 2008)

[b]Means within a column marked by the same letter are not statistically different at $p < 0.05$ according to least significant difference (n = 3)

[c]Least significance difference

Table 3 Estimated species-wise holding size required to produce wood for firewood and charcoal supply in Morogoro and for tobacco curing in Tabora, Tanzania

	Estimated farm size per household (ha)		
Tree species	Fuelwood[a]	Charcoal[a]	Tobacco[a]
Acacia auriculiformis	0.20	0.80	ND[b]
Acacia crassicarpa	0.09	0.34	0.34
Acacia julifera	0.15	0.61	0.39
Acacia leptocarpa	0.14	0.54	0.31
Acacia mangium	0.11	0.45	ND
Acacia nilotica	0.23	0.91	ND
Acacia polyacantha	0.15	0.59	ND
Gliricidia sepium	0.14	0.57	ND
Leucaena diversifolia	0.11	0.45	0.36

[a]Area estimated based on wood yield (Nyadzi et al. 2003; Kimaro et al. 2007, 2008) and volume or biomass of wood extracted annually from native forests and the population size (Luoga et al. 2000; Ramadhani et al. 2002). Where consumption rate was expressed in volume, wood basic density of the tested tree species (Table 1) was used to express biomass into volume
[b]*ND* Not determined because these species were not tested in Tabora

or through conversion (27%) of forests to agriculture where harvested wood is used as fuel (Makundi 2001). Wood extraction for agro-processing operations such as tobacco curing contributes 4–26% of deforestation in eastern and southern Africa (Sileshi et al. 2007). Our study, however, suggests that such high deforestation rates in Tanzania and other African countries could be reduced considerably through on farm wood supply using the rotational woodlot system (Table 2). Depending on tree species, wood from 1 ha of the rotational woodlot system was sufficient to satisfy fuelwood demand of a six member family for 7–16 years (Kimaro et al. 2007), assuming a household consumption rate of 10 kg week^{-1} (Biran et al. 2004). This demonstrates the high potential of planted tree fallows to reduce harvesting pressure on native forests.

Often, land availability is a limiting factor for widespread adoption of improved fallow technologies for both food and wood production. We estimated the land requirements for the rotational woodlot system to supply wood in order to meet the demand for fuelwood and charcoal at Kitulangalo area, Morogoro based on published annual consumption rates (Luoga et al. 2000, 2002). This analysis revealed that the 2,088 ha of forest lands cleared annually at Kitulaghalo could be avoided if each household allocate about 0.43–1.14 ha of farmland to rotational woodlot culture (Table 3). Similar analysis for rotational woodlot systems in Tabora district, western Tanzania indicated that 0.34–0.57 ha of land per household would be sufficient to produce wood to meet the demand for tobacco curing (Table 3), saving about 8,675 ha of forests annually (Ramadhani et al. 2002). This amount of farmland is comparable to the area (0.5–0.8 ha) under rotational woodlot culture in farmers' fields in the district (Ramadhani et al. 2002). Evidently, the rotational woodlot system holds high promise to satisfy domestic and commercial demands of wood since charcoal business in Kitulangalo (Luoga et al. 2000) and tobacco production in Tabora (Ramadhani et al. 2002) are the main economic activities of the residents there. Clearly, utilizing wood from this system can minimize degradation of native forests, especially in rural areas where wood is unsustainably extracted for energy supply (Luoga et al. 2002; Mwampamba 2007).

Wood Carbon

Species difference in wood C sequestration potential were also profound (Table 2). Carbon sequestered in wood of *A. crassicarpa* and *A. leptocarpa* (20.5–25.5 Mg C ha^{-1}), and *A. mangium* and *A. polyacantha* (18.0 and 18.9 Mg C ha^{-1}) were comparable to the wood C (19 Mg C ha^{-1}) in protected miombo forests (Williams et al. 2008; Zahabu 2008). Wood C accumulation rates also differed among tree fallows (2.3–5.1 Mg C ha^{-1} $year^{-1}$; Table 2) but were higher than the annual C increment (1.7–2.8 Mg C ha^{-1} $year^{-1}$) in the protected miombo woodlands of Morogoro (Zahabu 2008). Carbon sequestration rates reported in this study (Table 2) fall within the range of values (1.50–6.55 Mg C ha^{-1} $year^{-1}$) reported for tropical AFS (Nair et al. 2009). Based on the sequestration rates of tested species, it will take approximately 4–9 years for the tree fallows to recover wood C lost by converting native miombo forests containing 20.9 Mg C ha^{-1} (Zahabu 2008; Williams et al. 2008) to agriculture. This period is considerably shorter than the two to three decades required for re-growing miombo woodlands after cultivation (Williams et al. 2008), likely due to high productivity of these managed systems (4.6–10.2 Mg ha^{-1} $year^{-1}$; Table 2) compared to 0.04–2.91 ha^{-1} $year^{-1}$ reported for miombo forests (Kityo 2004). Overall, these results demonstrate the C sequestration potential of woodlot AFS. Although C in wood biomass can be released after fuelwood harvesting, it offsets CO_2 emissions from clearing local forests for woodfuel supply or from using fossil fuels. As noted in the previous section, on farm wood supply through rotational woodlot systems provides an alternative source of wood to forest dependent communities. In this way, planted tree fallows may help reduce forest degradation and address the *leakage* problem (the possibility of increased deforestation or CO_2 emissions) in areas outside forest based C sequestration projects.

Soil Organic Carbon

After a 5 year fallow period, organic C (21.6–25.6 Mg C ha^{-1}) in the top 0–15 cm soil depth under *A. nilotica*, *A. polyacantha*, and *A. mangium* was significantly higher than the organic C (13 Mg C ha^{-1}) in the continuously cropped soils (Table 4). These fallows also had more top soil C than that for the 0–10 cm depth in 6 year fallowed miombo soils (14.1 Mg C ha^{-1}), but were close to that of miombo woodland reserves (27.5 Mg C ha^{-1}; Walker and Desanker 2004). This rapid replenishment

Table 4 Soil organic carbon (0–15 cm soil depth) under planted tree fallows in Morogoro, Tanzania

Tree species	Soil organic C (Mg ha^{-1})
Acacia crassicarpa	15.8c[a]
Acacia mangium	25.6a
Acacia polyacantha	21.6ba
Gliricidia sepium	18.8bc
Acacia nilotica	22.7ba
Natural grass fallow	17.8bc
Continuous cropping	13.0c
LSD[b]	5.15
Probability	0.0078

[a]Means within a column followed by the same letter are not statistically different at $p<0.05$ according to Least significance difference (n=3)
[b]Least significance difference

of soil organic C reflects high organic matter addition through faster litter and root turnover processes during the fallow period, presumably due to the fast growth and intensive management of planted tree fallows. These results indicate that improved fallows can enhance soil C sequestration, especially in the degraded dry miombo soils where the buildup of soil C is very slow due combined effects of drought, fire, and/or termite activity (Williams et al. 2008).

The labile fraction of soil C under the rotational woodlot system, however, may be lost after wood harvesting for crop production, due to cultivation. When soils are disturbed during tree clearing and site preparation, faster rates of litter decomposition, and breakdown of soil organic matter may occur in the top soil. However, it has been shown that the stable fraction of soil organic matter associated with silt and clay is generally unaffected during the first 3 years, implying minimal soil disturbance by hand hoe cultivation (Solomon et al. 2000). Hence the rotational woodlot system may not adversely affect soil C pools because the recommended length of the post fallow cropping phase of this system is 3 years (Nyadzi et al. 2003; Kimaro et al. 2007). Long term studies examining changes of soil organic C over one rotation cycle would be appropriate to verify this premise.

Management and Policy Implications for Agroforestry Based C Sequestration

Both energy and forest policies in Tanzania recognize woodfuel as the main source of biomass fuel (URT 1992, 1998). However, the energy policy categorizes woodfuel as a non-renewable energy source because of unsustainable supply from local forests (URT 1992). As indicated earlier, on farm wood supply through agroforestry can sustainably produce wood for both domestic and small scale industrial operations in the country. However, the main limitations for fully utilizing this potential are limited labor and land availability. Many village afforestation programs that aimed at establishing woodlot plantations for fuelwood supply in Tanzania in the early 1970s, have failed on account of these constraints. Tree planting, weeding, and other management activities under this program were often neglected during peak cropping seasons because household labor was directed towards food crop production (Monela and Kihiyo 1999). The rotational woodlot system, however, addresses these problems as this system integrates wood and food crops in addition to providing environmental benefits such as C sequestration and making the system socially adoptable, economically feasible, and environment friendly. Cost benefit analyses revealed higher returns to labor ($ US 388/ha) and land ($US 2.67/workday) of up to six times in this system compared to the maize – natural fallow system (Franzel 2004). Similar analysis comparing the benefits of rotational woodlots and emerging small scale *Jatropha curcas* L. plantations in terms of bioenergy supply in the region also found higher returns as well as lower fuel production costs for the woodlot system (Wiskerke et al. 2010). Apparently, the high demand of wood for tobacco curing and scarcity of fuelwood due to extensive grazing and clearing of woodlands in western Tanzania could be the main reasons for this high profitability and hence adoption of the rotational woodlot system (Ramadhani et al. 2002).

The success of the rotational system can be partly linked to the tradition of establishing dry season grazing areas or fodder reserves known as *Ngitili* (Kamwenda 2002). *Ngitili* is a traditional silvopastoral system in which farmers in western Tanzania set aside an area of standing vegetation (grasses, trees, shrubs, and forbs) at the beginning of the rainy season for grazing during the dry season when pasture is depleted. Management of *Ngitili* is governed by customary rules and regulations set by a traditional assembly (*Dangashida*) and implemented by the village guards (*Sungusungu*) (Kamwenda 2002). In western Tanzania, tree fallow phase of the rotational woodlot system can also be managed as *Ngitili* to provide dry season grazing areas (Nyadzi et al. 2003), making the system part of the culture of livestock keeping communities in this region. Value additions through schemes of payment for ecosystem services such as REDD could promote adoption of rotational woodlot practices even in other semiarid areas that may not meet the socioeconomic conditions of western Tanzania. However, both REDD framework (URT 2009) and the agroforestry strategy (NASCO 2006) do not take full advantages of rotational woodlots for C sequestration in the country. This may be attributable to the limited research on the role of agroforestry in climate change mitigation and adaptation.

It is well known that tree based systems contribute to reductions in atmospheric CO_2 and offset CO_2 emissions through three main mechanisms, namely: C sequestration, C conservation, C substitution (Nair et al. 2009). This chapter illustrates that avoided deforestation through on farm wood supply can be a promising option for the conservation of C in biomass and soils in existing native forests, which are the main source of woodfuels in the developing countries. Clearing ~2,000–8,000 ha per year of native forests and woodlands in semiarid Tanzania could be avoided by adopting rotational woodlot system for wood supply alone, representing a substantial contribution to the country's efforts to reduce harvesting pressure on native forests.

Table 5 Drivers of deforestation and forest degradation in Tanzania[a]

Driver	Deforestation	Forest degradation
Shifting cultivation[b]	√	
Commercial farming e.g. biofuel, tobacco, sisal and tea[b]	√	√
Poor (lack) of land use plan	√	√
Forest fires	√	√
Over exploitation of forests[b]	√	√
Over grazing[b]	√	√
Mining e.g. minerals, salts and sand	√	√
Infrastructure development e.g. road and power lines	√	
Energy for domestic and industrial use[b]	√	√
Refugees – civil wars	√	√
Natural disasters including drought and floods	√	√
Weak law enforcement	√	√
Expansion of settlements	√	√

[a]Adapted from URT (2009)
[b]Causes of deforestation and degradation of forest that can be addressed by adopting the rotational woodlot system

Considering that woodfuel supply accounts for ~90% of domestic energy in Tanzania (Mwampamba 2007), this system holds considerable promise and provide a win-win alternative for meeting wood demand and enhancing environmental security. However, appropriate policy and programs to increase adoption of rotational woodlot systems and participation of small farmers in C trading schemes (e.g. REDD and AFOLU) aimed at mitigating the impacts of climate change by promoting sustainable forest and agricultural management practices to conserve tropical forests, are imperative.

It should be noted that forests and woodlands in most developing countries, especially in semiarid areas, are utilized heavily for woodfuel extraction and agricultural expansion (Tole 1998; Luoga et al. 2002). Hence implementing policy and programs, such as REDD, to conserve or protect native forests as a strategy to reduce atmospheric CO_2 emission can limit access to forest resources by local communities if no appropriate mechanisms are in place to provide meaningful alternatives. This in turn may adversely affect success of these initiatives either through failure of the forest dependent communities to comply with the regulations or by displacing deforestation and forest degradation activities to another area (*leakage*) in order to satisfy the woodfuel demand. One option to address these problems would be to promote AFS, such as rotational woodlots, which have the potential to supply wood for domestic use and small scale industrial operations like tobacco curing. As noted above, agroforestry is still not considered as a reliable source of woodfuel that can reduce harvesting pressure on native forests in the country; hence neglected in the current REDD programs. Yet the rotational woodlot system can address most of the drivers of deforestation and forest degradation, which are targeted by REDD initiatives (Table 5). This oversight underscores the need for further studies to examine the role of agroforestry in climate change mitigation and adaptation, and to develop appropriate mechanisms for accounting the C sequestered in AFS and other land uses. Current research to institutionalize *Ngitili* as community based forest management approaches for C sequestration under REDD programs in Tanzania is an attempt to fill this knowledge gap (TNRF 2010). This may provide a rationale for including rotational woodlot and other similar practices in the future.

Conclusions

Agroforestry systems utilizing fast growing tree species can considerably reduce forest degradation and rapidly (4–9 years) sequester atmospheric CO_2 at levels comparable to the natural miombo forest reserves in semiarid lands. Carbon accumulated in tree biomass and soils differed significantly among the tested fallows and the potentials of certain species (e.g., *A. crassicarpa, A. leptocarpa, A. mangium*) were comparable to that of the native forests. Although wood C can be released during combustion of harvested fuelwood, it represents amounts that offset increased CO_2 emissions from clearing local forests for wood supply or from using fossil fuels. Considering high dependency on woodfuel for domestic and commercial use by rural communities, this system holds promise to minimize degradation and CO_2 emissions. Additionally, the rotational woodlot system may not adversely affect soil organic C because stable fractions of soil organic matter are generally unaffected by cultivation and the short (1–3 years) post fallow cropping period recommended for this system. However, such high potential may be undermined in the absence of comprehensive mechanisms and policies to compensate for the avoided deforestation as well as a lack of mechanisms to limit free access to commercially extracted woodfuel from unprotected forests. Current efforts to formalize *Ngitili*, a traditional silvopastoral systems, as a community-based approach to sequester C under REDD programs in Tanzania will likely increase recognition of C sequestration potential of semiarid AFS in the tropics.

Acknowledgements The authors would like to thank anonymous reviewers for their constructive comments on the text. The authors also acknowledge support from research collaborators at Sokoine University of Agriculture, Faculty of Forestry (J. Msalilwa, E. Kafui, and F.T. Sogomba, N. Greyson, M. Hamisi, and N. Masanika) and at the University of Toronto (V.R. Timmer and Y. Teng) on earlier studies related to this chapter. Financial support from The African Forestry Research Network (AFORNET) is greatly appreciated.

References

Ali MS, Malimbwi RE, Iddi S (1997) Comparison of volume production, basic density and stem quality between *A. mangium* and *A. auriculiformis* grown in Zanzibar. J Trop For Sci 10:10–17

Anderson JM, Ingram JSI (1993) Tropical soil biology and fertility: a handbook of methods, 2nd edn. C.A.B. International, Wallingford, 221 p

Biran A, Abbot J, Mace R (2004) Families and firewood: a comparative analysis of the costs and benefits of children in firewood collection and use in two rural communities in Sub-Saharan Africa. Hum Ecol 32:1–25

Doran JC, Turnbull JW, Martensz PN, Thomson LAJ, Hall N (1997) Introduction to the species' digests. In: Doran JC, Turnbull JW (eds) Australian trees and shrubs: species for land rehabilitation and farm planting in the tropics. ACIAR Monograph No. 24, pp 89–384

FAO (2007) The state of the world's forests. Food and Agriculture Organization of the United Nations, Rome, 144 p. Available at http://www.fao.org/docrep/009/a0773e/a0773e00.HTM. Accessed 5 Oct 2010

Franzel S (2004) Financial analysis of agroforestry practices. In: Alvalapati JRR, Mercer DE (eds.) Valuing agroforestry systems. Kluwer Academic, The Netherland, pp 9–37

Funder M (2009) Reducing Emissions from Deforestation and Degradation (REDD): an overview of risks and opportunities for the poor. Danish Institute for International Studies (DIIS), Report No. 21, 64 p

Isaac ME, Gordon AM, Thevathasan N, Oppong SK, Quashie-Sam J (2005) Temporal changes in soil carbon and nitrogen in West African multistrata agroforestry systems: a chronosequence of pools and fluxes. Agroforest Syst 65:23–31

Jama BA, Mutegi JK, Njui AN (2008) Potential of improved fallows to increase household and regional fuelwood supply: evidence from western Kenya. Agroforest Syst 73:155–166

Kamwenda GJ (2002) Ngitili agrosilvipastoral systems in the United Republic of Tanzania. Unasylva 211:46–50

Kaonga M, Bayliss-Smith TP (2009) Carbon pools in tree biomass and the soil in improved fallows in eastern Zambia. Agroforest Syst 76:37–51

Kimaro AA, Timmer VR, Mugasha AG, Chamshama SOA, Kimaro DA (2007) Nutrient use efficiency and biomass production of tree species for rotational woodlot systems in Semiarid Morogoro, Tanzania. Agroforest Syst 71:175–184

Kimaro AA, Timmer VR, Chamshama SOA, Mugasha AG, Kimaro DA (2008) Differential response to tree fallows in rotational woodlot systems: post-fallow maize yield, nutrient uptake, and soil nutrients. Agric Ecosyst Environ 125:73–83

Kityo P (2004) Productivity and utilization of natural fuel wood resources: an evaluation of the current situation in some parts of Gaza Province, Mozambique. M.Sc. thesis, International Institute of Geo-Information Science and Earth Observation, Enschede, The Netherlands. 57 p. Available at http://www.itc.nl/library/papers_2004/msc/nrm/peter_kityo.pdf. Accessed 5 Oct 2010

Kwesiga F, Akinnifesi FK, Mafongoya PL, Mcdermott MH, Agumya A (2003) Agroforestry research and development in southern Africa during the 1990s: review and challenges ahead. Agroforest Syst 59:173–186

Lal R (2003) Offsetting global CO_2 emissions by restoration of degraded soils and intensification of world agriculture and forestry. Land Degrad Develop 14:309–322

Luhende R, Nyadzi GI, and Malimbwi RE (2006) Comparison of wood basic density and basal area of 5-year-old *Acacia crassicarpa*, *A. julifera*, *A. leptocarpa*, *Leucaena pallida* and *Senna siamea* in rotational woodlots trials in western Tabora, Tanzania. NFT News: Improvement and culture of nitrogen fixing trees. 9: 5–6. Available at http://www.iufro.org/science/divisions/division-2/20000/20800/20802/newsletters/. Accessed 10 Mar 2009

Luoga EJ, Witkowski ET, Balkwill K (2000) Economics of charcoal production in Miombo woodlands of eastern Tanzania: some hidden costs associated with commercialization of the resources. Ecol Econ 35:243–257

Luoga EJ, Witkowski ET, Balkwill K (2002) Harvested and standing wood stocks in protected and communal Miombo woodlands of eastern Tanzania. For Ecol Manag 164:15–30

Makundi WR (2001) Potential and cost of carbon sequestration in the Tanzanian forest sector. Mitig Adapt Strat Glob Chn 6:335–353

Malimbwi RE, Solberg B, Luoga E (1994) Estimation of biomass and volume in miombo woodland at Kitulangalo Forest Reserve, Tanzania. J Trop For Sci 7(2):230–242

Monela GC, Kihiyo VBMS (1999) Wood energy in Sub-Saharan Africa. In: Palo M, Uusivuori J (eds.) World forest society and environment. Kluwer Academic, The Netherlands, pp 153–160

Mountinho P, Santillli M, Schwartzman S, Rodriguues L (2005) Why ignore tropical deforestation? A proposal for including forests conservation in the Kyoto Protocol. Unasylva 56:27–30

Mwampamba TH (2007) Has the woodfuel crisis returned? Urban charcoal consumption in Tanzania and its implications to present and future forest availability. Energy Policy 35:4221–4234

Nair PKR, Kumar BM, Nair VD (2009) Agroforestry as a strategy for carbon sequestration. J Plant Nutr Soil Sci 172:10–23

NASCO (National Agroforestry Steering Committee) (2006) Popular version of National Agroforestry Strategy. Available at http://www.tafori.org/Read.Aspx?ID=100003. Accessed 2 July 2010

Negra C and Ashton R (2009) Roadmap for terrestrial carbon science: research needs for carbon management in agriculture, forestry and other land uses. The Terrestrial Carbon Group Project. Available at http://www.terrestrialcarbon.org/site/DefaultSite/filesystem/documents/Policy%20Brief%207_%20Roadmap%20for%20Terrestrial%20Carbon%20Science_ACh%20100304.pdf. Accessed 30 June 2010

Ngulube MR (1994) Evaluation of *Gliricidia sepium* provenances for alley cropping in Malawi. For Ecol Manag 64:191–198

Nyadzi GI, Otsyina RM, Banzi FM, Bakengesa SS, Gama BM, Mbwambo L, Asenga D (2003) Rotational woodlot technology in northwestern Tanzania: tree species and crop performance. Agroforest Syst 59:253–263

Orwa C, Mutua A, Kindt R, Jamnadass R, Anthony S (2009) Agroforestree Database: a tree reference and selection guide version 4.0. Available at http://www.worldagroforestry.org/sites/treedbs/treedatabases.asp. Accessed 10 Nov 2009

Palm C, Tomich T, Van Noordwijk M, Vosti S, Gockowski J, Alegre J, Verchot L (2004) Mitigating GHG emissions in the humid tropics: case studies from the Alternatives to Slash-and-Burn Program (ASB). Environ Dev Sust 6:145–162

Ramadhani T, Otsyina R, Franzel S (2002) Improving household income and reducing deforestation using rotational woodlots in Tabora district, Tanzania. Agr Ecosys Environ 89(3):229–239

Samek JH, Skole DL, Klinhom U, Butthep C, Navanugraha C, Uttaruk P, Laosuwan T (2011) Inpang carbon bank in northeast Thailand: a community effort in carbon trading from agroforestry projects. In: Kumar BM, Nair PKR (eds) Carbon sequestration potential of agroforestry systems: opportunities and challenges. Springer, Dordrecht, pp 263–280

SAS Institute Inc (2000) SAS version 8. SAS Institute Inc, Cary

Sileshi G, Akinnifesi FK, Ajayi OC, Chakeredza S, Kaonga M, Matakala PW (2007) Contribution of agroforestry to ecosystem services in the Miombo eco-region of eastern and southern Africa. Afr J Environm Sci Technol 4:68–80

Skutsch M, Zahabu E, Lovett J, McCall M, Singh SP, Trines E, Verplanke J, Karky BS, van Laake P, Banskota K, Basnet R, Ba L (2008) Dealing with emissions from dry forest degradation: a low cost community-based approach. Policy Paper no. 2: Kyoto: Think Global, Act Local project. Available at http://www.communitycarbonforestry.org/NewPublications/Policy%20note%20operationalising%20reduced%20degradation%20in%20REDD%20version%204.pdf. Accessed 30 Oct 2009

Solomon D, Lehmann J, Zech W (2000) Land use effects on soil organic matter properties of chromic Luvisols in semi-arid northern Tanzania: carbon, nitrogen, lignin and carbohydrates. Agr Ecosys Environ 78:203–213

Swallow B, van Noordwijk M, Dewi S, Murdiyarso D, White D, Gockowski J, Hyman G, Budidarsono S, Robiglio V, Meadu V, Ekadinata A, Agus F, Hairiah K, Mbile PN, Sonwa DJ, Weise S (2007) Opportunities for avoided deforestation with sustainable benefits. An interim report. ASB partnership for the Tropical Forest Margins, Nairobi, Kenya, 42 p. Available at http://www.asb.cgiar.org/PDFwebdocs/Report-on-Opportunities-for-Avoided-Deforestation-Sustainable-Benefits-web-low.pdf. Accessed 5 Oct 2010

Takimoto A, Nair PKR, Nair VD (2008) Carbon stock and sequestration potential of traditional and improved agroforestry systems in the West African Sahel. Agr Ecosys Environ 125:159–166

TNRF (Tanzania Natural Resource Forum) (2010) TZ REDD Newsletter. Issue 1 May 2010. Available at http://www.tnrf.org/files/E-INFO_REDD_Newsletter_Tanzania-1_May_2010.PDF. Accessed 2 July 2010

Tole S (1998) Sources of deforestation in tropical developing countries. Environm Manag 22:19–33

URT (United Republic of Tanzania) (1992) Tanzania energy policy. Available at http://www.tzonline.org/pdf/theenergypolicyoftanzania.pdf. Accessed 2 July 2010

URT (United Republic of Tanzania) (1998) Tanzania forest policy. Available at http://www.tzonline.org/pdf/nationalforestpolicy.pdf. Accessed 2 July 2010

URT (United Republic of Tanzania) (2009) National Framework for Reduced Emissions from Deforestation and Forest Degradation (REDD). Available at http://www.reddtz.org/images/pdf/redd%20framework%2009_new.pdf. Accessed 2 Oct 2009

van Noordwijk M, Suyamto DA, Lusiana B, Ekadinata A, Hairiah K (2008) Facilitating agroforestation of landscapes for sustainable benefits: tradeoffs between stocks and local development benefits in Indonesia according to the FALLOW model. Agr Ecosyst Environ 126:98–112

Verchot LV, van Noordwijik M, Kandji S, Tomich T, Ong C, Albrecht A, Mackensen J, Bantilan C, Anupama KV, Palm C (2007) Climate change: linking adaptation and mitigation through agroforestry. Mitig Adapt Strat Glob Chn 12:901–918

Walker SM, Desanker PV (2004) The impact of land use on soil carbon in Miombo woodlands of Malawi. For Ecol Manag 203:345–360

Wickens GE, Seif El Din, AG, Sita, G, Nahal, I (1995) Role of Acacia species in the rural economy of dry Africa and the Near East. FAO conservation guide 27. Available at http://www.fao.org/docrep/V5360E/v5360e0f.htm. Accessed 5 Oct 2010

Williams M, Ryan CM, Rees RM, Sambane E, Fernando J, Grace J (2008) Carbon sequestration and biodiversity of re-growing Miombo woodlands in Mozambique. For Ecol Manag 254:145–155

Wiskerke W, Dornburg T, Faaij APC, Malimbwi RE, Rubanza CDK (2010) Cost-benefit analysis of biomass energy supply option for rural smallholders in semiarid East Africa Shinyanga region in Tanzania. J Renew Sust Energy Rev 14:148–165

Zahabu EM (2008) Sinks and sources: a strategy to involve forest communities in Tanzania in global climate policy. Ph.D. thesis, University of Twente, The Netherlands, 235 p. Available at http://doc.utwente.nl/68918/. Accessed 5 Oct 2010

Silvopasture and Carbon Sequestration with Special Reference to the Brazilian Savanna (Cerrado)

P.K. Ramachandran Nair, Rafael G. Tonucci, Rasmo Garcia, and Vimala D. Nair

Abstract The Brazilian savanna, known as the Cerrado, extending over 200 million ha, is the largest neotropical savanna in the Americas. With its ongoing conversion to intensive agriculture since the 1960s, of which cultivated pastures for beef cattle production is a major form, this unique ecosystem is now considered threatened. Given the recognized role of trees in carbon (C) sequestration and greenhouse gas (GHG) mitigation, the silvopastoral system of tree plantation development on pasture lands is considered to be particularly relevant to this region. For the past two decades, eucalyptus-based silvopastoral systems have been established in the Cerrado region by growing agricultural crops (rice and soybean) in the first 2 years followed by *Brachiaria* forage and beef-cattle grazing from the third year of plantation establishment. Recent studies in a variety of situations indicate that agroforestry systems store higher amounts of C compared to single species cropping and grazing systems, both aboveground and belowground. The Brazilian savannas that have characteristically low aboveground C reserves hold considerable stocks of soil organic C, probably as a consequence of previous land use, the history of which is unknown. Most of this C is in a biodegradable form and is likely to be lost to the

P.K.R. Nair (✉)
Center for Subtropical Agroforestry, School of Forest Resources and Conservation,
University of Florida, P.O. Box 110410, Gainesville, FL 32611–0410, USA
e-mail: pknair@ufl.edu

R.G. Tonucci • R. Garcia
Animal Science Department, Federal University of Viçosa, P.H. Rolfs Avenue,
MG 36570-000, Viçosa, Brazil

Goats and Sheep, EMBRAPA, Sobral, Ceara State, Brazil
e-mail: rgtonucci@gmail.com; rgarcia@ufv.br

V.D. Nair
Soil and Water Science Department, University of Florida,
P.O. Box 110510, Gainesville, FL 32611, USA
e-mail: vdn@ufl.edu

B.M. Kumar and P.K.R. Nair (eds.), *Carbon Sequestration Potential of Agroforestry Systems: Opportunities and Challenges*, Advances in Agroforestry 8,
DOI 10.1007/978-94-007-1630-8_8, © Springer Science+Business Media B.V. 2011

atmosphere when the soil is disturbed during land conversion to agriculture and pasture. Adoption of sustainable land use systems such as silvopasture could reduce this potential hazard. Given the role of the Cerrado in the global C cycle and climatic change, these issues deserve well coordinated investigations.

Keywords Ecosystem degradation • *Eucalyptus* • Grasslands • GHG mitigation • Oxisols • Ruminants

Introduction

Silvopasture refers to the agroforestry practice of integrating trees in animal production systems. Broadly, there are two major forms of silvopasture: grazing and tree fodder systems. In grazing systems, cattle are allowed to graze on pasture under widely spaced or scattered trees, whereas in the tree fodder systems, the animals are stall-fed with fodder from trees or shrubs grown in blocks on farms (Nair 1993; Nair et al. 2008). The grazing system of silvopasture has recently gained prominence as an ecologically sustainable and environmentally desirable approach to managing degraded pasture lands in the industrialized countries (Mosquera-Losada et al. 2005; Garrett 2009). With the recent emphasis on the environmental impact of land use systems, the role of silvopasture and other agroforestry practices in mitigating climate change through carbon (C) sequestration has been a major area of research focus (Nair et al. 2010).

The silvopastoral system of tree plantation development on pasture lands and its role in C sequestration and greenhouse gas (GHG) mitigation are particularly relevant to Brazil, where land use changes in forestry and agricultural sectors are reported to be responsible for more than two-thirds of the GHG emissions (Comunicação Nacional 2004). For example, most of the agricultural areas in the vast Brazilian savanna, known as the Cerrado, are grasslands, of which at least 60% are in some stage of degradation (IBGE 2006). This paper presents an overview of the status of silvopasture in Brazil, and a summary of some recent studies on C sequestration in silvopastoral systems in Florida (USA), Minas Gerais (Brazil), and northwestern and Central Spain. With that background, we will present some perspectives on the GHG – primarily C dioxide (CO_2) – mitigation potential of silvopasture systems in the Brazilian Cerrado, and relate them to other similar ecosystems elsewhere.

Silvopasture in the Brazilian Savanna (Cerrado)

Brazil has a cattle population of 200 million on 100 million ha of cultivated pastures (IBGE 2006). Of these, beef production involves 180 million head, producing 8.0 Tg of meat per year. Growing national and international markets for meat and demand for better quality of meat necessitate important changes to a production system that relies largely on pasture, the production capacity of which has been depleted following years of exploitation. In the Cerrado region of Brazil, cultivated pastures cover about 49 million ha, supporting a herd of 40 million head representing more than 35% of the total Brazilian beef production. Thus, the region accounts for 35–40% of the beef cattle industry both in area and production.

The Cerrado

Savannas are a major component of the world's vegetation, covering one-sixth of the land surface, and accounting for 30% of the primary production of all terrestrial vegetation (Grace et al. 2006). In South America the savanna, mostly distributed in Brazil, Colombia, Venezuela, and Bolivia, feeds three of the major water basins: the Amazon, Paraguay, and São Francisco Rivers (Cochrane et al. 1985). The Brazilian savanna, called the Cerrado (Fig. 1), occurs manly in central Brazil in the states of Mato Grosso, Mato Grosso do Sul, Tocantins, Goias, and western parts of Minas Gerais, and extends over 200 million ha (Batlle-Bayer et al. 2010). The Cerrado is a wet savanna and it consists of a gradient of physiognomies, from grassland (called "campo limpo") to a sclerophylous forest (Cerradão), with over 10,000 species of plants, of which 45% are unique to the Cerrado.

The Cerrado region's typical climate is hot, semi-humid, with pronounced seasonality marked by a dry winter season from May through October. The annual rainfall ranges from 1,200 to 2,000 mm, 80–90% of which occurs during the summer (known, rightly, as the rainy) season between October and April. The mean annual temperature varies from 22°C in the south to 27°C in the north of the region (Bustamante et al. 2006). The soils are generally very old, deep, and inherently poor in nutrients such as phosphorus and calcium. They have high levels of aluminum and low levels of organic matter and pH. Oxisols and Entisols represent approximately 46% and 15% of the area, respectively (Reatto et al. 1998). Due to their low nutrient status and high acidity and aluminum concentration, soil organic matter (SOM) plays a particularly important role in the physical, chemical, and biological processes related to nutrient cycling, soil aggregation, and plant-water availability in the Cerrado (Resck 1998). The Cerrado trees have characteristic twisted trunks covered by a thick bark, and leaves, which are usually broad and rigid. Many herbaceous plants have extensive roots to store water and nutrients. The plant's thick bark and roots serve as adaptations for the periodic fires that sweep the Cerrado landscape. The adaptations protect the plants from destruction and make them capable of sprouting again after the fire.

The Cerrado region has been the focus of intense agricultural expansion since the 1960s, and a large area of native vegetation has been replaced by agriculture, cultivated pastures, and planted forests.[1] Satellite images showed that in 2002, 55% of

[1] EMBRAPA CERRADO (1999).

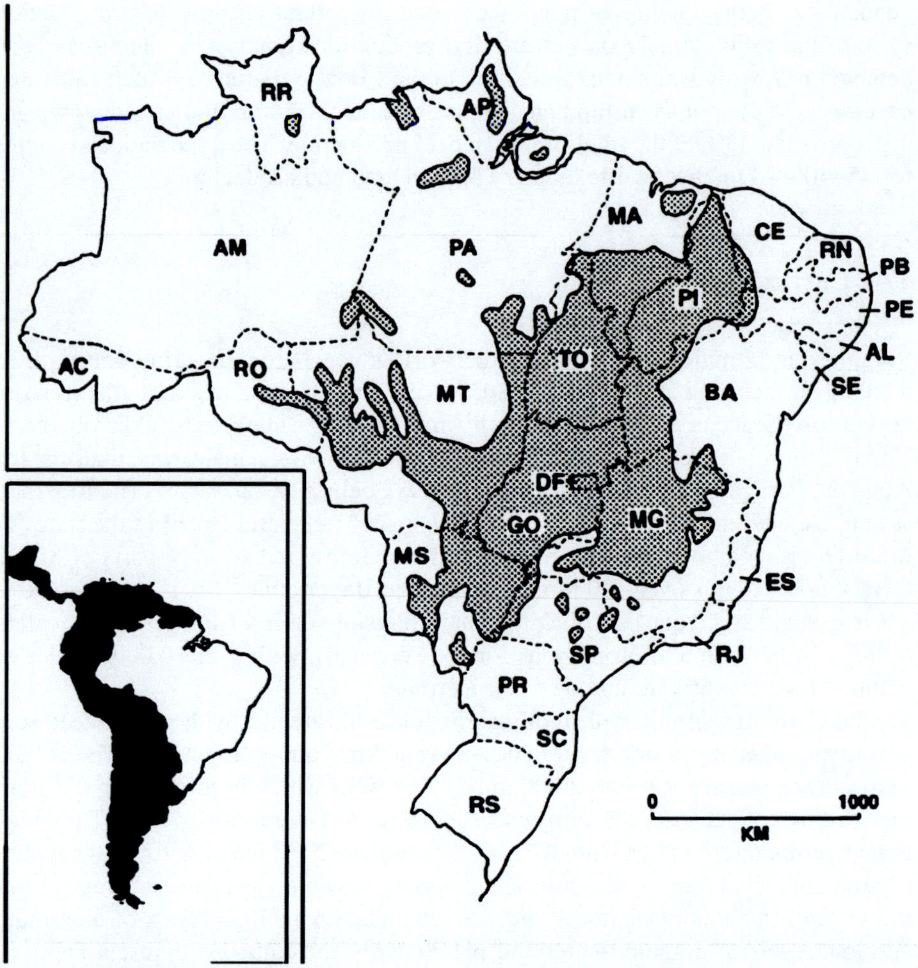

Fig. 1 Distribution of Cerrado vegetation in Brazil (letters are state abbreviations). Those referred to in connection with geographical patterns are: *DF* Federal District; *GO* Goiás; *MA* Maranhão; *MG* Minas Gerais; *MS* Mato Grosso do Sul; *MT* Mato Grosso; *PA* Pará; *TO* Tocantins

the Cerrado had already been transformed (Machado et al. 2004). During the period from 1975 to 1995, the area under crop cultivation in the Cerrado increased from 6.9 to 8.2 million ha (Bustamante et al. 2006). The major crops are soybean (*Glycine max* L. Merr), maize (*Zea mays* L.), rice (*Oryza sativa* L.), and beans (*Phaseolus vulgaris* L.). Soybean is the most important crop and it had its "boom" in the 1980s propelled by growing international demand for it. Cultivated pastures account for the largest agricultural expansion, mostly with the introduction of the African grass of the genus *Brachiaria*. Estimate of the pasture area in the Cerrado ranges from 35 to 50 million ha (Sano et al. 2000). Most of these cultivated pastures have, however, experienced some degree of degradation; they have lost, to some extent, their capacity to produce biomass due to deterioration of soil chemical, physical, and biological conditions.

Various types of landholdings and producers can be found in the Cerrado biome, ranging from large farms with areas of more than 20,000 ha and a variety of crop fields or cattle, to a large number of "small" farms with areas less than 100 ha. Planted forests are a relatively new land use system in this area that has gained popularity within the last decade; pasture lands are now being rapidly converted by interplanting with fast growing tree species. Large tracts of the Cerrado have also been planted to fast growing trees, especially eucalyptus hybrids (*Eucalyptus* spp.) and pines (*Pinus* spp.), which account for roughly two-thirds and one-third, respectively, of the approximately 5.5 million ha of planted forests in Brazil (ABRAF 2008). Most of these plantations have been planted over the small farmlands that used to raise cattle. This new development, motivated primarily by its monetary advantages, has brought up two major issues: the introduction of non-native tree species in the biome, and the decline – if not elimination – of the traditional activity of cattle rearing. Integrating cattle and trees as in silvopastoral systems offer the advantages of monetary benefits from planted forests and at the same time supports cattle rearing. There might be unexplored advantages via C sequestration too.

Silvopastoral Systems in the Cerrado

Silvopastoral systems were first established in the Cerrado region of Minas Gerais State about 20 years ago and the area under the practice has been increasing steadily since then. Accurate data on the spread of the system are not available; however, the current area under the practice is estimated to be about 14,000 ha (based on the authors' personal contact with local farms). Other areas of the Cerrado have also been cultivated with silvopastoral systems, mainly in the state of Mato Grosso do Sul. It is perceived (Dubé et al. 2000) that the establishment of silvopastoral systems can reduce the cost of establishment of the whole (beef + timber) system; furthermore, the additional income derived from the crops would be an economic incentive to tree plantation owners during the early years of plantation establishment.

The silvopastoral systems in the Cerrado are established by cultivating one or two annual crops in rows in between the widely-spaced tree rows. Crops such as rice and soybean are cultivated in the first and second year, respectively, after establishing eucalyptus (Fig. 2), the most common tree used in the system. It is planted at the spacing of 10 × 4 m or 8 × 4 m. Tree rows are aligned, as much as possible, in the east-west orientation in order to allow highest extent of light penetration to the understory grass between trees. Most planters limit the soil preparation for silvopastoral establishment to the minimum, mainly spot application of herbicides to kill weeds or any undesirable plant in the rows where the trees will be planted. This minimum soil preparation is important to avoid soil disturbance and oxidation of SOM. Soil moisture availability and mild temperature under trees create better

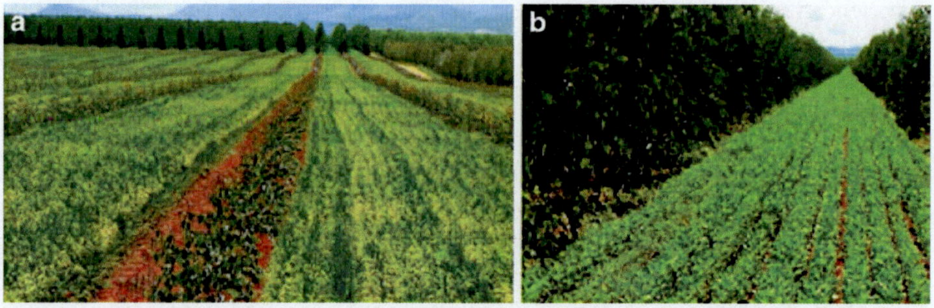

Fig. 2 During the early years of silvopastoral system establishment, agricultural crops are grown in between tree rows: (**a**) first year – rice planted after establishing eucalyptus; (**b**) second year – soybean planted after the rice was harvested. The pasture will be established in the third year after soybean harvest. Fazenda Riacho, Paracatu, Minas Gerais, Brazil

Fig. 3 A silvopastoral system of *Eucalyptus* spp. with *Brachiaria brizantha* (Hochst. Ex A. Rich.) Stapf, as the understory in Fazenda Riacho, Paracatu, Minas Gerais, Brazil. Note that the plants maintain their green color even in the peak dry season when the photo was taken

condition for mineralization of nitrogen (N) which contributes to improving and extending the forage quality in the dry season. In the third year, seeds of the grass *Brachiaria brizantha* (Hochst. Ex A. Rich.) Stapf, are sown to constitute the understory (Fig. 3). Sixty days after sowing the grass seeds, beef cattle are stocked in the area for grazing (Fig. 4).

In spite of the steady increase in the area under silvopasture in Brazil, information about beef cattle production under these systems is relatively scanty. A pioneering

Fig. 4 Beef cattle in the silvopastoral system of eucalyptus and *Brachiaria brizantha* (Hochst. Ex A. Rich.) Stapf, in Fazenda Riacho, Paracatu, Minas Gerais, Brazil

example of the system, which contributed to studies and popularization of this technology, is the "Fazenda Riacho" (Riacho Farm), an agroforestry unit of the Votorantim Siderurgia group, located in the Cerrado region of Minas Gerais. Bernardino et al. (2011) studying the beef cattle performance in the silvopastoral system in this farm reported that fertilizing the understory grass (*B. brizantha*) with N and potassium (K) resulted in increase in higher grass dry matter production and higher meat production per ha; on average, the increase in the animal live weight gain (LWG) was directly proportional to the fertilizer rates used. Fertilization with N and K is considered important in the establishment and management of silvopastoral system in the Brazilian Cerrado (Andrade et al. 2001; Bernardino 2008). Other studies are currently under way to evaluate the effects of doses and sources of N on the productivity of the understory in silvopastoral systems.

An alternative to the use of inorganic fertilizers is the introduction of forage legumes to constitute part of the understory in silvopastoral systems. Legumes can add N to the system and thus reduce the cost of N input as well as the environmental hazard associated with fertilizer N. They can also enhance the forage quality, resulting in better cattle performance. Paciullo et al. (2004, personal communication)[2] evaluated the weight gain of dairy heifer grazing a silvopastoral system with three

[2] Paciullo DSP, Aroeira LJM, Viana AF, Malaquias JD, Rodrigues NM, Carvalho CAB, Costa FJN, and Verneque RS (2004) Desempenho de novilhas mestiças Europeu x Zebu, mantidas em sistemas silvipastoril ou em monocultura de Braquiária. In: *Reunião Anual da Sociedade Brasileira de Zootecnia*, Campo Grande, SBZ, CD-ROM.

different tree species and an understory of *Brachiaria decumbens* Stapf. and *Stylosanthes guianensis* (Aubl.) Sw. and a pasture with the same grass species. They reported that, while no differences were found between silvopastoral system and grassland in the rainy season, a 40% gain in the heifer weight was noticed under silvopasture in the dry season. Alvim et al. (2005) also found a better weight gain in the dry season for heifers grazing in the understory of a silvopastoral system when compared to *B. decumbens* pasture.

Several studies have indicated the potential of silvopastoral system in beef-cattle production in the state of Rio Grande do Sul. Silva (1998) evaluated the effect of two densities of *Eucalyptus saligna* Sm. plantations, spaced 2×3 m and 2×6 m (1,666 and 833 trees ha^{-1}) and three forage offers (6.0, 9.6, and 13.0%) on the beef cattle performance, and found that the highest LWG per ha 215 kg, in the medium (9.6%) forage offer and lowest tree density (833 trees ha^{-1}). Furthermore, Silva et al. (2001) studied the animal performance, stocking rate, and the residual forage in a silvopastoral system with acacia negra (*Acacia mearnsii* De Wild.) spaced 2×3 m and 2×5 m and two understory species (*B. brizantha* cv. Marandu and *Panicum maximum* Jacq. cv. Gatton). The best results for LWG, animal gain per ha, and stocking rate were obtained with lower tree density, for both understory species. Lucas (2004) studying a silvopastoral system with acacia negra at a stand density of 500 trees ha^{-1} and understory of *P. maximum* cv. Gatton, established for 8 years and grazed during 445 days, found a total LWG of 747 kg ha^{-1} (average of 1.8 kg^{-1} ha^{-1} grazing day^{-1}). These results show a high contrast with the average LWG productivity of 50 kg^{-1} ha^{-1} year^{-1} of the traditional extensive grazing based on native pasture in Rio Grande do Sul State.

Research Results on Carbon Sequestration in Silvopastoral Systems

During the past few years, the University of Florida (UF) Center for Subtropical Agroforestry (CSTAF) has been involved in soil C sequestration studies under a range of agroforestry systems and related land use systems (Nair et al. 2010). The overall objectives were to quantify soil organic matter (SOC) accumulation and sequestration in different types of agroforestry systems in a variety of ecological and geographical conditions, determine C storage in different soil fractions up to at least 1-m depth, and quantify, wherever possible, C contribution by C_3 and C_4 plants (~ trees and herbaceous plants) using natural C isotopic differences between the two groups. The studies included silvopastoral systems in three countries, under different agroecological conditions (Table 1). Detailed descriptions of climate and soil conditions, land use systems, and their management are reported in specific papers published from each study. Briefly, the Florida sites included a silvopasture with slash pine (*Pinus elliottii* Engelm.) and adjacent treeless bahiagrass pasture (*Paspalum notatum* Flüggé). In Spain, two silvopastoral systems were studied: a simulated silvopasture with pine (*Pinus radiata* D. Don) or birch

Table 1 Site- and system details of the University of Florida, Center for Subtropical Agroforestry, research sites for carbon sequestration studies under silvopastoral systems

Location; coordinates	Climate (m.a.p, mm; mean temp. range, °C)	Silvopasture system	Land uses	Age (# years since establishment)
Florida, USA; 28° to 29° N; 81° to 83° W	Humid subtropical; 1330; −3 to 28	Slash pine (*Pinus elliottii* Engelm.) + bahia-grass (*Paspalum notatum* Flüggé)	Pasture Silvopasture Pasture Silvopasture	50 12 55 14
Central Spain; 39° 59′ N; 6° 6′ W	Subhumid mediterranean; 600; 8–26	Dehesa: Cork oak (*Quercus suber* L.) silvopasture	Cork oak	80
Northwestern Spain; 43° 9′ N; 7° 30′ W	Humid Atlantic; 1200; 5.8–18	Eur. birch (*Betula pendula* Roth.) with orchard grass, *Dactylis glomerata* L.	Pasture Silvopasture	 15
Minas Gerais, Brazil; 17° 36′ S; 46° 42′ W	Cerrado: Subhumid tropical; 1350; 20–30	*Eucalyptus* spp. with understory of *Brachiaria* spp. (fodder grass)	Forest Silvopasture Pasture	 14

Source: Nair et al. (2010)
m.a.p mean annual precipitation

(*Betula pendula* Roth) with *Dactylis* spp./*Trifolium* spp. in between, with an adjacent treeless pasture on Inceptisols in northwestern Spain; and a traditional dehesa silvopastoral system with cork oak trees (*Quercus suber* L.) on Alfisols in central Spain. In the dehesa system, total C stock was determined near (2 m) and away (15 m) from the tree. The study sites in Minas Gerais, Brazil, included a eucalyptus silvopasture system (*Eucalyptus* spp. with understory of *Brachiaria* spp. as fodder grass) compared with a pasture system and an adjacent forest stand. The soil orders of the study sites included Spodosols and Ultisols (both in Florida, USA), Inceptisols (northwestern Spain), Alfisols (central Spain), and Oxisols (Minas Gerais, Brazil). At each location, soils were sampled to at least 1-m depth from four to six layers (sampling depths) according to replicated experimental design procedures. All soil samples from the different sites were fractionated into three different aggregate-size fractions [macro (2,000–250 μm), micro (250–53 μm), and silt- and clay- sized fractions (<53 μm)], and the C content in each fraction was determined by dry combustion using an automated C analyzer (Thermo Finnegan Flash EA 1112 NC; Thermo Fisher Scientific Inc. Waltham, MA, USA).

The total SOC varied considerably within the different systems to a meter depth suggesting differences in C sequestration potential that reflects climatic conditions, soil types, and the plant species (Fig. 5). The highest SOC stock to a meter depth was in the Oxisols of Brazil (~ 400 Mg ha^{-1}: Tonucci et al. 2011) and the lowest SOC was in the sparsely tree-dominated locations of the dehesa system in central Spain (Mean: 31 Mg ha^{-1}; Howlett 2009).

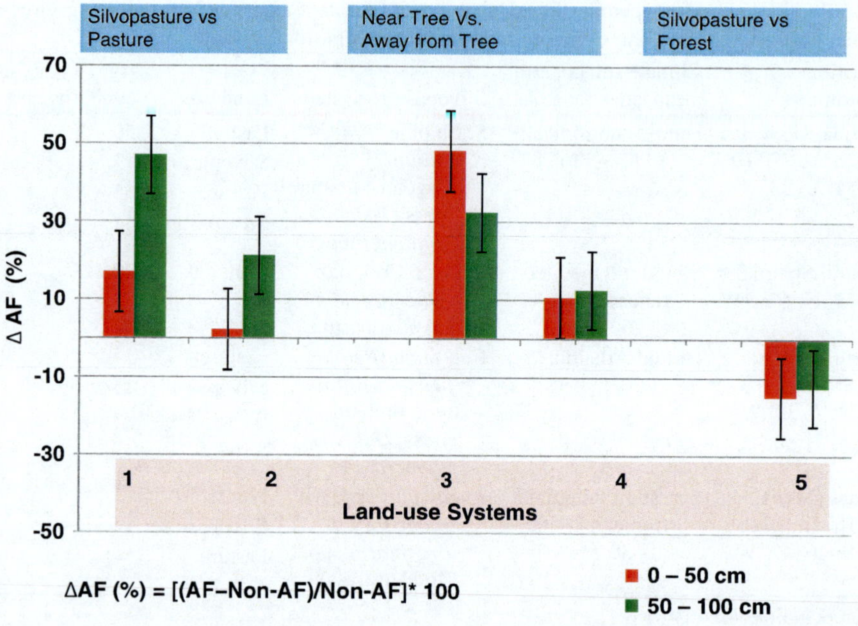

	Systems; age (# years since AF system installation)	Location	Soil Order
1	Pine + pasture vs. treeless pasture; 30 y	Florida, USA	Ultisols
2	Pasture under birch trees vs. treeless pasture; 15 y	Northern Spain	Inceptisols
3	Under tree vs. away from trees (Dehesa); 80 y	Central Spain	Alfisols
4	Under trees vs. away from trees; Parkland system; >50 y	Segou, Mali	Alfisols
5	Brachiaria + Eucalyptus vs. Treeless forage stand; 30 y	MG, Brazil	Oxisols

Fig. 5 Changes in soil C stock under different agroforestry (Silvopasture) vs. non-AF systems. ΔAF (%) = [(AF – Non-AF) / Non-AF]* 100 (Adapted from Nair et al. 2010)

In the Florida study, silvopastures had greater amounts of SOC stored within a meter soil profile compared to adjacent treeless pastures (Fig. 6). Using stable C isotope signatures, Haile et al. (2010) showed that C in the deeper soil profile was derived from the tree component, i.e. the slash pine of the silvopastoral system. Further, the relatively stable C fraction (<53 μm) was found to be derived from the tree component (Haile et al. 2010). In the study in central and northwestern Spain, the traditional dehesa (cork oak) silvopastoral system with sparse tree density had lower SOC in the whole soil compared to the managed silvopasture system with

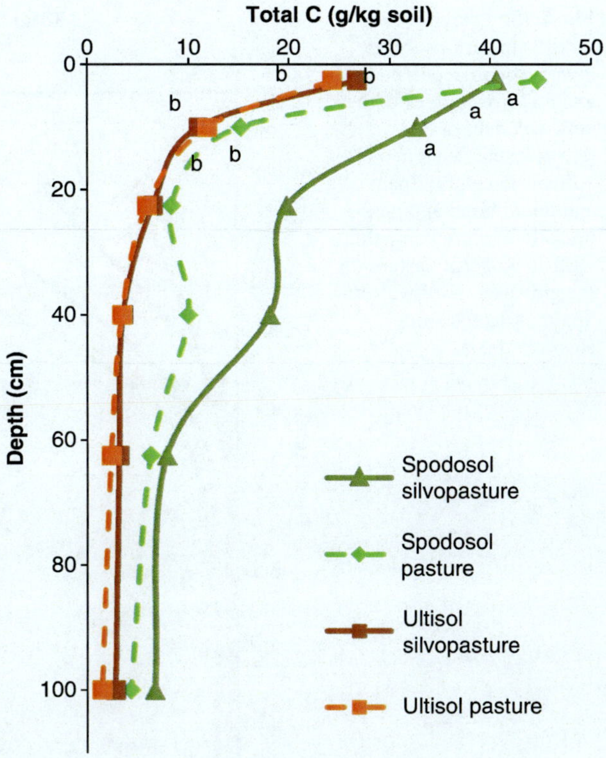

Fig. 6 Soil carbon stock (g kg^{-1}) in various soil layers to 1-m depth in adjacent pasture and silvopasture systems in two soil orders at two Florida locations (Adapted from Haile et al. 2010)

higher tree density (Fig. 7). Within the dehesa system of central Spain, the soil near the tree, compared to that away from the tree, stored more C. In pasture system in Minas Gerais, the AFS (eucalyptus-based silvopasture) had the highest SOC content in the macro-sized- and silt + clay- sized fractions compared with the forest- and pasture soils (Fig. 8).

In addition to the above (silvopastoral) studies, similar studies were conducted in three other countries: in the homegardens in Kerala, India (Saha et al. 2009, 2010); parkland- and other systems in Mali, West Africa (Takimoto et al. 2008, 2009); and shaded cacao systems in comparison with adjacent natural forest in southeast Bahia, Brazil (Gama-Rodrigues et al. 2010). The results from these multi-location (five-country) studies showed that:

1. The amount of C stored in soils depends on soil qualities, especially silt + clay content
2. Tree-based systems, compared to treeless systems, store more C in deeper soil layers under comparable conditions
3. Higher SOC content is associated with higher species richness and tree density
4. Soil near the tree, compared to away from the tree, stores more C

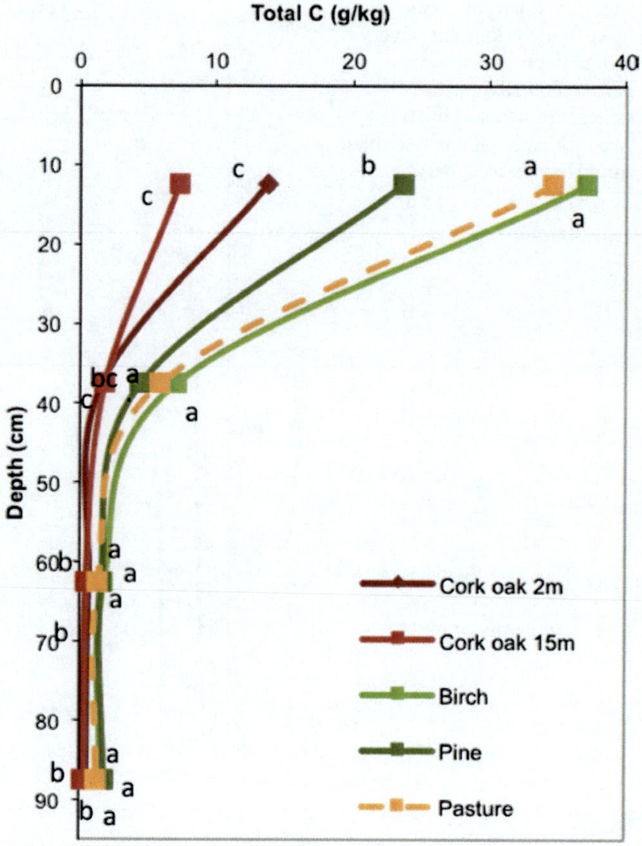

Fig. 7 Soil carbon stock (g kg^{-1}) in various soil layers to 1-m depth, near and away from trees in the cork-oak dehesa (silvopasture) system on Alfisols in central Spain, and under birch and pine tree silvopasture compared with an adjacent pasture on Inceptisols in northwestern Spain (Adapted from Howlett 2009)

Furthermore, C_3 plants (trees) were found to contribute to more stable C in the soil than C_4 plants (grasses) in deeper soil profiles in Florida soils (Spodosols and Ultisols); but this was not the case in the Cerrado soils (Oxisols).

In spite of the limitations of the separate studies upon which this analysis is based, the results show the intrinsic differences and enormous variations in soil C stock among different soils orders and land use systems. Although the studies at the different locations were not designed specifically to compare C stock across soil orders under similar ecological and management conditions, a general trend of higher C stock in soils containing higher amounts of clay and silt was evident. Among the agroforestry and other systems studied, the differences in C stock are up to more than 100-fold. While much of these differences can be attributed to the type of land use systems and ecological regions, there are clear differences among land use systems within the same ecological regions and soil orders. In general, tree-based agricultural systems, compared to treeless systems, store more C. Furthermore, land use history of the site seems to have a major and overriding role in determining the amount of C stored in the soils, such that the previous land use history of a site has the most effect than any other factor in determining the C content in that soil.

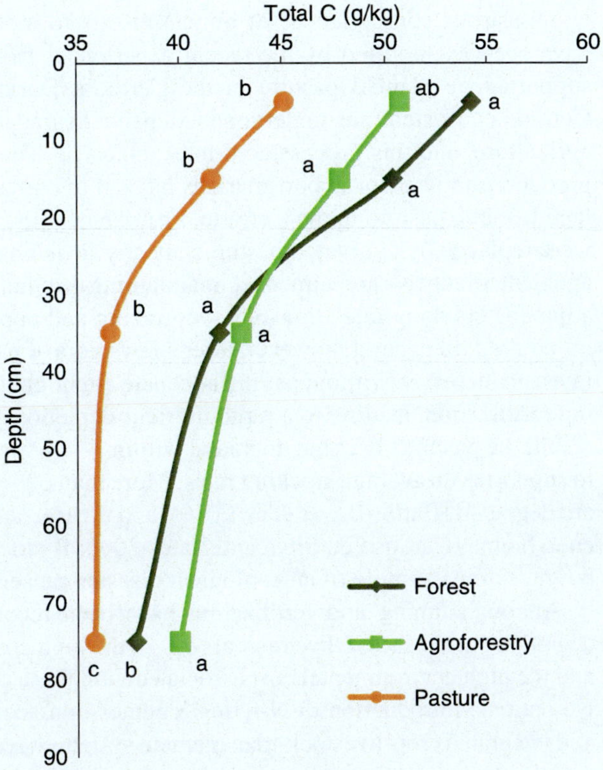

Fig. 8 Soil carbon stock (g kg^{-1}) in various soil layers to 1-m depth in adjacent forest, pasture and silvopasture systems on Oxisols at Minas Gerais, Brazil (Adapted from Tonucci et al. 2011)

In summary, available results indicate that AFS store higher amounts of C, compared to single species cropping and grazing systems, in both aboveground and belowground compartments of the system. The C sequestration potential of AFS seems especially significant in the soil, particularly in soil depths below 50 cm (Nair et al. 2010). The extent of C sequestration will, obviously, depend on a number of site-specific factors as well as system management.

Some Perspectives on Silvopasture and Carbon Sequestration in the Cerrado

Grasslands that cover nearly three billion hectare globally, with roughly two-thirds in the tropics and one-third in the temperate region, constitute a major ecosystem of the world. Being semiarid lands, they are resource limited, especially in N and water. Silvopastoralism is a major land use system in the savannas, with extensive grazing under dispersed stands of indigenous trees in the vast African savanna.

Disastrous consequences of human efforts to modify these fragile ecosystems have been exemplified by the recent experiences from intensive cattle production supported by planted pasture in the Cerrado. During the past few decades, the Cerrado ecosystem has undergone extensive degradation because of conversion to agriculture, and this process continues unabated. The increasing demand for dairy products and beef for export markets has led to conversion of large areas of native land to cattle pasture in the Cerrado. From the 1970s, the native savanna grasslands were replaced by cultivated pastures, mostly *Brachiaria* species. Best pasture management practices are aimed at maintaining productivity and full soil cover by adjusting stocking rates to avoid overgrazing and application of fertilizer and lime as required. In general, however, after a few years of pasture establishment, stocking rates are increased without paying adequate attention to pasture management including fertilization, leading to a rapid nutrient-depletion and pasture degradation. As a result, the pastures become degraded within 3–4 years to the extent of being unable to support even average stocking rates. More than 60% of the pastures in the Cerrado are degraded (Batlle-Bayer et al. 2010), and the Cerrado is now regarded as a threatened biome (Cardoso da Silva and Bates 2002; Boddey et al. 2004): an unfortunate but revealing example of an ecological disaster caused by human intervention.

Although liming and fertilization have been recommended for reclaiming the degraded pastures (de Oliveira et al. 2004), the high cost of production and transport, and the high environmental cost associated with these practices make them unattractive options. Introduction of N-fixing legumes in association with improved grasses, and integrated crop-livestock-management systems have been proposed for sustaining grassland productivity in the Cerrado region, but have not yet been widely adopted, partly because the legume+grass mixture of understory could not be sustained for long. Worldwide, improved grassland management (e.g., application of fertilizers, adapted stocking rates, introduction of legumes and irrigation) is reported to have the potential to lead to a significant soil C sink (Conant et al. 2001). However, such results have to be viewed with caution, because many of them have used the degraded pasture system as 'baseline' for the comparison and not the native Cerrado ecosystem.

The Brazilian savannas have small aboveground C reserves compared to forest biomes; but their soils hold considerable stocks of organic C. Bustamante et al. (2006) estimated that soils of the Cerrado region contained an average stock of 117 (range: 100–174) Mg C ha^{-1} (for native Cerrados). In a review of changes in organic C stocks upon land use conversion in the Cerrado, Battle-Bayer (2010) cited reports of SOC stocks ranging from 123 to 209 Mg C ha^{-1} from different locations in the Cerrado. Our own studies (Tonucci et al. 2011) have shown much higher stock of C in the Cerrado soils (Table 2, Fig. 8). It has also been suggested that the land use history of the site could have a major influence on C stock and distribution of C in different size-fractions under in different land use systems. The Cerrado has only recently (four decades) been opened up for conversion to agricultural/livestock/forestry purposes. Unfortunately no information could be obtained about the previous site history dating back to 200 years or more. A preliminary evaluation using ^{14}C dating technique in soils of the forest, pasture and silvopasture sites used in the study sites of Tonucci et al. (2011) suggested the possibility that the Cerrado region

Table 2 Total soil organic carbon storage to one-meter depth in soils in silvopastoral systems at various study sites

Study location	Silvopastoral system	Land use	Soil: Order	Soil: pH	Soil: Bulk density (Mg m^{-3})	Soil organic carbon stock to 1-m depth (Mg ha^{-1})	Reference
Florida, USA; Humid subtropical	Slash pine (*Pinus elliottii* Engelm.) + bahiagrass (*Paspalum notatum* Flüggé)	Pasture	Spodosols	5.5	1.5	66.7	Haile et al. (2008, 2010)
		Silvopasture		5.4	1.5	102	
		Pasture	Ultisols	6.2	1.7	30.8	
		Silvopasture		5.7	1.6	37.3	
Central Spain; Subhumid Mediterranean	Dehesa: Cork oak (*Quercus suber* L.) silvopasture	Away from tree (15 m)	Alfisols	4.1	1.3	26.5	Howlett (2009)
		Near tree (2 m)		4.0	1.3	50.2	
Northwestern Spain; Humid Atlantic	Silvopasture with birch (*Betula pendula* Roth.)	Pasture	Inceptisols	4.2	1.5	133	Howlett et al. (2011)
		Silvopasture		4.2	1.4	150	
Minas Gerais, Brazil; Cerrado: Subhumid tropical	*Eucalyptus* spp. with understory of *Brachiaria* spp. (fodder grass)	Forest	Oxisols	5.1	1.0	353	Tonucci et al. (2011)
		Pasture		5.4	1.2	408	
		Silvopasture		5.2	1.0	353	

See Table 1 for additional site description

had previously been under a "high-C-storing" system for a long time (Tonucci et al., personal communication: September 2010). The $\delta^{14}C$ values for various soil depth classes under different land use systems studied by Tonucci et al. (unpublished data) suggest that the organic matter in the surface soil of these systems was of recent addition whereas the C in the lower soil layers had been stored for much longer time periods. These results, although preliminary, suggest that the Cerrado soils stock high amounts of C derived possibly from previous land use, the details of which are unknown. The bottom line is that the Cerrado biome has a high stock of C in soil, probably as a consequence of previous land use; most of this C is in a biodegradable form, and could be lost to the atmosphere with soil disturbance. Adoption of sustainable land use systems such as silvopasture could reduce this potential hazard. These concerns call for well coordinated and detailed investigations on this important issue.

Conclusions

Our studies, though limited, suggest that if sustainable silvopastoral systems could be developed as alternatives to conversion of forest lands to support animal production, the high levels of C footprint of animal production in developing countries could be reduced considerably. Between the two forms of savanna conversions – to produce grass and grain – the former, however, is a lesser evil environmentally, and grass-fed beef, which is far more efficient in overall energy use than grain-fed beef, would leave a lesser C footprint than intensive grain production systems including the grain-based beef. Thus, shifting from input-intensive pasture- and grain production to environment-friendly silvopastoral systems could reduce GHG emission, promote C sequestration in soils, and enhance the soil's resilience by increasing the SOC pool.

Although the Cerrado is a unique ecosystem, many of these projections and perspectives could be applicable to other savanna regions too. The extent to which these results are applicable in the savannas of other continents such as Africa and Asia is unclear, because of not only the differences in soils and other ecological conditions, but also the vast differences in management systems. In most parts of the African and Asian savannas, silvopastoralism consists mostly of extensive animal grazing in open lands with scattered trees with practically no land use intensification involving fertilizers and such external inputs, unlike in the fertilized and management-intensive silvopasture in the Cerrado. Nevertheless, results from the extensive dehesa silvopasture system of Spain as well as other studies from the Parklands (extensive, open grazing) system in Mali, West Africa, indicate the positive role of trees in SOC build-up. Given that globally the savanna ecosystem covers a sixth of the total land area and account for a third of total plant production, the role of savanna silvopastoral systems in global C sequestration and climate change mitigation deserve serious investigation. With the increased awareness of the role of savannas, the Cerrado in particular, in the global C cycle and climate change, the international community has a stake in such efforts.

References

ABRAF (2008) Associação Brasileira de Produtores de Florestas Plantadas. Censo Florestal de 2008. São Paulo, SP, 90 p. Available at http://www.abraflor.org.br/estatisticas/ABRAF08-EN.pdf. Accessed Jan 2011

Alvim MJ, Carvalho MM, Botrel MA, Pires MFA (2005) Ganho de peso vivo de novilhas leiteiras em sistemas silvipastoril e em pastagem de *Brachiaria decumbens*. In: Reunión de la Associación Latinoamericana de Producción Animal 19, Tampico, Brazil, pp 20–22

Andrade CMS, Garcia R, Couto L (2001) Fatores limitantes ao crescimento do capimtanzânia em um sistema agrossilvipastoril com eucalipto, na região dos cerrados de Minas Gerais. Rev Bras Zootecn 30:1178–1185

Batlle-Bayer L, Batjes NH, Bindraban PS (2010) Changes in organic carbon stocks upon land use conversion in the Brazilian Cerrado: a review. Agric Ecosyst Environ 137:47–58

Bernardino FS, Tonucci RG, Garcia R, Neves JCL, Rocha GC (2011) Produção de forragem e desempenho de novilhos de corte em um sistema silvipastoril: efeito de doses de nitrogênio e oferta de forragem. Rev Bras Zootecn (in press)

Boddey RM, Macedo R, Tarre´ RM, Ferreira E, de Oliveira OC, de Rezende C, Cantarutti RB, Pereira JM, Alves BJR, Urquiaga S (2004) Nitrogen cycling in *Brachiaria* pastures: the key to understanding the process of pasture decline. Agric Ecosyst Environ 103:389–403

Bustamante MMC, Corbeels M, Scopel E, Roscoe R (2006) Soil carbon storage and sequestration potential in the Cerrado region of Brazil. In: Lal R, Cerri CC, Bernoux M, Etchevers J, Cerri CEP (eds) Carbon sequestration in soils of latin America. Food Products Press, Binghamton, pp 28–43

Cardoso da Silva JM, Bates JM (2002) Biogeographic patterns and conservation in the South American cerrado: a tropical savanna hotspot. Bioscience 52:225–234

Cochrane TT, Azevedo LG, Thomas D, Madeira Neto J, Adamoli J, Verdesio JJ (1985) Land use and productive potential of American savannas. In: Tothill J, Mott JC (eds) International savanna symposium 1984. Australian Academy Science, Canberra, pp 114–124

Comunicação Nacional (2004) Comunicação nacional inicial do Brasil à conservação – quadro das nações unidas sobre mudanças do clima. Coordenação-Geral de Mudanças Globais de Clima. Ministério da Ciência e Tecnologia, Brasília, Brasil, 74 p

Conant RT, Paustian K, Elliott ET (2001) Grassland management and conversion intro grassland: effects on soil carbon. Ecol Appl 11:343–355

de Oliveira OC, de Oliveira IP, Alves BJR, Urquiaga S, Boddey RM (2004) Chemical and biological indicators of decline/degradation of *Brachiaria* pastures in the Brazilian Cerrado. Agric Ecosyst Environ 103:289–300

Dubé F, Couto L, Garcia R, Araújo GAA, Leite HG (2000) AvaliaçãoEconômica de um sistema agroflorestal com eucalyptus sp. No Noroeste de Minas Gerais: o caso da Companhia Mineira de Metais. Rev Árvore 24:437–443

EMBRAPA CERRADO (1999) Embrapa Cerrados: conhecimentos, tecnologias e compromisso ambiental. Planaltina, 34 p

Gama-Rodrigues EF, Nair PKR, Nair VD, Gama-Rodrigues AC, Baligar VC, Machado RCR (2010) Carbon storage in soil-size fractions under cacao agroforestry systems in Bahia, Brazil. Environ Manag 45:274–283

Garrett HE (2009) North American agroforestry: an integrated science and practice, 2nd edn. American Society of Agronomy, Madison, 369 p

Grace J, José JS, Meir P, Miranda HS, Montes RA (2006) Productivity and carbon flux of tropical savannas. J Biogeogr 33:387–400

Haile SG, Nair PKR, Nair VD (2008) Carbon storage of different soil-size fractions in Florida silvopastoral systems. J Environ Qual 37:1789–1797

Haile SG, Nair VD, Nair PKR (2010) Contribution of trees to soil carbon sequestration in silvopastoral systems of Florida. Glob Change Biol 16:427–438

Howlett D (2009) Environmental amelioration potential of silvopastoral agroforestry systems in Spain: soil carbon sequestration and phosphorus retention. Ph.D. Dissertation, University of Florida, Gainesville

Howlett DS, Mosquera-Losada M-R, Nair PKR, Nair VD, Rigueiro-Rodríguez A (2011) Soil carbon storage in silvopastoral systems and a treeless pasture in northwestern Spain. J Environ Qual. doi:10.2134/jeq2010.0145

IBGE (2006) Censo Agropecuário 2006: resultados preliminares. Instituto Brasileiro de Geagrafia e Estatística, Rio de Janeiro. Available at http://www.ibge.gov.br/home/estatistica/economia/agropecuaria/censoagro/2006/agropecuario.pdf. Accessed Jan 2011

Lucas NM (2004) Desempenho animal em sistemas silvipastoril com acácia-negra (*Acácia mearnssi* De Wild.) e rendimento de matéria seca de cultivares de *Panicum maximum* sob dois regimes de luz solar. In: Zootecnia. Universidade Federal do Rio Grande do Sul, Santa Maria, Brazil, 127 p

Machado RB, Neto MBR, Pereira PGP, Caldas EF, Gonçalves DA, Santos NS, Tabor K, Steininger M (2004) Estimativas de perda da área do Cerrado brasileiro. Conservação Internacional, Brasília, DF p. 26. Available at http://www.conservation.org.br/arquivos/RelatDesmatamCerrado.pdf. Accessed 20 Oct 2010

Mosquera-Losada MR, McAdam J, Rigueiro-Rodriguez A (eds) (2005) Silvopastoralism and sustainable land management. CAB International, Wallingford, UK, 418 p

Nair PKR (1993) An introduction to agroforestry. Kluwer Academic, The Netherlands, 499 p

Nair PKR, Gordon AM, Mosquera-Losada MR (2008) Agroforestry. In: Jorgensen SE, Faith BD (eds) Encyclopedia of ecology. Elsevier, Oxford, pp 101–110

Nair PKR, Nair VD, Kumar BM, Showalter JM (2010) Carbon sequestration in agroforestry systems. Adv Agron 108:237–307

Reatto A, Correia JR, Spera ST (1998) Solos do Bioma Cerrado: aspectos pedologicos. In: Sano SM, Almeida SP (eds) Cerrado: Ambiente e Flora. EMBRAPA-CPAC, Planaltina, pp 47–87

Resck DVS (1998) Agricultural intensification systems and their impact on soil and water quality in the cerrado of Brazil. In: Lal R (ed) Soil quality and agricultural sustainability. Ann Arbor Press, Chelsea, pp 288–300

Saha SK, Nair PKR, Nair VD, Kumar BM (2009) Soil carbon stock in relation to plant diversity of homegarden systems in Kerala, India. Agroforest Syst 76:56–65

Saha SK, Nair PKR, Nair VD, Kumar BM (2010) Carbon storage in relation to soil size-fractions under some tropical tree-based land use systems. Plant Soil 328:433–446

Sano EE, Barcellos AO, Bezerra HS (2000) Assessing the spatial distribution of cultivated pasture in the Brazilian savanna. Pasturas Trop 22:2–15

Silva JLS (1998) Produtividade de componentes de um sistema silvipastoril constituído por *Eucalyptus saligna e pastagens cultivadas e nativas no Rio Grande do Sul*. In: Zootecnia. Universidade Federal de Viçosa, Viçosa, 178 p

Silva JLS, Saibro JC, Castilhos ZMS (2001) Situação da pesaquisa e utilização de sistemas silvipastoris no Rio Grande do Sul. In: Carvalho MM, Alvim MJ, Carneiro JC (eds) Sistemas Agroflorestais pecuários: opção de sustentabilidade para áreas tropicais e subtropicais. FAO, Juiz de Fora, pp 257–283

Takimoto A, Nair PKR, Nair VD (2008) Carbon stock and sequestration potential of traditional and improved agroforestry systems in the West African Sahel. Agric Ecosyst Environ 125:159–166

Takimoto A, Nair VD, Nair PKR (2009) Soil carbon sequestration potential of agroforestry practices in the West African Sahel. Agroforest Syst 76:11–25

Tonucci RG, Nair PKR, Nair VD, Garcia R, Bernardino FS (2011) Soil carbon storage in silvopasture and related land use systems in the Brazilian Cerrado. J Environ Qual. doi:10.2134/jeq2010.0162

Biomass and Carbon Accumulation in Land Use Systems of Claveria, the Philippines

Shushan Ghirmai Brakas and Jens B. Aune

Abstract This study was undertaken to assess the standing biomass and carbon (C) accumulation in the smallholder farming systems of Claveria, the Philippines. Nineteen land use types were identified and the age and standing biomass assessed by field measurements and the use of allometric equations. Aboveground C stock varied from 2.9 to 234 Mg ha^{-1}. The highest stock was observed in the preserved forest followed by homegardens whereas the lowest was observed in the grasslands. In general, C accumulation in aboveground biomass decreased with increasing tree diversity. The highest rate of C accumulation was found in mango (*Mangifera indica* L.) plantation (17.9 Mg ha^{-1} year^{-1}) followed by banana (*Musa* spp.) + fruit trees (13.6 Mg ha^{-1} year^{-1}). Low (<1 Mg ha^{-1} year^{-1}) rates of C accumulation were observed in coconut (*Cocos nucifera* L.) plantations, coconut + banana, bush fallow, and grasslands. Agroforestry systems like homegardens and corn (*Zea mays* L.) + timber and fruit trees can have both high rates of C accumulation and high tree diversity, implying the synergy between C accumulation and maintenance of tree diversity.

Keywords Carbon stock • Carbon accumulation • Homegardens • Tree diversity

Introduction

Land use and land use changes affect the exchange of greenhouse gases between terrestrial ecosystems and the atmosphere. Forest clearing by burning and conversion to agriculture and pasture causes large carbon (C) fluxes into the atmosphere

S.G. Brakas • J.B. Aune (✉)
Department of International Environment and Development Studies, Noragric, Norwegian University of Life Sciences, P.O. Box 5003, 1432, Aas, Norway
e-mail: shushanfree@yahoo.com; jens.aune@umb.no

B.M. Kumar and P.K.R. Nair (eds.), *Carbon Sequestration Potential of Agroforestry Systems: Opportunities and Challenges*, Advances in Agroforestry 8,
DOI 10.1007/978-94-007-1630-8_9,

(Nabuurs et al. 2007). Carbon dioxide (CO_2) emissions from land use and land use changes, predominantly from forested areas, account for 33% of global CO_2 emissions between 1850 and 1998 (Bolin and Sukumar 2000).

The establishment of Clean Development Mechanisms (CDM) as one of the instruments in the Kyoto Protocol to control emissions of greenhouse gases (UNFCCC 1998) has led to an increased interest in reducing CO_2 in the atmosphere through forest based C sequestration (CS) projects. The most significant increases in CO_2 sequestration can be achieved by moving from lower biomass land use systems such as grasslands, agricultural fallows, and permanent shrub lands to forest based land use systems such as natural forests, forest plantations, and agroforestry (Roshetko et al. 2002). Potential mechanisms to reduce C losses and increase C sinks include forest management by protecting and conserving the C pools of the existing forests (Brown 1996). Forests sequester more than 92% of the world's C and between 20 and 100 times more C per hectare than agricultural lands (Andrasko 1990). Slowing deforestation, augmenting afforestation, and intensifying silviculture can significantly contribute to the conservation or sequestration of significant quantities of terrestrial C (Dixon et al. 1993). Alternative forest management systems such as fuelwood plantations and woodlots may also have the potential to sequester C (Brown et al. 1993; Kimaro et al. 2011; Quinkenstein et al. 2011) and agroforestry is particularly relevant in this respect. Benefits that agroforestry systems provide in addition to C sequestration are increased food production, improved nutritional quality of food, fodder, improved soil fertility, timber, and build the asset base on the farm (World Agroforestry Centre 2010a). Nair et al. (2009) showed that the estimates of C sequestration potential in agroforestry systems are highly variable ranging from 0.29 to 15.21 Mg ha^{-1} $year^{-1}$. The objective of this paper is to assess the aboveground C stock and the annual rate of C accumulation in the smallholder land use systems of Claveria, the Philippines and to study the relationship between tree diversity, C stock, and C accumulation rates. This information is important in order to identify land use systems that can both contribute to C sequestration and at the same time preserve species richness. The smallholder agroforestry systems of Claveria could provide an attractive environment for C investment through CDM projects.

Materials and Methods

Study Area

The study was conducted in the southwestern part of Claveria located in the province of Misamis Oriental in the northern part of the Mindanao region, the Philippines (8°38′ N; 124°55′ E; 300–800 m altitude; Fig. 1). It was selected as a representative area where considerable forest and grasslands areas have been converted into

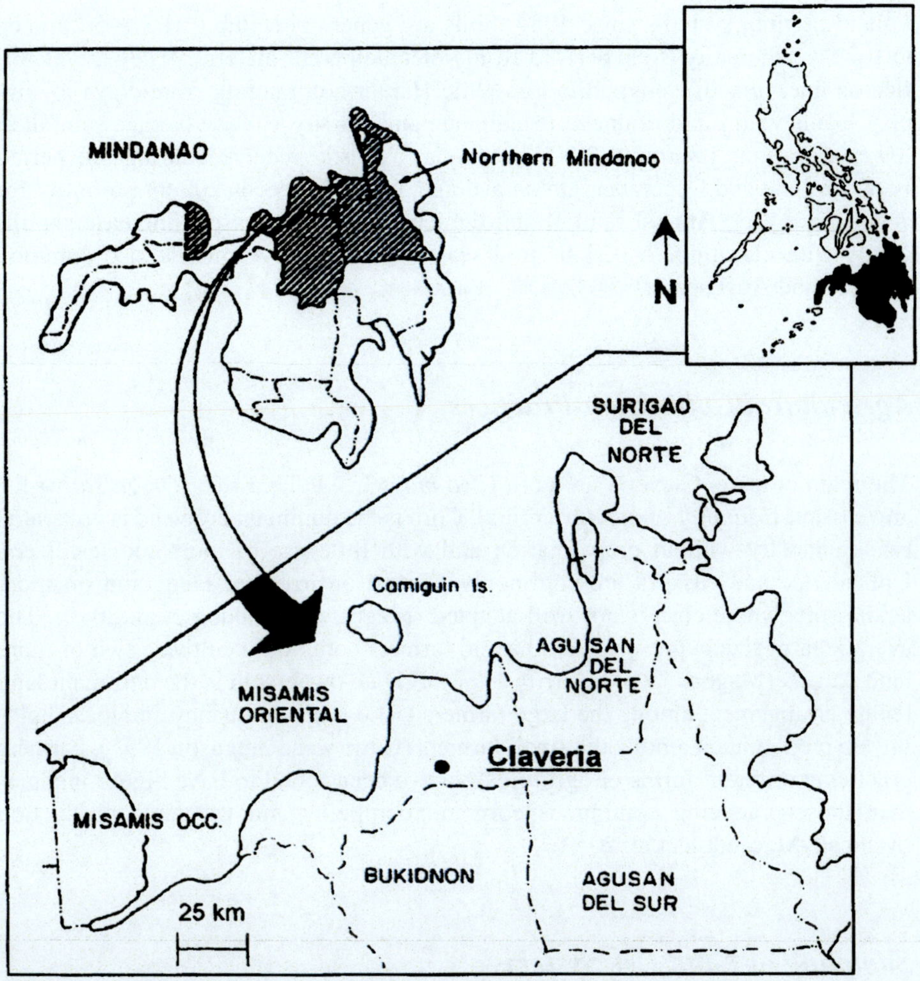

Fig. 1 Location map of the study area (*inset*: Map of the Philippines) (Source: Stark 2000)

agroforestry. In 1949, grasslands (59%) and forests (14%) were the dominant land uses and cultivated land occupied 9% of the area. From 1949 to 1967, however, settlements by small farmers accelerated. Consequently, cultivated area doubled to 20% and grasslands decreased to less than 50%, while the forested areas remained unchanged. Between 1967 and 1988, croplands doubled again to 41% of the total land area. Perennial croplands (mainly coffee, *Coffea* spp.) covered 4% of the land in 1967, but increased to 30% by 1988. Only 1% of the area remained forested in 1988 and the grasslands were reduced to 18% (Garrity and Augustin 1995). The study site in Claveria also represents one of the research sites of the World Agroforestry Centre (ICRAF). It has been the site of intensive work on sustainable upland farming systems since 1984. Soils are generally acidic (pH 3.9–5.2), deep (>1 m) weathered oxisols, derived from volcanic parent material. They are classified as fine, mixed, isohyperthermic Ultic Haplorthox ranging from clays to silty clay loams with rapid drainage (Magbanua and Garrity 1988). Average rainfall at lower elevations is around 2,000 mm year^{-1}, mostly received during the period between June and December. On an average, rainfall exceeds 200 mm month^{-1} for 7 months a year. Annual rainfall and the length of rainy seasons increase significantly with elevation. A serious mid season drought in August is also common (Garrity and Augustin 1995).

Agricultural/Cultivation Practices

The main crops in Claveria are corn (*Zea mays* L.), upland rice (*Oryza sativa* L.) and cassava (*Manihot esculenta* Crantz). Corn is the dominant crop and is cultivated twice annually without crop rotation and with little use of inorganic fertilizers. Upland rice and cassava are commonly planted on irrigated lands and on more acidic soils where corn is not well adapted (Magcale-Macandog et al. 2010). The average farm size is presently 3.0 ha and farmers commonly cultivate two or more land parcels (Magbanua and Garrity 1988). Private ownership is the dominant land tenure arrangement among the large farmers (>3.0 ha), but tenancy and lease holding are predominant among the small farmers (Garrity and Augustin 1995). Farmers practicing different forms of agroforestry have been found to have higher incomes than those practicing continuous corn monocropping and vegetable production (Magcale-Macandog et al. 2010).

Sampling of Land Use Systems

Field visits were conducted to identify the land use systems in the study area. In addition, a workshop was conducted in which the land care facilitators at the World Agroforestry Centre identified the prominent land use types in this area. Land use systems varied from systems with high tree diversity to those with low tree diversity; the salient attributes of which are described in Table 1. Major products included timber, fruits, cereals, and fodder. Within land use system variations in species composition and age of trees were profound.

Aboveground biomass was measured in 19 land use systems (Table 1) in order to assess the C stock and the rate of C accumulation in the identified land use systems. The term C accumulation used in this paper should not be equated with C sequestration as the latter depends on the amount of recalcitrant C remaining at the end of tree rotation and the final use of the tree products in addition to accumulation of C in aboveground biomass (Montagnini and Nair 2004).

Table 1 Salient attributes of the land use systems of Claveria, the Philippines

Land use-system	Average age (year)	Average no. of tree sp./plot
Preserved forest[a]	>100	20
Natural forest[b]	>100	22.8
Multistrata agroforest[c]	38.8	28.5
Homegarden[d]	21	12.7
Coffee plantation[e]	30	2.3
Coconut plantation[f]	21.7	2.8
Woodlot[e]	10.5	1.5
Mango plantation[e]	7.8	1.5
Banana plantation[e]	1.5	1
Fallow with indigenous and fruit trees[g]	15	3.5
Corn coffee[h]	30	1
Corn + timber and fruit trees[h]	9.3	8.3
Corn + timber trees[h]	8.3	1.7
Corn + mango[h]	6	2
Corn + banana[h]	1.5	1.7
Corn monocropping[h]	1	1
Coconu + banana[h]	21.7	2
Banana + fruit trees[i]	4.7	2
Grassland[j]	50	1
Bush fallow[k]	25	n/a

[a]Located in the neighboring municipality (*Initao*), a fenced and strictly guarded forest, which excludes the local people. Common trees are *Aglaia* spp., *Dehaasia triandra* Merr., *Euphoria didyma* Blanco., *Litchi philippinensis* (Radlk.) Leenh., *Mallotussp.*, *Microdesmis casearifolia* Planch., *Millettia brachycarpa* Merr., *Pterospermum* sp. and *Ziziphus* spp.

[b]Mature forests of more than 100 years age consisting of indigenous tree species such as *Bixa orellana* L., *Shorea negrosensis* Foxw., *Lithocarpus llanosii* (A.DC.) Rehder, *Cinnamomum mercadoi* Vidal, *Euphoria didyma* Blanco., *Pentacme contorta* Merr. et Rolfe, *Musa textilis* Nee., *Litsea segregata* Elmer, *Headaphne* sp., *Syzygium brevistylum, Aglaia alternifolia, Ziziphus hutchinsonii,* and *Podocarpus brevifolius* (Stapf) Foxw. Illegal logging is practiced in the area

[c]Complex agroforestry system with no particular arrangement in the distribution of tree species. It resembles the natural forest and is mostly situated along the creeks. It is a species-rich system composed of indigenous trees, timber trees, fruit trees, coconut trees, coffee, and banana scattered in the area. Most common indigenous trees: *Artocarpus ovata* Blanco, *Cassia spectabilis* DC., *Sandoricum koetjape* Merr., *Sandoricum vidalli* Merr., *Artocarpus blancoi* (Elmer) Merr., *Artocarpus odoratissimus* Blanco, *Bixa orellana* L., *Litsea philippinensis* Merr., *Pterocarpus indicus* Willd., *Artocarpus rubrovenia* Warb., *Bactris gasipaes* H.B.K., and *Shorea almon* Foxw

[d]A common tree based system usually located near the households. It is composed of a variety of fruit trees (e.g. *Artocarpus heterophyllus* Lam., *Garcinia mangostana* L., *Lansium domesticum* Corr., *Mangifera indica* L., *Annona muricata* L., *Theobroma cacao* L.), timber trees, coconut, banana and coffee, *mainly* grown for home consumption and some commercial production

[e]Consists of single species planted in rows. The woodlots consist of timber trees such as *Gmelina arborea* Roxb., *Eucalyptus deglupta* Blume, and *Acacia mangium* Willd.

[f]Coconut is the most widespread tree in the study area. In a coconut plantation the coconut trees are planted in rows at a distance of 10 × 10 m from each other. It is also found scattered in farms or on borders of farms

[g]Land use systems established on lands left fallow for more than 15 years. These lands were planted with corn and cassava before they were left fallow allowing indigenous trees to grow. In most of such land use systems, the land was left undisturbed for several years and fruit trees were planted at a later stage

[h]Corn based systems are systems have corn as the main crop and are intercropped with timber or fruit trees, timber and fruit trees, banana, coconut, or coffee (scattered in contour and in alleys). These land use systems have a wide variation in spacing of crops and trees

[i]Banana is intercropped with mango, jack fruit, and the like

[j]Imperata grasslands for pasture

[k]Intended for soil fertility restoration

The farms included in the study were mainly smallholder agroforestry systems dominated by tree crops with few monoculture plantation systems. They included trees on farms, trees on farm boundaries, and crop fallow rotations. Measurements of biomass were also made in natural forests and in a forest reserve. Since all natural forest cover in the study area was converted to cultivated lands, to get biomass and C data for natural forests, measurements were made in the nearby forests. Two samples were taken in the same municipality where Claveria is situated (Tagmaray in Malitbog and Abacahan in Mat-i) and one in the neighbouring Bukidnon municipality. Three farms were randomly selected for each land use system in order to represent the area assigned to that land use system. For banana (*Musa* spp.) and mango (*Mangifera indica* L.) plantations, only two farms each were selected.

Sampling Protocol for Standing Tree Biomass

A total of 55 farms were sampled for the 19 land use systems. Live and dead trees, logs, and understory vegetation were sampled on each farm. The protocol developed by the Alternatives to Slash and Burn project (Hairiah et al. 2001) was used in this study. Two quadrats of 200 m^2 (40×5 m) were selected for each farm. A 40 m line was laid out within each quadrat and trees with a diameter at breast height (DBH) >5 cm located within 2.5 m on each side of the line were sampled. For plots which contained trees above 50 cm in diameter, a 20×100 m area was sampled and all trees with a DBH >30 cm were measured. The diameter of each tree within the plot was measured at 1.3 m above the soil surface except for trees with trunk irregularities. In such cases, height was measured above the irregular part of the stem. For trees branching below the measurement height, all branches >5 cm were measured at 1.3 m aboveground and the diameters were summed (ΣD^2) to get an equivalent diameter.

Sampling Protocol for Understory Vegetation and Dead Trees on the Ground

Understory biomass was collected from four 1 m^2 sub-quadrats randomly placed within each of the two 40×5 m quadrats. All aboveground vegetation other than trees with DBH >5 cm was harvested at ground level and weighed. A random subsample of this vegetation was weighed fresh in the field and again after oven drying. Lengths of the dead trees on the ground (necromass) within each quadrat, middle diameters of the logs as well as botanical names were recorded.

Assessing the C Stock

The weights of aboveground non-woody vegetation and dead trees on the ground for each quadrat were summed and divided by the sampling area. Trees and understory vegetation were assumed to contain 45% C of their biomass (Schroth et al. 2002). Tree biomass (*W*, dry weight) was estimated using the allometric equation (Ketterings et al. 2001) on the basis of stem diameter (*D*) at 1.3 m above the ground.

$$W = 0.11pD^{2+c} \tag{1}$$

Where *p* is the wood density and the coefficient *c* is based on the allometric relationship between tree height (*H*) and diameter (default value for $c = 0.62$). The data on wood density was extracted from a wood density database created by ICRAF (World Agroforestry Centre 2010b).

For pruned coffee (*Coffea arabica* L.), bamboo and banana (*Musa* sp.) the following equations (van Noordwijk et al. 2002) were used.

$$\text{Pruned coffee: } 0.281 * D^{2.06} \tag{2}$$

$$\text{Bamboo} : 0.131 * D^{2.28} \tag{3}$$

$$\text{Banana} : 0.03 * D^{2.13} \tag{4}$$

Where $D =$ diameter at breast height (cm)

Total dry weight of the understory vegetation (Hairiah et al. 2001) was calculated as:

$$Total\ dry\ weight\ (kg\ /\ m^2) = \frac{Total\ \ fresh\ weight\ (kg) * Subsample\ dry\ weight\ (g)}{Subsample\ \ fresh\ weight\ (g) * Sample\ area\ (m^2)}. \tag{5}$$

The biomass of unbranched cylindrical trees such as coconut (*Cocos nucifera* L.) and dead trees on the ground was calculated with an equation based on cylinder volume while assuming a density of 0.4 g cm^{-3} (Hairiah et al. 2001):

$$Biomass = \pi\, D^2 hs\ /\ 40 \tag{6}$$

Where biomass is expressed in kg, $D =$ tree diameter (cm), $h =$ length (cm), and $s =$ density (g cm^{-3}). Carbon in the roots and the topsoil (0–20 cm) were not included in this study due to financial and time constraints. Carbon accumulation in aboveground biomass for each land use system was assessed by measuring the C stock in each plot and by dividing by the number of years since establishment of the plot. Only plots with known year of establishment were assessed. The age of trees was obtained using local informants. Calculation of standard deviation and analysis of variance were undertaken in order to test the results statistically (Microsoft Excel, Minitab).

Results

Aboveground Carbon Stocks

The aboveground C stocks calculated for the 19 land use systems varied significantly (2.9–234 Mg C ha^{-1}; Table 2). The highest C stock was observed in the preserved forest followed by homegardens, whereas the lowest was in the grasslands. Forest, homegarden, mango plantation, multistrata agroforest, and coffee plantation had C stocks >100 Mg ha^{-1}. On the other hand, corn + mango, corn + timber trees, coconut plantation, coconut + banana, bush fallow, corn + banana, corn monocropping, and grasslands had relatively low C stocks (<20 Mg ha^{-1}). Corn + coffee, banana + fruit trees, banana plantations, fallow with indigenous and fruit trees, corn + timber and fruit trees, and woodlots were found to contain C stocks within the range of 40–100 Mg ha^{-1}.

Aboveground C accumulation rates (Table 2) also varied significantly among the land use systems. Highest rate of C accumulation was found in the mango

Table 2 Carbon stock and rate of carbon sequestration in different land use systems of Claveria, the Philippines

Land use system	Above ground carbon stock (Mg ha^{-1})	Standard error	Carbon accumulation rate (Mg ha^{-1} $year^{-1}$)	Standard error
Preserved forest	234.5	N/A	2.3	N/A
Homegarden	159.7	59.9	9.4	4.0
Multistrata agroforest	155.8	19.0	4.1	1.2
Natural forest	147.5	43.5	1.5	0.2
Mango plantation	118.9	0.6	17.9	6.0
Coffee plantation	112.3	21.5	5.3	2.6
Corn coffee	85.3	18.5	2.8	0.6
Banana + fruit trees	72.9	48.1	13.6	7.6
Fallow with indigenous and fruit trees	56.7	36.6	3.2	1.6
Corn + timber/fruit trees	53.5	22.3	7.8	5.1
Woodlot	40.6	13.3	5.7	2.2
Corn + timber trees	18.8	9.8	3.0	1.2
Corn + mango	16.7	6.3	3.1	0.8
Coconut plantation	14.3	8.5	0.9	0.4
Coconut + banana	11.5	0.7	0.5	0.1
Banana plantation	7.2	1.5	6.2	2.4
Bush fallow	5.4	0.6	0.3	0.1
Corn + banana	3.9	1.4	3.7	1.6
Corn monocropping	3.4	0.0	3.4	0.0
Grassland	2.9	0.2	0.1	0.0
Mean	66.1		4.7	
	$F=5.63$		$F=2.46$	
	$p<0001$		$p<0.05$	

plantation (17.9 Mg ha^{-1} $year^{-1}$) followed by banana+fruit trees (13.6 Mg ha^{-1} $year^{1}$). A low (<1 Mg ha^{-1} $year^{-1}$) rate of C accumulation was noted for the coconut plantations, coconut+banana, bush fallow and grasslands. Homegardens, corn+timber and fruit trees, corn+timber trees, banana plantation, woodlots, coffee plantation, and multistrata agroforest were intermediate (4–10 Mg C ha^{-1} $year^{-1}$).

Biodiversity in Relation to C Stock and C Accumulation

The number of tree species in each plot of the sampled land use system was plotted against C stock and annual aboveground C accumulation. In general, high aboveground C stocks were found in the land use systems with high tree diversity (preserved forest, homegarden and natural forest; Fig. 2). Corn+timber and corn+fruit trees also were found to contain a fairly high C stocks and number of tree species. Conversely, the land use systems with few tree species had low C stocks (e.g., corn, corn+banana, banana plantation, coconut+banana, corn+timber, and corn+mango). Two land use systems with C accumulation above 10 Mg ha^{-1} $year^{-1}$ (mango plantation and banana+fruit trees) also had low tree diversity. However, homegarden and corn+timber and fruit trees were able to combine high C accumulation rates (7–10 Mg C ha^{-1} $year^{-1}$) with high rates of tree diversity (8–12 tree species per plot).

Discussion

There is a tendency that C stock is decreasing with increasing tree diversity, after an initial increase (Fig. 2). Despite the negative trend at high tree diversity, this study has shown that smallholder agroforestry systems in Claveria, The Philippines have high C accumulation rates in aboveground biomass. Land use systems such as mango plantations, multistrata agroforest, homegarden, banana+fruit trees, corn+timber trees, corn+timber and fruit trees, coffee plantations, and woodlots showed C accumulation rates (>4 Mg ha^{-1} $year^{-1}$; Table 2). This shows that agroforestry systems that have high tree diversity may also have high C accumulation rates. Tree-crop systems sequestered C at a rate higher than those containing only annual crops or grasslands, which has limited accumulation of C. Therefore, significant quantities of C can be sequestered by moving away from grasslands, bush fallows, and agricultural fallows to tree based systems like agroforests and forest plantations. This is consistent with the findings of Tomich et al. (2002). Annual crops will only accumulate carbon through the roots and retention of crop residues, whereas the tree crops will accumulate carbon through roots, litter, and aboveground biomass. The C accumulation rates found at Claveria are higher than those reported by Pandey (2002): 2–4 Mg ha^{-1} $year^{-1}$ for complex agroforests and 7–9 Mg ha^{-1} $year^{-1}$ in agroforestry systems with one dominant species. Montagnini and Nair (2004) reported

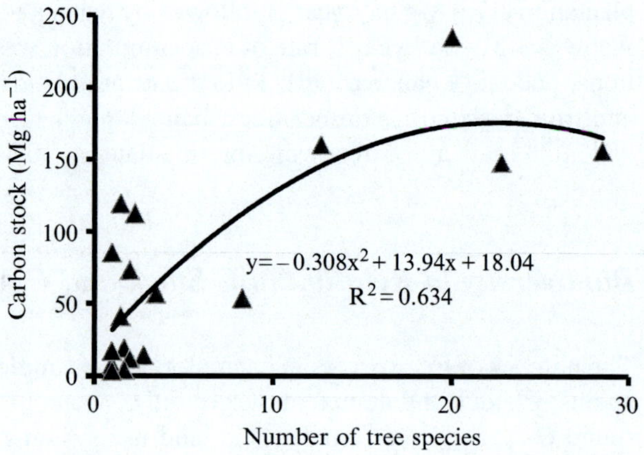

Fig. 2 Relationship between carbon stock and number of tree species in the land use systems of Claveria, the Philippines

C accumulation in tropical smallholder agroforestry to be in the range of 1.5–3.5 Mg ha^{-1} year^{-1}.

The results on C stock and C accumulation rates found in this study must be interpreted with caution, because the C stock and the C accumulation rates are dependent on the age of the plants, plant density, soil fertility of the site, rainfall and other factors. Old plantation/stands will have high C stocks, but low C accumulation rates since they have reached maturity while young plantations will have low C stock, but higher accumulation rates since the plantation will be in an active growth phase. It is difficult to account for these factors unless controlled experiments are conducted.

In addition to the accumulation of high average C stock, agroforestry systems have several advantages over monocultures. The monocropping systems (banana, mango, coconut, woodlots) are mostly for commercial purposes, while the agroforestry systems include crops for household consumption. Agroforestry also may provide a viable combination of C storage with minimal negative effects on food production (Pandey 2002). High and long term biomass accumulation with early generation of income from annual and semi-perennial intercrops is a characteristic feature of agroforestry systems. In addition, they allow for long term accumulation of capital in large sized trees and would provide more complete canopy cover than certain tree crop monocultures (Schroth et al. 2002). There is less risk in practicing agroforestry than monocropping with respect to climatic disasters such as floods and drought, market fluctuations, and pest/disease attacks. From Claveria itself it has been previously reported that agroforestry systems increase food security and provide additional income to farmers (Magcale-Macandog et al. 2010).

Implications for CDM

The results showed that man made forests (homegardens and multistrata agroforests) accumulated higher C stocks than natural forests (Table 2). However, the present C stock for the natural forests (147.5 Mg ha^{-1}) is lower than that reported by Brown (1996) who found that moist tropical forests contain between 155 and 187 Mg C ha^{-1}. This could perhaps be due to the illegal logging of the large trees in the government-owned forestlands. Overall, mango plantations, homegarden, multistrata agroforestry, and corn + timber trees could be good candidates for CDM projects due to their moderate to high C accumulation rates. Homegardens, multistrata agroforestry, and corn + timber and fruit trees also have the potential to maintain high tree diversity.

The landscape in Claveria is a mosaic of different land use systems. Furthermore, there is considerable heterogeneity within each land use system depending on planting density, age of plantation, species diversity, landscape characteristics, and soil quality. This is probably the major reason why it is difficult to establish CDM projects in Claveria as the monitoring costs are usually very high. It may be possible to make the land use systems more homogeneous through deliberate actions (e.g., uniform plant density, stand age, and management). However, such a course of action is not advisable, as the farmers will lose flexibility with regard to land management and income regeneration opportunities. An alternative option for C management in the Philippines may be to preserve the remaining natural forests in a Reduced Emission for Deforestation and Forest Degradation (REDD) scheme (C conservation: UN REDD Programme 2010). The C stock of the natural forest was estimated to be 147 Mg C ha^{-1}, but the rate of C sequestration was fairly low (1.5 Mg C ha^{-1} $year^{-1}$), which is consistent with previous reports (Sampson and Sedjo 1997). As Cairns and Meganck (1994) stated, integrated forest management including land use planning focused on preservation of primary forests, intensified use of non-timber resources, agroforestry applications, and selective plantation forestry may help to sequester C and meet the needs of local people.

Conclusions

This paper shows that certain land use systems in Claveria are characterized by a high potential for C accumulation in aboveground biomass. For instance, mango plantations and banana + fruit trees had an annual aboveground C accumulation rates above 10 Mg C ha^{-1} $year^{-1}$, albeit having lower tree diversity. In general C accumulation rates decreased with increasing tree diversity, but there were also land use systems that combined high tree diversity with high C accumulation (e.g., homegarden and corn + timber and fruit trees). This shows that there is not necessarily any contradiction between high C accumulation rates and high tree diversity. However, a major obstacle to C project in the area is the heterogeneous land use systems making it costly to establish CDM projects and accurately monitor the C accumulation rates. Preserved forests contained the highest number of species and had the highest stock of C. Preserving forest through C payments, therefore, may be an option for saving the forests in the Philippines.

Acknowledgements This research was funded by the World Agroforestry Centre. We thank the field assistants Ronillo Sungkit, Hipolito Lumansoc, Gil Arsenal and Ephren Pagalan for their genuine efforts in making the data collection process a success; Dr. Manolo Bertomeu for his insights about the study area and providing important information about the landscape and the trees. Dr. Meine van Noordwijk is also acknowledged for his advice during the initial stages of the data collection.

References

Andrasko K (1990) Global warming and forests: an overview of current knowledge. Unasylva 41:3–11

Bolin B, Sukumar R (2000) Global perspective. In: Watson RT, Noble IR, Bolin B, Ravindranath NH, Verardo DJ, Dokken DJ (eds) IPCC special report on land use, land use change and forestry. Cambridge University Press, London, pp 21–51

Brown S (1996) Mitigation potential of carbon dioxide emissions by management of forests in Asia. Ambio 25(4):273–278

Brown S, Hall CAS, Knabe W, Raich J, Trexler MC, Woomer P (1993) Tropical forests: their past, present, and potential future role in the terrestrial carbon budget. Water Air Soil Pollut 70:71–94

Cairns MA, Meganck RA (1994) Carbon sequestration, biological diversity, and sustainable development: integrated forest management. Environ Manag 18(1):13–22

Dixon RK, Winjum JK, Schroeder PE (1993) Conservation and sequestration of carbon: the potential of forest and agroforest management practices. Glob Environ Chn 2:159–173

Garrity DP, Augustin PC (1995) Historical land use evolution in a tropical acid upland agroecosystem. Agr Ecosyst Environ 53:83–95

Hairiah K, Sitompul SM, van Noordwijk M, Palm C (2001) Methods for sampling carbon stocks above and below ground, ASB lecture note 4B. International Center for Research in Agroforestry (ICRAF), Bogor, Indonesia. Available at http://www.worldagroforestrycentre.org/Sea/Products/Training/Materials/lecture%20notes/ASB-LecNotes/ASBLecNote%204B.pdf. Accessed 23 Aug 2010)

Ketterings QM, Coe R, van Noordwijk M, Ambagau Y, Palm CA (2001) Reducing uncertainty in the use of allometric biomass equations for predicting aboveground tree biomass in mixed secondary forests. For Ecol Manag 146:199–209

Kimaro AA, Isaac ME, Chamshama SAO (2011) Carbon pools in tree biomass and soils under rotational woodlot systems in eastern Tanzania. In: Kumar BM, Nair PKR (eds) Carbon sequestration potential of agroforestry systems: opportunities and challenges. Springer, Dordrecht, pp 129–143

Magbanua RD, Garrity DP (1988) Acid upland agro-ecosystems: a micro level analysis of the Claveria research site. In: Proceedings of the 1988 acid upland research design workshop. International Rice Research Institute, Los Banos, pp 1–20

Magcale-Macandog DB, Rañola FM, Rañola RF, Ani PAB, Vidal NB (2010) Enhancing the food security of upland farming households through agroforestry in Claveria, Misamis Oriental, Philippines. Agroforest Syst 79:327–342

Montagnini F, Nair PKR (2004) Carbon sequestration: an underexploited environmental benefit of agroforestry systems. Agroforest Syst 61–2:281–295

Nabuurs GJ, Masera O, Andrasko K, Benitez-Ponce P, Boer R, Dutschke M, Elsiddig E, Ford-Robertson J, Frumhoff P, Karjalainen T, Krankina O, Kurz WA, Matsumoto M, Oyhantcabal W, Ravindranath NH, Sanz Sanchez MJ and Zhang X (2007) Forestry. In climate change 2007: mitigation. In: Metz B, Davidson OR, Bosch PR, Dave R, Meyer LA (eds) Contribution of working group III to the fourth assessment report of the Intergovernmental Panel on Climate Change. Cambridge University Press, Cambridge, UK/New York, pp 541–584

Nair PKR, Kumar BM, Nair VD (2009) Agroforestry as a strategy for carbon sequestration. J Plant Nutr Soil Sci 172:10–23

Pandey DN (2002) Carbon sequestration in agroforestry systems. Climate Policy 2:367–377

Quinkenstein A, Böhm C, Matos E, Freese D, Hüttl FR (2011) Assessing the carbon sequestration in short rotation coppices of *Robinia pseudoacacia* on marginal sites in northeast Germany. In: Kumar BM, Nair PKR (eds) Carbon sequestration potential of agroforestry systems: opportunities and challenges. Springer, Dordrecht, pp 201–216

Roshetko JM, Delaney M, Hairiah K, Purnomosidhi P (2002) Carbon stocks in Indonesian homegarden systems: can smallholder systems be targeted for increased carbon storage? Am J Altern Agric 17:138–148

Sampson RN, Sedjo RA (1997) Economics of carbon sequestration in forestry: an overview. Crit Rev Environ Sci Technol 27:S1–S8

Schroth G, D'Angelo SA, Teixeira WG, Haag D, Lieberei R (2002) Conversion of agroforestry and monoculture plantations in Amazonia: consequences for biomass, litter and soil carbon stocks after 7 years. For Ecol Manag 163:131–150

Stark M (2000) Soil management strategies to sustain continuous crop production between vegetative contour strips on humid tropical hillsides. Technology development and dissemination based on farmer's adaptive experimentation in the Philippines. Southeast Asia Regional Research Program, International Centre for Research in Agroforestry: Bogor, 225 p. Abstract available at http://www.worldagroforestrycentre.org/sea/Publications/searchpub.asp?publishid=310. Accessed Oct 2010

Tomich TP, de Foresta H, Dennis R, Ketterings Q, Murdiyarso D, Palm C, Stolle F, Suyanto, van Noordwijk M (2002) Carbon offsets for conservation and development in Indonesia. Am J Alt Agric 17:125–137

UN REDD programme (2010). About REDD +. Available at http://www.un-redd.org/AboutREDD/tabid/582/Default.aspx. Accessed 25 Aug 2010

UNFCCC (1998) Kyoto Protocol to the United Nations Framework Convention. Available at http://unfccc.int/resource/docs/convkp/kpeng.pdf. Accessed 27 Aug 2010

van Noordwijk M, Rahayu S, Hairiah K, Wulan YC, Farida A, Verbist B (2002) Carbon stock assessment for a forest-to-coffee conversion landscape in Sumber-Jaya (Lampung, Indonesia): from allometric equations to land use change analysis. Sci China C 45(Supp):75–86

World Agroforestry Centre (2010a). Agroforestry and our role. Available at http://www.worldagroforestrycentre.org/about_us/our_role_in_agroforestry. Accessed 23 Oct 2010

World Agroforestry Center (2010b) Wood density database. Available at http://www.worldagroforestry.org/sea/Products/AFDbases/WD/. Accessed 23 Aug 2010

Part II
Agrobiodiversity and Tree Management

Linking Carbon, Biodiversity and Livelihoods Near Forest Margins: The Role of Agroforestry

Götz Schroth, Maria do Socorro Souza da Mota, Terry Hills, Lorena Soto-Pinto, Iwan Wijayanto, Candra Wirawan Arief, and Yatziri Zepeda

Abstract Agroforestry systems distinguish themselves from other forms of agriculture through their ability to store higher amounts of carbon (C) in their biomass, and often also to conserve more biodiversity. However, in both regards they are generally inferior to forests. Therefore, the impact of agroforestry practices on landscape C stocks and biodiversity needs to be analyzed both in terms of the interactions between agroforestry and forest, which may be positive or negative, and in terms of the conservation of C and biodiversity in the farming systems themselves. This paper argues that in forest frontier situations, the most important characteristic of land use systems in terms of C and biodiversity conservation is to be "land-sparing" (i.e. minimizing forest conversion), which requires a certain level of intensification.

G. Schroth (✉)
Mars Incorporated, Santarém, Pará, Brazil

Federal University of Western Pará, Santarém, Pará, Brazil
e-mail: goetz.schroth@effem.com

M.S.S. da Mota
Federal University of Western Pará, Santarém, Pará, Brazil
e-mail: msmota13@ig.com.br

T. Hills
Conservation International, Cairns, Australia
e-mail: t.hills@conservation.org

L. Soto-Pinto
El Colegio de la Frontera Sur (ECOSUR), San Cristobal de las Casas, Chiapas, México
e-mail: lsoto@ecosur.mx

I. Wijayanto • C.W. Arief
Conservation International-Indonesia, Jakarta, Indonesia
e-mail: i.wijayanto@conservation.org; candra.arief@gmail.com

Y. Zepeda
Conservation International-Mexico, Tuxtla Gutierrez, México
e-mail: yatzirizepeda@gmail.com

B.M. Kumar and P.K.R. Nair (eds.), *Carbon Sequestration Potential of Agroforestry Systems: Opportunities and Challenges*, Advances in Agroforestry 8,
DOI 10.1007/978-94-007-1630-8_10,

In land use mosaics, on the other hand, where natural habitat has already been reduced to small fragments, land use practices should also be biodiversity-friendly and have high levels of C storage to complement those in natural vegetation. Agroforestry has a role to play in both situations by making land use more sustainable and by making inhabited reserves ecologically and economically more viable. The paper presents three case studies where different sets of incentives are used to provide communities with the means to conserve C and biodiversity on their land and adjacent forest. In the Sierra Madre de Chiapas, Mexico, C trading is combined with shade coffee (*Coffea* sp.) production to conserve and increase tree cover on farm land in biosphere reserves. In North Sumatra, Indonesia, coffee-growing communities receive technical and marketing support and assistance with legalizing their land tenure situation as incentives to stop forest conversion for coffee, with a prospect of adding C trading later. In the central Brazilian Amazon, communities reforest their land in an extractive reserve and offer reforestation credits on a local market while laying the basis for a more tree-based reserve economy. In all three cases, the bundling of various forms of incentives is meant to increase the resilience of the respective approach to market and policy changes. Approaches like these would benefit from a better integration of agricultural and forest policies.

Keywords Amazon • Biosphere reserve • Environmental service rewards • Extractive reserve • North Sumatra • Sierra Madre de Chiapas

Introduction

Agroforestry systems are distinct from other forms of agriculture in their ability to store higher amounts of carbon (C) in the above- and belowground biomass and soils (Montagnini and Nair 2004; Nair et al. 2010). Similar characteristics – substantial and preferably complex and multi-layered canopies formed by native tree species, reduced levels of disturbance, and high levels of litter and soil organic matter – are also basic ingredients of land use systems that harbor elevated levels of biodiversity in vegetation, litter, and soil (McNeely and Schroth 2006; Schroth and Harvey 2007). Therefore, agroforestry systems (AFS) and especially their most complex, forest-like forms termed "agroforests" (Michon and de Foresta 1999; Schroth et al. 2004a) often combine higher C stocks with higher biodiversity compared to simpler-structured land use systems based on annual crops or sole stands of tree crops. However, the C stocks of AFS are generally lower than those of the natural forests of their respective site. For example, C stocks in cacao (*Theobroma cacao* L.) agroforests in southern Cameroon were 62% of those of mature forest at the same site, and were also significantly less than those of old secondary forests (Kotto-Same et al. 1997). The same is true for biodiversity. Although extensively managed agroforests may harbor a large number of native plant and animal species including certain endangered and endemic species, other more strictly forest-dependent and slow-growing species will avoid them or be progressively eliminated through hunting, weeding, and lack of reproduction, ceding their place to more "weedy" and commercially useful species (Siebert 2002; Sonwa et al. 2007; Cassano et al. 2009). Therefore, when reflecting upon the role of agroforestry systems (or land use in general) in conserving C stocks and biodiversity, it is necessary to consider the wider landscape with its dynamic patterns of forest, agroforestry, agriculture, and other land uses, rather than just the (agroforestry) plot or farm.

Per-area yields of agricultural crops, including tree crops, are generally highest in intensively managed, simply structured (e.g. little or unshaded) production systems, although the life cycle of tree crops is often shorter under such conditions (Beer et al. 1998). Biodiversity and C stocks in production systems, on the other hand, are generally higher in complex structured, diversified, and less intensively managed systems, including complex agroforests (Michon and de Foresta 1999; Rice and Greenberg 2000; Schroth et al. 2004a). It has therefore been asked under what conditions either "wildlife-friendly" but less productive land use practices (such as complex agroforests) or more intensive practices whose higher per-area yields make it easier to "spare" land for habitat conservation would lead to higher overall biodiversity in a given landscape (Green et al. 2005). These authors suggested that if intensification of land use would lead to proportionally greater biodiversity loss than yield increase, it would be more efficient to practice agriculture and biodiversity conservation in separate areas, i.e. follow a "land-sparing" agricultural strategy. If, on the other hand, yield increase through intensification would lead to proportionally smaller losses of biodiversity, then a "wildlife-friendly" agricultural strategy where production and conservation are integrated in the same area (as is the case in complex agroforests) would lead to an overall better conservation outcome for the landscape as a whole. The same principles can be readily applied to C stocks. Although this approach is theoretically appealing, in practice the situation rarely presents itself in such a clear-cut manner, because some wildlife and plant species that are sensitive to even small levels of disturbance require forest habitat, while other species (including certain rare and endemic species) may even do better in somewhat disturbed areas including agroforests (Cassano et al. 2009; Oliveira et al. 2011).

Another way to consider the relative importance that should be given to biodiversity and C conservation either on-farm (i.e. "wildlife-friendly farming") or off-farm (i.e. through "land-sparing" agriculture combined with forest set-asides) within a landscape-wide conservation and development strategy is to distinguish between two types of landscapes: (1) areas where agriculture advances into a forest frontier (e.g. the Amazon, Central Africa or parts of Indonesia), and (2) the more "advanced" stage of landscape transformation of agriculture-forest mosaics where the frontier has "closed" and the landscape is composed of interspersed patches of agriculture or agroforestry with some remnants of natural forest (Chomitz et al. 2006). In the "frontier" case, a primary goal of a "biodiversity and climate-friendly" agricultural development strategy must be to minimize forest conversion, therefore "land-sparing" technologies that generate high yields and farmer incomes in a sustainable manner from a relatively small area of land, combined with effective forest conservation policies should be prioritized (Ewers et al. 2009). This requires agricultural intensification, e.g., through productive planting material and inputs to maintain soil fertility, and may include the use of agroforestry practices for income diversification and increased soil conservation (Schroth and da Mota 2007). Gockowski and Sonwa (2011) analyzed land use scenarios based on different cacao production technologies in West Africa where much forest has been lost to low-producing cacao production systems over the last half-century (Ruf and Schroth 2004). They estimated that, had intensification technologies, including intensively managed cacao-timber agroforests, supported by effective forest conservation policies, legislative reforms, and functioning input and credit markets been systematically pursued from the outset, the same total amount of cacao could have been produced (and income generated) on a smaller area of land. Consequently, over 21,000 km² of deforestation and 1.4 billion Mg CO_2 emissions could have been avoided, while at the same time preserving these countries' valuable timber and non-timber forest resources.

In mosaic landscapes, on the other hand, the size and number of forest fragments may already be too much reduced to conserve healthy populations and assemblages of the regional fauna and flora, especially of naturally rare and wide-ranging species. Therefore, in addition to the need to conserve the remaining patches of forest habitat, relatively more emphasis should be placed on creating or maintaining on-farm habitat and corridors compared to agricultural frontier situations, i.e., a "wildlife-friendly" strategy should be pursued. For example, in southern Bahia, Brazil, shade cacao systems, locally called *cabruca*, play an important role in the conservation of substantial C stocks (Gama-Rodrigues et al. 2011) as well as a large number of endemic plant and animal species in a landscape where natural forest cover has been reduced to less than 10% of its original extent (Faria et al. 2007; Cassano et al. 2009; Oliveira et al. 2011). In both phases of landscape transformation through agricultural expansion, therefore, agroforestry can play an important role in maximizing biodiversity and C conservation, as will be illustrated in the case studies later in this chapter.

Unfortunately, a common situation in tropical land use is quite the opposite of what was outlined above. In frontier situations, where land prices are low and prices of agricultural inputs needed for intensification are high, land use is often wasteful in terms of land and forest consumption rather than "land-sparing" (Barbier 2005). Once the frontier has closed, land becomes more expensive and agrochemical inputs cheaper, and so a greater emphasis is placed on intensification precisely when "wildlife-friendly" land uses are most needed to complement the dwindling natural habitat. There is, however, some reason for hope that this situation could change in the future. As the case studies below will show, C and biodiversity conservation are locally already becoming sources of income for tropical farmers, complementing income from agricultural production itself, and such opportunities could expand through several mechanisms:

1. A number of certification systems recognize practices that correlate with biodiversity conservation both at the farm level (e.g., shade use in tree crops, maintenance of riparian buffer strips, and on-farm forest reserves) and to some extent at the landscape level (e.g., prohibition of deforestation). Although not specifically designed for that purpose, these practices also impact favorably on C stocks.

Furthermore, some certifiers (such as the Rainforest Alliance: www.ra.org) are working to integrate C sequestration explicitly into their standards. Since the demand for certified agricultural commodities is increasing rapidly on the global markets, environmental certification is a way through which farmers may benefit from biodiversity and C conservation through increased market access, price premiums and also the technical support that often comes with certification programs.

2. While environmental certification of smallholder tropical farmers is well established in Latin America and rapidly advancing in Africa and Asia (Neilson 2008), access for smallholder farmers to markets for C credits from afforestation/reforestation projects that reward high C and biodiversity agroforestry practices has advanced more slowly. This is due to the significant technical and administrative obstacles and transaction costs that are inherent in the development of C projects (Torres et al. 2010; Brown et al. 2011). Examples of agroforestry projects where these obstacles have to some extent been overcome are presented below and in other chapters of this volume.
3. High C and biodiversity production systems may also be rewarded indirectly by opening additional market opportunities for farm timber and non-timber products for local, national, and potentially international markets (Sonwa et al. 2007; Gockowski et al. 2010).

A well established approach to the simultaneous pursuit of livelihood development and the conservation of biodiversity and ecosystem services, including C stocks, is the creation of specifically managed areas such as sustainable use reserves, including biosphere reserves, where land use options are regulated by management plans distinguishing various use and non-use zones and prohibiting deforestation and certain forms of land use that are considered unsustainable or destructive. In return, the traditional, legal inhabitants have access to certain forms of government support such as secure land tenure, housing, and special government or externally funded projects. This type of legally inhabited, sustainably managed areas is particularly well established in Latin America. Despite land use restrictions and their (partial) focus on forest conservation, some reserves produce significant amounts of agricultural commodities. For example, the biosphere reserves of the Sierra Madre de Chiapas in southern Mexico that are discussed in the first case study are one of the most important production areas of Arabica coffee in Mexico, while the Tapajós-Arapiuns Extractive Reserve that is presented in the third case study hosts substantial areas of community rubber agroforests (Schroth et al. 2003) although many of these are now temporarily abandoned awaiting an increase in rubber prices and better market access. The role that agroforestry can play in the conservation of biodiversity and C stocks at a landscape scale by increasing the economic and ecological viability of biosphere and sustainable use reserves has not received much attention, although these reserves offer a unique institutional framework for integrating conservation and development objectives and could offer relatively easy opportunities for the labeling of “sustainable landscapes” as a form of distinguishing their products on regional and global markets (Ghazoul et al. 2009).

In the following, we present case studies from ongoing projects from a mosaic landscape in Mexico and forest frontier landscapes in Sumatra and the Brazilian Amazon where agroforestry practices contribute to preserving the biodiversity, C stocks and other ecosystem services both directly on farms and indirectly through their interaction with natural forest and their contribution to the sustainable livelihoods of their inhabitants.

Case Study 1 – Sierra Madre de Chiapas, Mexico

In the first case study, we discuss a project that uses agroforestry practices to connect smallholder farmers in several biosphere reserves in southern Mexico to voluntary C markets, thereby reinforcing and complementing existing incentives to use sustainable and biodiversity-friendly land use methods and helping to protect the integrity of the reserves. The Sierra Madre de Chiapas is a mountain chain covering about 1.8 million ha in southern Mexico that runs parallel to the Pacific coast (Fig. 1). The region is recognized for its biodiversity and provides important watershed services to the surrounding lowlands, especially the narrow but agriculturally important coastal plain and the central valley with the state capital Tuxtla Gutierrez. The higher elevations of the Sierra Madre are included in a system of biosphere and forest reserves that host over 27,000 inhabitants (Schroth et al. 2009).

The Sierra Madre is an important production area of high-quality Arabica coffee and many of its inhabitants make a living as coffee growers, especially at elevations upward of 600 m and in the southern and more humid parts of the mountains, while cattle and annual crops such as maize (*Zea mays* L.), cultivated in slash-and-burn systems, are more important in the drier north and at lower elevations. Since the mid 1990s, the US-based non-profit organization Conservation International (www.conservation.org) had been working with coffee farmers especially in the buffer zone of El Triunfo Biosphere Reserve, providing technical assistance in agricultural best practices, such as the use of diversified coffee shade and the conservation of forest, with the objective of harmonizing farming and biodiversity conservation. While initially working with several private sector partners, the program received a significant boost in the late 1990s through a partnership with Starbucks Coffee Company which sourced coffee from participating farmers and created its Organic Shade Grown Mexico brand (Perez-Aleman and Sandilands 2008). By the mid-2000s the program involved about 900 farmers and had some notable successes: participating farmers "earned 20% more per ha [compared to non-participants]; nine out of ten families were able to make improvements to their homes; 72% reported being able to consume meat more than once every 10 days, compared to only 50% for non-participants" (Perez-Aleman and Sandilands 2008). However, the cost of the technical assistance to the farmers and the dependency on external funding made it difficult to sustain and further scale up the program, therefore additional incentive and funding mechanisms were needed.

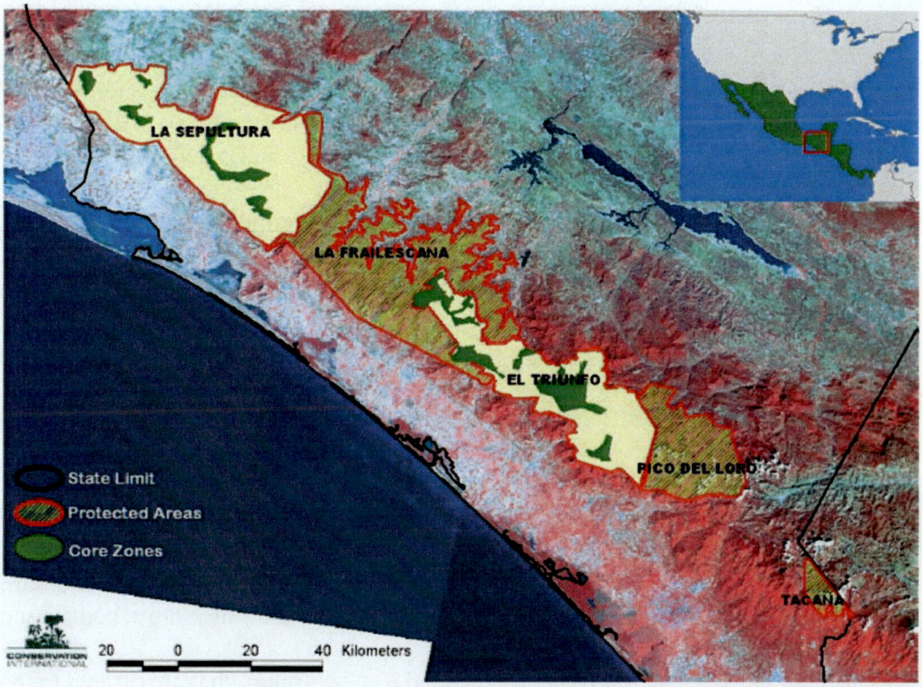

Fig. 1 The protected areas system of the Sierra Madre de Chiapas, Mexico (From Schroth et al. 2009)

Practices such as the use of complex shade canopies in coffee and the conservation of forest, beside protecting biodiversity, also increase (or maintain) the C stocks of farms and landscapes. Therefore, in the mid 2000s, when C markets were slowly emerging as an option to generate additional farmer income, it seemed logical to pursue the integration of the coffee program with C sequestration, thereby diversifying the incentives offered to farmers for C and biodiversity friendly land use practices. A second reason for pursuing this integration was that coffee growing, although a very important land use in the higher parts of the Sierra Madre, is by no means the only one. Highland farmers also grow annual crops such as maize and beans (*Phaseolus* sp.) in fallow rotations, raise small numbers of cattle, and extract forest products such as *xate* palm leaves (*Chamaedorea* sp.) that are used in decoration. These other land uses, which become relatively more important towards the lower and northern parts of the Sierra Madre, were practically not affected by the coffee program, and this reduced its potential to impact the landscape as a whole. For example, conservation and C benefits obtained in the coffee farms could partly be offset by fire use in pasture areas or clearing of secondary forest for food crops by non-participating farmers. This was particularly evident in the drought year 1998 when a total area of 37,336 ha or 22% of La Sepultura Biosphere Reserve was affected by wildfires, including 20% crown fires, that destroyed forest and coffee farms (Schroth et al. 2009).

In 2007, Conservation International therefore partnered with a local NGO, Ambio, which managed the *Scolel Té* (Tzeltal-Mayan for "The growing tree") program in other parts of Chiapas and Oaxaca, already benefiting several hundred families. This program, which is described in detail in this volume by Ruiz-De-Oña-Plaza et al. (2011), connects small farmers to voluntary C markets through a participatory, community based land use planning process where farmers can choose among a number of "modules" (such as planting additional shade trees in coffee, establishing live fences, reforestation of pasture, fire damaged or landslide areas, etc.) that generate C benefits and implement them with the help and under the monitoring of the program (Torres et al. 2010; Zepeda et al. 2010). The *Scolel Té* program began in 1994 as an academic project by the research and higher education institution *El Colegio de la Frontera Sur* (ECOSUR) in San Cristobal de las Casas, in partnership with the University of Edinburgh and Pajal Yakaktic, a farmer organization based in Chiapas. At that time, a participatory diagnosis and design study was carried out to select the main technical interventions and determine the associated costs as well as the social organization required for their implementation. As a result, the cooperative Ambio was created. Starting in 1997, Ambio negotiated C credits from the project on the voluntary market using the "Plan Vivo" standard (www.planvivo.org) that had been developed for *Scolel Té* but is now being used globally (Torres et al. 2010).

The *Scolel Té* approach is flexible. The land use "modules" that are offered to the communities are treated as initial suggestions and are later often modified and adapted to the farmers' specific needs, for example by maintaining colonizer trees among planted trees or grazing cattle under trees. Another important advantage of the approach is that it leads to early payments to farmers, with installments in years 1, 2, 3, 5 and 8 subject to the results of a continuous monitoring and technical support program, while other methodologies often involve a delay of many years between project design and implementation, credit sales, and actual payments to the land users which is a disincentive to small farmers (Torres et al. 2010; Ruiz-De-Oña-Plaza et al. 2011).

For the pilot project integrating Conservation International's coffee program with the *Scolel Té* C sequestration program, initially eight coffee communities were chosen. Although costly in logistical terms, the eight pilot communities were widely spread across the Sierra Madre so that the approach could be tested under a range of biophysical and socioeconomic site conditions. Areas where the reserve administration (the National Commission of Protected Areas) perceived a high risk of land use change either from forest to agriculture or from coffee to annual food crops were prioritized. Through a participatory process, the communities were familiarized with the Plan Vivo methodology and the various land use modules that would generate C credits. Within the first 2 years of the project, 144 farmers participated in the capacity building process of which finally 54 planted trees on a total of 57.25 ha of land, opting mostly (83% of the area) for live fences in pastures or sometimes in coffee. Live fences are an agroforestry technique that can easily be integrated into the local farming systems without negatively affecting crop or pasture yields, thus presenting low opportunity costs, but can sequester non-negligible amounts of C

(28–54 Mg per ha depending on the site; Torres et al. 2010) and generate significant positive impacts in terms of biodiversity conservation (Harvey et al. 2004). On the other hand, and in line with expectation, farmers opted less frequently (14.8% of the area) for increasing the shade canopies in their coffee farms because they already used very dense shade and further increasing it could have augmented disease pressure and compromised the coffee yields. The remaining 2.2% of the area were used for improved fallow plantings. Based on these choices and estimated growth rates, the C income of the participating farmers was estimated at USD 295 over the next 5 years. In 2009, the program forward-sold the first C credits.

Building on this initial pilot phase, the project is now being scaled up to 19 communities in the Sierra Madre, while more farmers are joining in already participating communities. By end 2010, an additional 176 farmers had committed to planting 376 ha, with an even stronger preference for live fences (90% of the committed area). With the expansion of the program, it is hoped that eventually a critical mass will be reached where further growth will be less dependent on external funding. This scaling-up in the field needs to go hand in hand with scaling-up of marketing efforts for the C credits, as well as the sustainably produced coffee, to avoid future bottlenecks. Experience will show if there are synergies on the market in advertising both sustainably grown commodities and C credits with a strong social component from the same landscape, and if this will eventually lead to the recognition of the Sierra Madre as a "sustainable origin" or "sustainable landscape" to help distinguish its products in an increasingly crowded marketplace for certified or otherwise "special" products.

The design process of this project integrating conservation agriculture with C trading also revealed strong links between climate change mitigation and adaptation and the role that agroforestry practices can play in both (Schroth et al. 2009). Beside the predicted increase in temperatures, which may negatively affect coffee quality and thus its value on the market, the increasing risk of extreme climate events, including rainstorms as well as droughts, is a particular concern for a region that has been severely affected in the past by hurricanes and wildfires (Schroth et al. 2009). Significantly, recent research has shown that complex vegetation, such as forest and shade coffee, reduces the vulnerability of farmland to landslides (Philpott et al. 2008). On the other hand, fire management plans and reforestation of pasture land, where many wildfires originate, increase C sequestration and reduce the risk of accidental C losses during drought years (Schroth et al. 2009).

Case Study 2 – North Sumatra, Indonesia

This case study is focused on illustrating the practical role of agroforestry in improving livelihoods and reducing deforestation in the context of the Indonesian coffee sector. The study is being undertaken in the highlands of North Sumatra, Indonesia. Sumatra is the third largest island in Indonesia, measuring 1,800 by 400 km. It contains an extraordinary wealth of natural resources and habitat diversity (Whitten et al. 2000), which are crucial for maintaining the welfare of the island's 50 million people. Dairi district with its capital Sidikalang is one of the key coffee growing areas of North Sumatra, with an annual production of 9,300 Mg on about 20,000 ha, out of about 80,000 ha under coffee for all of North Sumatra. It is situated adjacent to Lake Toba, the largest volcanic lake on earth, location of two hydropower plants (Asahan and Lae Renum) that are critical for Sumatra's power supply, and a key freshwater biodiversity area (Fig. 2).

The margins of Lake Toba are largely deforested and sedimentation and agricultural runoff are impacting the ecology of the lake. The sedimentation rates are estimated at 1–3 cm per year according to the Indonesian Environment Ministry.[1] Only the western margin is still under protection forest, most of which is in Dairi district (Fig. 2). According to the district administration, Dairi has lost 60% of the vegetation of its water catchments due to deforestation. This often involved the encroachment of protection forests, including for coffee farming. In quantitative terms, this deforestation is relatively insignificant compared to the deforestation that has affected Sumatra's lowlands over the past 15 years. However, considering the already small remaining area and ecologically sensitive role of the protection forest in the Lake Toba watershed, further forest loss in this area is clearly a concern. Furthermore, although the conversion of production or protection forest into agricultural land often happened many years ago, the production of coffee (and other agricultural products) on encroached land is formally illegal and, according to field interviews, subjects the farmers to occasional fines. With an increasing emphasis on traceability in the global (and Indonesian) coffee industry, it also complicates the access to higher-paying specialty coffee markets for these communities (Arifin et al. 2008; Neilson 2008). In fact, the project presented here was partly motivated by the concerns of Starbucks coffee buyers who had witnessed forest conversion for coffee farms in this important coffee supplying region (C. Jordan, 2005, personal communication to G.S.).

Based on field surveys in 2005 in communities along the forest boundary and interviews with government officials in Dairi district by Conservation International and its local partner, the fair-trade company ForesTrade-Indonesia, the project focused on four coffee communities, Barisan Nauli, Sileu-leu Parsaroan, Pagambiran, and Perjuangan (Fig. 2). As in other similar studies of the causes of forest encroachment and seemingly unsustainable use of forest resources in Sumatra (Arifin et al. 2008; McCarthy 2006), the interviews revealed a complex set of factors driving the encroachment. These included an influx of migrants, their allocation of land at the forest boundary by resident relatives possibly with the intention of expanding the agricultural area of the communities, lack of clarity about the exact location of the legal boundary between agricultural and protection forest land among the communities, and lack of enforcement of forest protection laws by the authorities. They also included lack of technical support and agrochemical inputs to help farmers establish coffee farms on degraded grassland – of which large areas are available – instead of

[1] Jakarta Post, 15 May 2010.

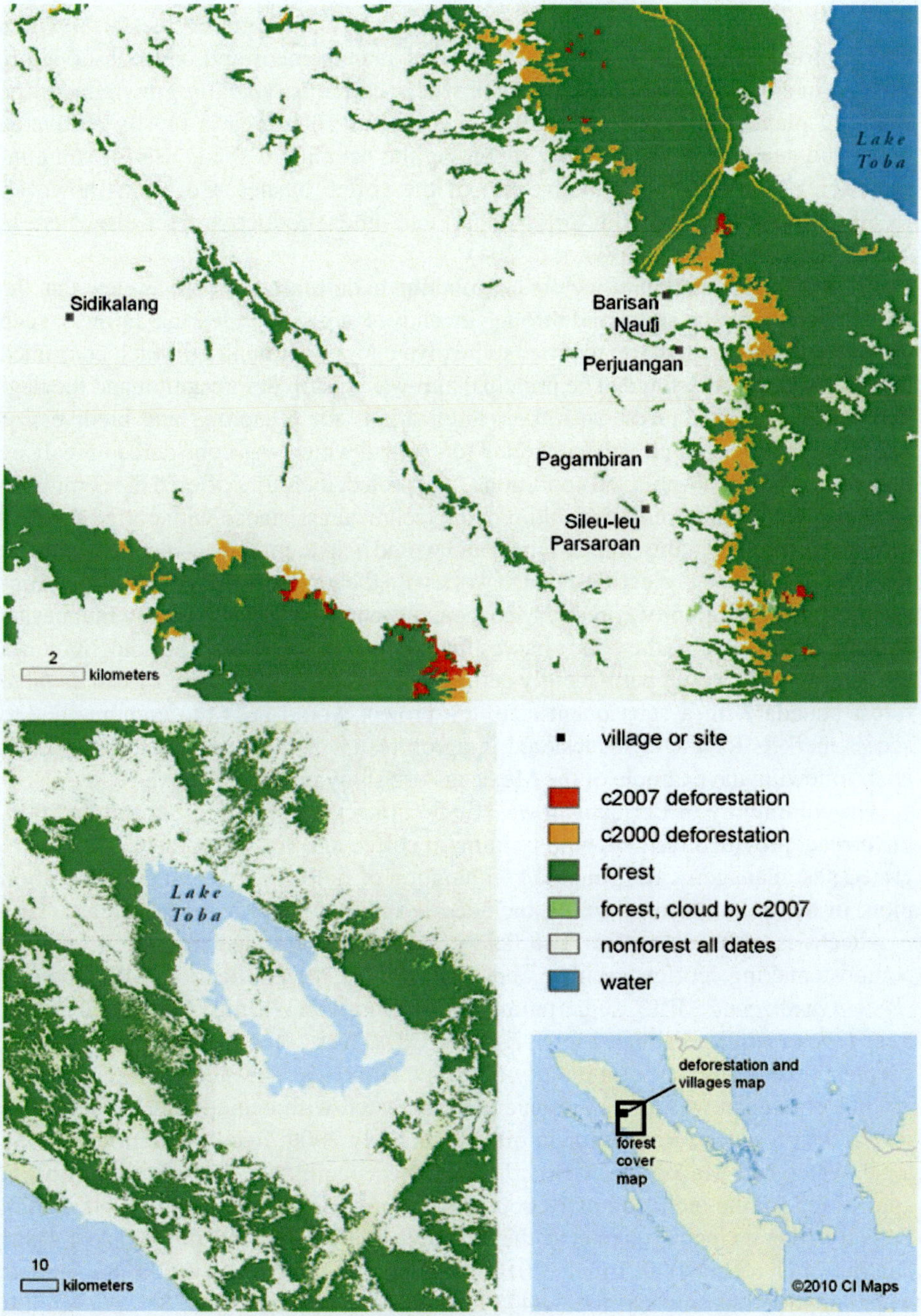

Fig. 2 Coffee communities on the boundary of a forest protecting the watershed of Lake Toba in North Sumatra, Indonesia (map by Kellee Koenig)

the more fertile forest soil, and to maintain and rehabilitate coffee farms when coffee productivity has decreased for lack of management and soil conservation. Finally, in contrast to neighboring Aceh where coffee is typically grown under the shade of planted legume trees, coffee in the Lake Toba area is mostly cultivated under full sun conditions, thereby foregoing the benefits of shade use for soil conservation and a longer productive life of the coffee bushes. Coffee is, however, locally grown under benzoin (*Styrax* spp.) and other productive trees elsewhere in North Sumatra (Garcia Fernandez 2004).

From the constellation of factors contributing to deforestation, it was clear that the problem could not be addressed through incentives targeting individual farmers, such as conventional certification, but had to involve entire communities which controlled the access to the forestland. The principal aim was to stop deforestation and the degradation of standing forest with their implications for C storage and biodiversity, rather than reforesting already converted forest land which was considered unrealistic under the given socio-political conditions. The project therefore offered the communities an agreement whereby it would provide technical assistance with coffee agroforestry practices and quality improvement and would help them access specialty markets for their coffee. The project would also work with the government to include the communities in a Community Forestry Management scheme as established by Indonesian law that would legalize the coffee harvesting though not the sale of the land. In return, the communities would permanently demarcate and monitor a jointly agreed *de facto* forest boundary. In a subsequent step, the project would help the communities to access markets for the C sequestered in agroforestry plantings on the former forest land, following the example of the Mexican case study presented above.

The community of Perjuangan was the first to accept the offer. Starting in 2006, the project provided farmers with training in coffee agroforestry practices and integrated pest management through a combination of field schools and demonstration plots in the communities, which included a strong focus on the advantages and practicalities of shade coffee. The field schools also provided training in organic compost making, coffee pruning and management, diversification with avocado (*Persea americana* Mill.), sugar palm (*Arenga pinnata* (Wurmb) Merr.) and timber trees (*Toona sureni* (Blume) Merr., *T. sinensis* (A. Juss.) Roem.), post harvest management, marketing, and community-managed nurseries. The *de facto* forest boundary of Perjuangan was agreed upon and demarcated with cement blocks and a tree row in the presence of government officials in early 2009. In collaboration with the local Watershed Management Board, the project facilitated the submission of an application for the inclusion of 10,000 ha of former forest land into the Community Forestry Management scheme to the Minister of Forestry and the Head of Dairi District in October 2009. In late 2010, these community conservation agreements were supported by approximately 475 households in Pagambiran, 340 households in Sileu-leu Parsaoran, around 370 households in Perjuangan, and 24 members of the local farmer group in Barisan Nauli, which has a population of approximately 2000. In 2010, the 280 farmers of Sileu-leu Parsaoran managed to arrange their first sale of coffee to a major exporter, demonstrating their increased capacity to produce a quality product, and connect that product with the international market.

The expected declaration of the converted forest land as "community forest" will allow the communities not only to legally harvest their coffee, but also to directly benefit from C revenues that would be generated through agroforestry plantings and forest conservation on this land. However, to enable the C revenues to flow, there are still legal barriers since the legal framework of C trading in Indonesia does not yet permit smallholder participation in the global voluntary C markets, though promising advances in the respective policy discussions have been made, for example through the publication of the draft "National Strategy for the Reduction of Emissions from Deforestation and Forest Degradation (Nastra REDD plus)" in 2010.

Case Study 3 – Tapajós-Arapiuns Extractive Reserve, Brazilian Amazon

In this section we discuss the contribution of agroforestry and a specific Brazilian market for reforestation credits to increasing the ecological and economic viability, thereby helping to conserve its forest C stocks and biodiversity. The study area is the Tapajós-Arapiuns Extractive Reserve in western Pará State, Brazilian Amazon. Similar to the previous case study, C trading is not (yet) a component of the project which uses another environmental service market to create incentives and rewards for practices that, while mainly targeting sustainable timber and non-timber supplies, also conserve landscape C stocks and biodiversity.

Protected areas, including sustainable use reserves and indigenous lands, are widely recognized for their contribution to reducing deforestation and forest degradation in the Amazon (Nepstad et al. 2006). They play a key role in the conservation of the biodiversity and C stocks of this largest of tropical forests. Extractive reserves as a specific form of inhabited protected areas were created as a response to the conflicts between traditional rubber tappers and expanding cattle ranches in the western Brazilian Amazon during the 1980s (Cardoso 2002). The first extractive reserve was created in 1990 and today there are more than 11 million ha of extractive reserves and more than ten million ha of (closely related) sustainable development reserves in the Brazilian Amazon (ISA 2007), forming large-scale corridors with other forms of protected areas and indigenous lands (Fig. 3). Ruiz-Perez et al. (2005) demonstrated the effectiveness of the Alto Juruá Extractive Reserve in the western Amazonian state of Acre in reducing deforestation rates compared to surrounding areas.

The concept of extractive reserves was criticized early on based on the commonly held view that extractivism in species-rich tropical forest is rarely a way out of poverty (Homma 1993). However, the economic basis of "extractive reserves" is not always extractivism, but may be family agriculture and agroforestry complemented by fishing and some hunting. Such is the case in the Tapajós-Arapiuns Extractive Reserve, an area of 650,000 ha with approximately 20,000 inhabitants in about 70 communities located mostly on the banks of the Tapajós and Arapiuns

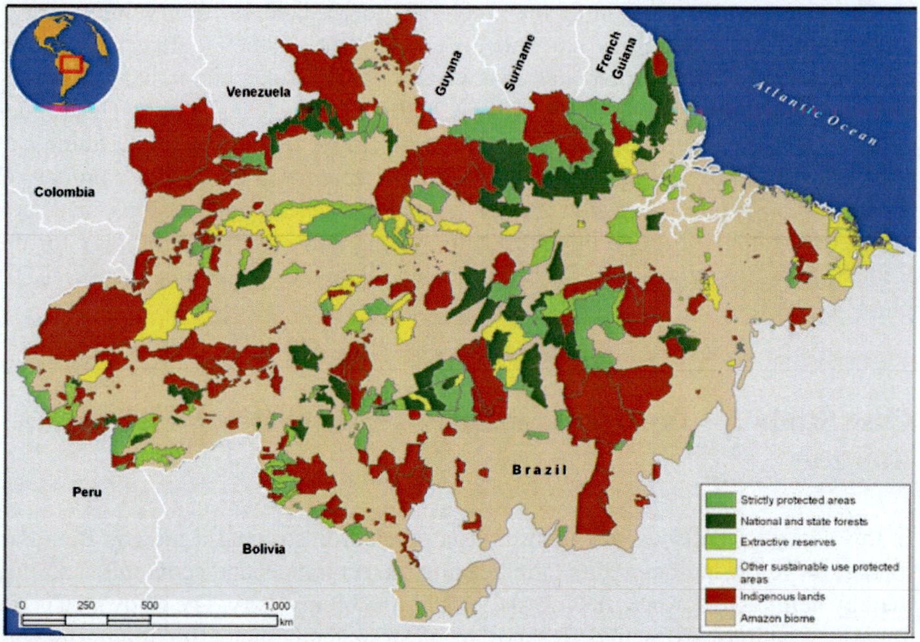

Fig. 3 Protected areas in the Brazilian Amazon. National forests, state forests and extractive reserves are different types of sustainable use protected areas (map by Luis Barbosa)

rivers in western Pará state (Fig. 4). The area was protected in 1998 after a history of conflict between communities and logging companies (Menton 2003).

The Tapajós valley has a strong agroforestry tradition, reaching back at least to the early twentieth century (Schroth et al. 2003). Farmers used to plant rubber seeds (*Hevea brasiliensis* H.B.K. M.-Arg.) in their slash-and-burn plots together with their staple cassava (*Manihot esculenta* Crantz) and some other food crops. After the second cassava harvest, the rubber saplings were tended for a few years until they could cope with the evolving fallow vegetation. The plot was then abandoned until the rubber trees had reached sufficient size for tapping at an age of 7–14 years (the earlier dates indicating that very small trees were tapped). At that time, paths connecting the rubber trees were cleared in these plantations turned secondary forests, and the trees were tapped about twice per week during the rainy season and allowed to rest during the dry season when the latex flow is reduced. Unlike in the very similar and better documented Indonesian rubber agroforests (Michon and de Foresta 1999), neither detailed biodiversity nor C studies have been carried out in the traditional rubber agroforests of the Tapajós. However, these agroforests can reach an age of more than 50 years, are structurally complex, and are responsible for the almost continuous tree cover of the banks of the Tapajós and Arapiuns rivers, merging further inland into secondary and old-growth forests (Fig. 4). They store significant amounts of C (a sample of eight rubber agroforests on the eastern river bank

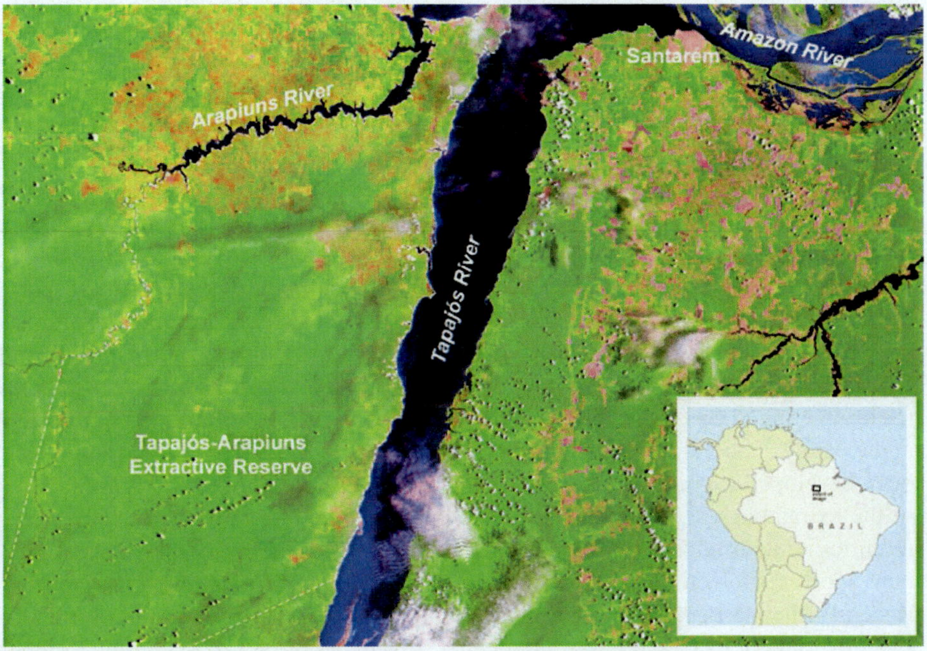

Fig. 4 Location of the Tapajós-Arapiuns Extractive Reserve on the left bank of the Tapajós River in western Pará state, Brazilian Amazon. On the satellite image, the more disturbed "agroforestry zones" along the river banks can be distinguished from the relatively intact interior of the reserve. The contrast to the unprotected area south of the city of Santarém is clearly visible (map by Kellee Koenig)

had estimated C stocks in the aboveground tree biomass of 143 (S.D.= 63) Mg ha^{-1}; G. Schroth, unpublished data) and provide wildlife populations of the adjacent forest with seasonal food resources such as rubber seeds and fruits from interspersed fruit trees (Schroth et al. 2003). However, their main potential from a landscape point of view in this "forest frontier situation" would be to offer a sustainable source of income to their owners, thereby increasing the economic viability and ecological integrity of the reserve as a whole. Unfortunately, in recent years low and fluctuating prices of rubber latex and sometimes even difficulties to sell latex in the region have meant that most rubber agroforests at the Tapajós are now abandoned and the cultivation of cassava in slash-and-burn systems for self-consumption and sale of roasted flour (*farinha*) on the market in Santarém has become the main activity and source of income of the reserve inhabitants. This strong reliance on slash-and-burn agriculture in a seasonally fire-prone region is considered undesirable by the reserve administration. With the growing human population, it risks driving an inland expansion of the agricultural frontier within the reserve, and to potentially increase deforestation, human-wildlife conflict, and hunting (Carvalho and Pezzuti 2010). Furthermore, community meetings revealed that cassava growing and roasting are considered much heavier and unhealthier work by the often elderly reserve inhabitants, compared to rubber tapping in the shade.

These traditional rubber agroforests are being planted from seeds, and although some trees in a plantation can be very productive, most trees are of low productivity (Schroth et al. 2004b). A government sponsored attempt in the late 1990s to provide the communities with grafted saplings has been unsuccessful owing to high mortality, apparently related to logistical problems in seedling distribution and insufficient technical assistance (species of several other tree crop species suffered a similar fate). However, even in the surroundings of Belterra and Aramanaí on the eastern river bank, where farmers had, until recently, ready access to grafting technology from commercial rubber plantations, grafted rubber plantings are the exception rather than the rule, suggesting that the use of seeds to establish rubber agroforests is a conscious strategy to reduce costs and risks, both ecological ones from fire and drought, and economic ones from unpredictable markets and government policies (Schroth et al. 2003). A technological package to improve the productivity of new and existing rubber agroforests without interfering with their "low input logic" has been developed (Schroth et al. 2004b) and presented to more than 20 communities in the extractive reserve in 2004. But despite considerable interest from the communities it had limited impact given that with the low rubber prices of the time, few new agroforests were established and most existing agroforests were not tapped. Recently, the market price of natural rubber has increased and, in addition, the federal government of Brazil has announced a subsidized price for Amazonian rubber (Brasil 2010) which could dramatically change this situation, but this subsidy has not yet become available to the communities in the reserve at the time of writing.

Parallel to these efforts to revitalize rubber agroforestry, a new agroforestry practice was introduced in cooperation with the reserve administration over the past few years. In Brazil, the federal "forest supply" (*fomento florestal*) law of 1996 requires that companies that use wood from unsustainable sources (such as forest conversion) replant a proportional number of trees, with the long-term objective of creating a closed cycle of wood use and planting, thereby reducing the pressure on natural forests. In subsequent directives, a ratio of eight trees per cubic meter of wood has been established. In 2003, the environmental authorities decided to implement this legislation systematically in the Amazon and needed suppliers of reforestation credits that could be purchased by those wood consumers that had no own land or technical capacity to conduct reforestation operations to offset their consumption. The idea thus arose to technically and legally enable community organizations in the Tapajós-Arapiuns Extractive Reserve to reforest with native trees areas of reserve land that had been deforested through slash-and-burn agriculture or earlier logging and sell reforestation credits (not trees) to wood consuming companies in the region. This seemed to be a "win-win" approach that would allow the communities over the short term to earn additional income from credit sales and on the longer term to redirect their activities from slash-and-burn agriculture to the commercialization of non-timber and timber products from individually owned and registered single or mixed species plantations of commercial tree species. The environmental authorities, on the other hand, would increase their supply of credits that they could leverage to compel wood consuming companies to comply with the legislation while at the same time addressing the problem of insufficient land use options in the reserve.

The *fomento florestal* law did not explicitly mention the possibility of communities reforesting federal land (such as extractive reserves), but a request for clarification from the project to the Directorate of Forests in Brasília obtained a positive response, as long as the reforestation activity was not in conflict with the management plan of the reserve. Following an onerous administrative process that took well over a year and involved government agencies at the region, state, and federal levels, a previously existing association of five communities, which had been chosen for this project for its dynamic leadership, critical mass of members interested in reforestation, and existence of a (dysfunctional) community nursery, received the official authorization to administer reforestation projects in the reserve under this law. This was to our knowledge the first time that such an authorization was obtained by a community based organization in an extractive reserve in Brazil. While this administrative process was ongoing, inhabitants of the five communities as well as some neighboring communities produced timber tree seedlings in the community nursery and planted them in their slash-and-burn plots, as they previously used to do with rubber seeds and seedlings. In 2006 and 2007, the community association completed three credit sales totaling about R$ 25,000 (~USD 15,000) the returns of which were distributed to the participants, used to produce more seedlings in the community nurseries (which mostly employed women from the communities) and to cover the costs of administration and technical assistance. These sales increased substantially the demand for the new agroforestry practice among the participating and other communities in the reserve. In 2008, the project won support from the World Bank through its annual Development Marketplace competition which allowed the project to be scaled up to presently over 350 families from 46 communities, including some in very remote parts of the reserve that had rarely been reached by earlier projects. Although by choice of the communities all plantings were individually owned, the seedlings were produced in community nurseries and so community organization and technical support to communal work absorbed a large share of the project's resources. At the time of writing, the main challenge of the project was that in the course of the decentralization of the Brazilian forest administration, the responsibility for the implementation of the *fomento florestal* law had been shifted from the federal to state agencies, temporarily interrupting the community organizations' access to the reforestation credit market. This problem was being addressed through discussions with government agencies at different levels. While the outcome of these discussions is difficult to predict, it should be noted that the credit sales, while an important encouragement to the communities, are only part of the benefits they receive, the more important long-term ones being the creation of a basis for a reserve economy founded, once again, on tree products.

This project, like those described in the previous case studies, has the dual objectives of conserving the forest resources of the extractive reserve with its C stocks and biodiversity, and to improve the livelihoods of its inhabitants in a sustainable way. Although among its objectives are forest conservation, the partial substitution of slash-and-burn agriculture with tree based land uses and reforestation, the focus of the project is not on the (international) trading of C credits, but rather on an emerging domestic market for reforestation credits that is little known even in

Brazil. While the *fomento florestal* law and its directives established rigorous accreditation criteria and procedures for those who wish to offer reforestation credits, resulting in a complex administrative process that could not be managed by communities without competent external support, the process is still easier and faster than that involved in the development of most C trading projects. Given the potential size of the market (eight trees per cubic meter of unsustainably extracted wood, a broad category covering all wood derived from forest conversion and logging or fuelwood harvesting without sustainable management plan), the mainstreaming of the approach in Brazilian reserve management and forestry policy and practice would result in a significant boost to agroforestry and community forestry as components in the management of sustainable use reserves, thereby protecting their C and biodiversity resources and the livelihoods of their inhabitants. Moreover, due to the lower opportunity costs of land in reserve compared to non-reserve areas, reforestation credits are among the very few products for which sustainably managed reserves have a comparative advantage on the market.

Conclusions

The case studies have shown a number of ways how agroforestry can contribute to linking C storage through forest conservation and reforestation, with their (assumed) benefits for biodiversity conservation, and livelihoods improvement. The sequestration of C in the tree biomass of diverse and structurally complex land use systems is only one such role, although the one that has received the most attention in the literature. It is most important in mosaic landscapes, where natural forests have been reduced to fragments, thereby increasing the relative contribution of farm land to landscape C stocks and biodiversity. The C and biodiversity rich shade-cacao systems (*cabruca*) that make up the majority of the "forest" cover in the extremely biodiversity rich landscape of southeastern Bahia, Brazil, and the shade coffee systems of the Sierra Madre de Chiapas, Mexico, are prime examples for traditional agroforestry practices that combine environmental and livelihood functions. In forest frontier situations, on the other hand, the highest priority in a strategy linking agricultural development with climate change mitigation and biodiversity conservation must be to minimize the need for forest conversion by enabling land users to obtain an adequate and sustainable income from their land, thereby supporting direct forest conservation policies, while the C storage and biodiversity in the farming system itself, although highly desirable, are of secondary importance.

The three case studies have shown that agroforestry can play a key role in stabilizing land use mosaics (as in the Sierra Madre de Chiapas) and forest frontiers (as in North Sumatra and the Amazon) by linking land use policies with incentives based on commodity markets and various types of environmental service markets. Case studies 1 and 3 emphasized inhabited, sustainable use protected areas as a key policy tool for stabilizing forest landscapes that has been widely used in Latin America and that provides a useful framework for such mixed incentives. While in case study 1, C sequestered in agroforestry systems was directly traded, case study 3 illustrated that other types of environmental service markets or rewards, including ones operating on the sub-national or local level, may also result in positive outcomes in terms of C and biodiversity conservation and additional income for land users. Case study 2 from Indonesia where land conflicts are commonplace showed that the regularization of land tenure and use rights of forest boundary communities was a key topic in the stabilization of the forest frontier. It is being addressed as part of a package that also involves the production of an agricultural commodity (Arabica coffee) in agroforestry systems for specialty markets and sets the stage for a subsequent inclusion of C trading after clarification of the legal framework.

While all three case studies describe work in progress, several key lessons emerge. The first lesson is that the combination of different incentives can result in better and more lasting outcomes in terms of land use change than single types of incentives, such as C trading or certification, provided that the institutions to coordinate such complex incentive mechanisms are in place. Different incentives for similar land use practices (such as C and biodiversity conservation) can add up and reinforce each other. Also, through the bundling of several types of incentives, different groups of land users within the same landscape, such as coffee and cattle producers or recent migrants and established farmers, can be targeted simultaneously. The bundling of incentives also reduces the risk of losing past achievements and the trust of communities if one incentive becomes temporarily unavailable or less attractive, e.g., through policy changes or price fluctuations in commodity or environmental service markets. And finally, one incentive, such as the regularization of the land tenure situation, may be a precondition for other incentives, such as the participation in premium commodity and environmental service markets. This lesson implies that the standard requirement of additionality in land based C projects (i.e. the requirement that the land use change that a project intends to bring about would not happen in the absence of the C payment) should be carefully balanced with the goal of project sustainability, which will usually be greater if the intended land use change does not depend on C payments alone.

The second major lesson is that when designing incentive programs for agroforestry practices, their impacts must be considered at the scale of the landscape and must include plot or farm level effects as well as interactions between land use and natural vegetation with its typically high C stocks and biodiversity. Especially, land use systems cannot be considered "biodiversity- (or carbon-) friendly" if their low yields and the low incomes they generate drive additional conversion of natural habitat. Where the conversion of natural habitat for agriculture is unavoidable (and permitted by law), the primary preoccupation must be that it results in efficiently used, sustainable, and productive land use systems so that the need for further conversion is minimized. Contributing to these objectives is agroforestry's most important role in C and biodiversity conservation at the agriculture-forest frontier.

Finally, agricultural development policies and forest conservation policies should be better integrated to avoid contradictions and achieve locally, nationally, and globally desirable outcomes in terms of development and environmental conservation. The probability that this lesson will be learned and applied in current frontier development regions, such as Central Africa and the Amazon, increases with the development of markets for C sequestration and other environmental services and the growing emphasis on sustainable production practices in international agricultural commodity markets.

Acknowledgements Project work mentioned in this paper was funded by grants of Starbucks Coffee Company to Conservation International and the World Bank Development Marketplace to the Association of the Rural Extractivist Producers of the Left Margin of the Tapajós (APRUSPEBRAS). We thank Luis Barbosa for Fig. 3 and Kellee Koenig for Figs. 2 and 4. Jeff Sayer and Jeff McNeely made valuable comments on an earlier draft of the paper.

References

Arifin B, Geddes R, Ismono H, Neilson J, Pritchard B (2008) Farming at Indonesia's Forest Frontier: understanding incentives for smallholders. Australian National University, Canberra, 14 p. Available at http://www.aigrp.anu.edu.au/docs/projects/1017/neilson_brief.pdf. Accessed Jan 2011

Barbier EB (2005) Natural resources and economic development. Cambridge University Press, Cambridge, 410 p

Beer J, Muschler R, Kass D, Somarriba E (1998) Shade management in coffee and cacao plantations. Agroforest Syst 38:139–164

Brasil (2010) Plano Agrícola e Pecuário 2010–2011. Ministério da Agricultura, Pecuária e Abastecimento, Brasilia, 48 p. Available at http://www.irga.rs.gov.br/uploads/anexos/1290195013Planos_Agricolas.pdf. Accessed Jan 2011

Brown DR, Dettmann P, Rinaudo T, Tefera H (2011) Poverty alleviation and environmental restoration using the Clean Development Mechanism – a case study from Humbo Ethiopia. Environ Manage (in press)

Cardoso CAS (2002) Extractive reserves in Brazilian Amazonia: local resource management and the global political economy. Ashgate, Aldershot, 274 p

Carvalho EAR and Pezzuti JCB (2010) Hunting of jaguars and pumas in the Tapajós-Arapiuns Extractive Reserve, Brazilian Amazonia. Oryx 44:610–612

Cassano CR, Schroth G, Faria D, Delabie JHC, Bede L (2009) Landscape and farm scale management to enhance biodiversity conservation in the cocoa producing region of southern Bahia, Brazil. Biodivers Conserv 18:577–603

Chomitz KM, Buys P, DeLuca G, Thomas TS, Wertz-Kanounnikoff S (2006) At Loggerheads? Agricultural expansion, poverty reduction, and environment in the Tropical Forests. World Bank Policy Research Report.Washington, DC. Available at www.worldbank.org/tropicalforestreport. Accessed Jan 2011

Ewers RM, Scharlemann JPW, Balmford A, Green RE (2009) Do increases in agricultural yield spare land for nature? Glob Change Biol 15:1716–1726

Faria D, Paciencia MLB, Dixo MBO, Laps RR, Baumgarten J (2007) Ferns, frogs, lizards, birds and bats in forest fragments and shade cacao plantations in two contrasting landscapes in the Atlantic forest, Brazil. Biodivers Conserv 16:2335–2357

Gama-Rodrigues EF, Gama-Rodrigues AC, Nair PKR (2011) Soil carbon sequestration in cacao agroforestry systems: a case study from Bahia, Brazil. In: Kumar BM, Nair PKR (eds.) Carbon sequestration potential of agroforestry systems: opportunities and challenges. Springer, Dordrecht, pp 85–99

Garcia Fernandez C (2004) *Benzoin,* a resin produced by *Styrax* trees in North Sumatra Province, Indonesia. In: Kusters K and Belcher B (eds) Forest products, livelihoods and conservation: case studies of non-timber forest product systems, vol 1: Asia. Center for International Forestry Research, Bogor, pp 151–168

Ghazoul J, Garcia C, Kushalappa CG (2009) Landscape labelling: a concept for next-generation payment for ecosystem service schemes. For Ecol Manag 258:1889–1895

Gockowski J, Sonwa DJ (2011) Cocoa intensification scenarios and their predicted impact on climate change, biodiversity conservation, and rural livelihoods in the Guinea rainforest of West Africa. Environ Manage (in press)

Gockowski J, Tchatat M, Dondjang J-P, Hietet G, Fouda T (2010) An empirical analysis of the biodiversity and economic returns to cocoa agroforests in Southern Cameroon. J Sustain For 29:638–670

Green RE, Cornell SJ, Scharlemann JPW, Balmford A (2005) Farming and the fate of wild nature. Science 307:550–557

Harvey CA, Tucker NIJ, Estrada A (2004) Live fences, isolated trees, and windbreaks: tools for conserving biodiversity in fragmented tropical landscapes. In: Schroth G, Fonseca GAB, Harvey CA, Gascon C, Vasconcelos HL, Izac A-MN (eds) Agroforestry and biodiversity conservation in Tropical Landscapes. Island Press, Washington, pp 261–289

Homma AKO (1993) Extrativismo Vegetal na Amazônia – Limites e Oportunidades. Embrapa-SPI, Brasilia, 202 p

ISA (2007) Amazônia Brasileira – areas protegidas (map). Instituto Socioambiental, São Paulo. Available at http://www.socioambiental.org/esp/desmatamento/site/. Accessed Jan 2011

Kotto-Same J, Woomer PL, Appolinaire M, Zapfack L (1997) Carbon dynamics in slash-and-burn agriculture and land use alternatives of the humid forest zone in Cameroon. Agric Ecosyst Environ 65:245–256

McCarthy JF (2006) The fourth circle – a political ecology of Sumatra's rainforest frontier. Stanford University Press, Stanford, 353 p

McNeely JA, Schroth G (2006) Agroforestry and biodiversity conservation - traditional practices, present dynamics, and lessons for the future. Biodivers Conserv 15:549–554

Menton MC (2003) Effects of logging on non-timber forest product extraction in the Brazilian Amazon: community perceptions of change. Int For Rev 5:97–105

Michon G, de Foresta H (1999) Agro-forests: incorporating a forest vision in agroforestry. In: Buck LE, Lassoie JP, Fernandes ECM (eds.) Agroforestry in sustainable agricultural systems. Lewis Publishers, Boca Raton, pp 381–406

Montagnini F, Nair PKR (2004) Carbon sequestration: an underexploited environmental benefit of agroforestry systems. Agroforest Syst 61:281–295

Nair PKR, Nair VD, Kumar BM, Showalter JM (2010) Carbon sequestration in agroforestry systems. Adv Agron 108:237–307

Neilson J (2008) Global private regulation and value-chain restructuring in Indonesian smallholder coffee systems. World Dev 36:1607–1622

Nepstad DC, Schwartzman S, Bamberger B, Santilli M, Ray D, Schlesinger P, Lefebvre P, Alencar A, Prinz E, Fiske G, Rolla A (2006) Inhibition of Amazon deforestation and fire by parks and indigenous lands. Conserv Biol 20:65–73

Oliveira LC, Neves LG, Raboy BE, Dietz JM (2011) Abundance of jackfruit (*Artocarpus heterophyllus*) affects group characteristics and use of space by golden-headed lion tamarins (*Leontopithecus chrysomelas*) in cabruca agroforest. Environ Manag (in press)

Perez-Aleman P, Sandilands M (2008) Building value at the top and the bottom of the global supply chain: MNC-NGO partnerships. Calif Manag Rev 51:24–49

Philpott SM, Lin BB, Jha S, Brines SJ (2008) A multi-scale assessment of hurricane impacts on agricultural landscapes based on land use and topographic features. Agric Ecosyst Environ 128:12–20

Rice RA, Greenberg R (2000) Cacao cultivation and the conservation of biological diversity. Ambio 29:167–173

Ruf F, Schroth G (2004) Chocolate forests and monocultures - an historical review of cocoa growing and its conflicting role in tropical deforestation and forest conservation. In: Schroth G,

Fonseca GAB, Harvey CA, Gascon C, Vasconcelos HL, Izac A-MN (eds) Agroforestry and biodiversity conservation in Tropical Landscapes. Island Press, Washington, pp 107–134

Ruiz-De-Oña-Plaza C, Soto-Pinto L, Paladino S, Morales F, Esquivel E (2011) Constructing public policy in a participatory manner: from local carbon sequestration projects to network governance in Chiapas, Mexico. In: Kumar BM, Nair PKR (eds.) Carbon sequestration potential of agroforestry systems: opportunities and challenges. Springer, Dordrecht, pp 247–262

Ruiz-Perez M, Almeida M, Dewi S, Costa EML, Pantoja MC, Puntodewo A, Postigo AA, Andrade AG (2005) Conservation and development in Amazonian extractive reserves: the case of Alto Jurua. Ambio 34:218–223

Schroth G, da Mota MSS (2007) Tropical agroforestry. In: Scherr SJ, McNeely JA (eds.) Farming with nature – the science and practice of ecoagriculture. Island Press, Washington, pp 103–120

Schroth G, Harvey CA (2007) Biodiversity conservation in cocoa production landscapes: an overview. Biodivers Conserv 16:2237–2244

Schroth G, Coutinho P, Moraes VHF, Albernaz AKM (2003) Rubber agroforests at the Tapajós river, Brazilian Amazon - environmentally benign land use systems in an old forest frontier region. Agric Ecosyst Environ 97:151–165

Schroth G, Harvey CA, Vincent G (2004a) Complex agroforests - their structure, diversity, and potential role in landscape conservation. In: Schroth G, Fonseca GAB, Harvey CA, Gascon C, Vasconcelos HL, Izac A-MN (eds.) Agroforestry and biodiversity conservation in Tropical Landscapes. Island Press, Washington, DC, pp 227–260

Schroth G, Moraes VHF, da Mota MSS (2004b) Increasing the profitability of traditional, planted rubber agroforests at the Tapajós river, Brazilian Amazon. Agric Ecosyst Environ 102:319–339

Schroth G, Laderach P, Dempewolf J, Philpott SM, Haggar JP, Eakin H, Castillejos T, Garcia-Moreno J, Soto-Pinto L, Hernandez R, Eitzinger A, Ramirez-Villegas J (2009) Towards a climate change adaptation strategy for coffee communities and ecosystems in the Sierra Madre de Chiapas, Mexico. Mitig Adapt Strat Glob Change 14:605–625

Siebert S (2002) From shade- to sun-grown perennial crops in Sulawesi, Indonesia: implications for biodiversity conservation and soil fertility. Biodivers Conserv 11:1889–1902

Sonwa DJ, Nkongmeneck BA, Weise SF, Tchatat M, Adesina AA, Janssens MJJ (2007) Diversity of plants in cocoa agroforests in the humid forest zone of Southern Cameroon. Biodivers Conserv 16:2385–2400

Torres AB, Marchant R, Lovett JC, Smart JCR, Tipper R (2010) Analysis of the carbon sequestration costs of afforestation and reforestation agroforestry practices and the use of cost curves to evaluate their potential for implementation of climate change mitigation. Ecol Econ 69:469–477

Whitten T, Damanir SJ, Anwar J, Hisyam N (2000) The ecology of Sumatra. Periplus, Singapore, 478 p

Zepeda Y, Schroth G, Hernandez R (2010) Complementary environmental service reward programmes for sustainable mosaic landscapes in the Sierra Madre de Chiapas, Mexico. Mt Forum Bull 2010:80–82

Assessing the Carbon Sequestration in Short Rotation Coppices of *Robinia pseudoacacia* L. on Marginal Sites in Northeast Germany

Ansgar Quinkenstein, Christian Böhm, Eduardo da Silva Matos, Dirk Freese, and Reinhard F. Hüttl

Abstract The assessment of the carbon (C) sequestration potential of different land use systems is receiving increasing attention within the European Union forced by aspects of optimum humus content of soils and the debate on climate change. Short rotation coppice crops (SRC) emerge as a promising land use option both for bioenergy production and C sequestration. In this study, C storage in the biomass and the soil under four SRC systems of *Robinia pseudoacacia* L. was investigated. The plantations were established on reclamation sites in the mining district of Lower Lusatia in 1995, 2005, 2006, and 2007. Samples were collected in the winter of 2007 and 2009. Average aboveground dry matter (DM) production ranged from 0.04 to 9.5 Mg ha^{-1} $year^{-1}$ for 1–14 years of growth, respectively. Total stocks of soil organic carbon (SOC) at 0–60 cm depth after 2 and 14 years of growth were 22.2 ± 11.3 and 106.0 ± 11.7 Mg ha^{-1}, respectively. Interpreting the data as a false chronosequence, the average rate of soil C sequestration in the 0–60 cm layer was 7.0 Mg ha^{-1} $year^{-1}$. Hot water extractable carbon (HWC) that represents the labile fraction of SOC was highest in the oldest plantation (1.4 Mg ha^{-1} for the 0–30 cm layer). The relative proportion of HWC in SOC, however, did not change substantially between different aged SRC, indicating that with time, because of increasing

A. Quinkenstein (✉) • C. Böhm • E. da Silva Matos • D. Freese
Soil Protection and Recultivation, Brandenburg University of Technology, Konrad-Wachsmann-Allee 6, D-03046 Cottbus, Germany
e-mail: quinkenstein@tu-cottbus.de; boehmc@tu-cottbus.de; eduardo.matos@embrapa.br; freese@tu-cottbus.de

R.F. Hüttl
Chair of Soil Protection and Recultivation, Brandenburg University of Technology, Konrad-Wachsmann-Allee 6, D-03046 Cottbus, Germany

Helmholtz Centre Potsdam – GFZ German Research Centre for Geosciences, Telegrafenberg, G 309, D-14473 Potsdam, Germany
e-mail: Reinhard.Huettl@gfz-potsdam.de

B.M. Kumar and P.K.R. Nair (eds.), *Carbon Sequestration Potential of Agroforestry Systems: Opportunities and Challenges*, Advances in Agroforestry 8,
DOI 10.1007/978-94-007-1630-8_11,

stocks, C became increasingly stabilized within the soils. Overall, plantations of *R. pseudoacacia* seem to be a promising land use option for post-mining areas due to their high capacity for sequestering C within biomass as well as a high potential to increase soil C stocks on marginal sites.

Keywords Bioenergy • False chronosequence • Hot water extractable carbon • Post-mining area • Soil carbon stock

Introduction

With the publication of its Fourth Assessment Report, the Intergovernmental Panel on Climate Change (IPCC) emphasized that climate change is one of the most challenging problems presently facing mankind (IPCC 2007). Important strategies for mitigating the negative consequences of rising atmospheric concentrations of greenhouse gases are the development of advanced techniques for sequestering carbon (C) and the substitution of fossil fuels by renewable energy resources to satisfy the demand for energy, heat, and fuel without emitting additional carbon dioxide (CO_2) into the atmosphere. Utilization of woody biomass for bioenergy production is an important component in realizing these aims. Within the woody biomass option, short rotation coppice (SRC) plantations for biomass production feature several environmental advantages, store C, and come increasingly into play on agricultural set aside areas, agricultural marginal lands, and reclamation sites of post-mining landscapes.

Short rotation coppice systems usually consist of fast growing trees such as willow (*Salix* spp. L.), poplar (*Populus* spp. L.) or black locust (*Robinia pseudoacacia* L.) used as woody energy crops. These trees are planted at high densities of up to 12,000 (and sometimes even more) plants per hectare and in a planting pattern that allows for mechanical operations. Following 2–6 years of growth, the sprouts are harvested with specialized machinery during the winter season when the plants are dormant. Because the trees are able to resprout after cutting, several rotations can be taken before yield declines, and the plants need to be replaced. Depending on plant productivity, the SRC plantation can be run for 20–30 years. The key determinants of SRC productivity are water and nutrient availability, weed and pest control, light, and temperature. On fertile sites in the temperate region, annual growth rates between 4 and 14 Mg DM ha^{-1} $year^{-1}$ have been reported for willow and poplar (Mitchell et al. 1999; Scholz and Ellerbrock 2002). On marginal sites with low water availability, however, poplar and willow show comparatively poor growth (Grünewald et al. 2007).

The district of Lower Lusatia in northeast Germany is characterized by ongoing opencast lignite mining activities and the accruing reclamation sites are increasingly used for cultivating woody energy crops. Experiments on reclamation sites have reported growth rates between 1 and 6 Mg DM ha^{-1} $year^{-1}$ for poplar and willow (Bungart and Hüttl 2004; Grünewald et al. 2007). Higher and more stable growth rates of 4–9.5 Mg DM ha^{-1} $year^{-1}$ were found for the pioneer species, *R. pseudoacacia* (Grünewald et al. 2007, 2009). This leguminous tree is known for its ability to fix nitrogen, tolerance to water stress, and adaptation to well aerated, light soils (Rédei et al. 2008). *R. pseudoacacia* is native to southeastern North America and has been introduced to Europe during the seventeenth century. It grows remarkably fast and copes well with infertile and acidic soils, produces nutrient rich, well decomposable litter and dense wood of high quality, which is not only useful for bioenergy production but is also suitable for fuel, wood fiber, timber, and poles. The species is also suitable for forage and beekeeping purposes (Rédei et al. 2008). For these reasons, *R. pseudoacacia* is generally preferred for SRC plantations on marginal sites in the post-mining landscapes within the region (Grünewald et al. 2007).

Within SRC systems, C is sequestered in different components of the plantations. Large amounts of C accumulate within the living biomass – both above- and below-ground. A substantial share of the C compounds, produced during the processes of photosynthesis, is stored in the shoots, which are frequently removed and typically used for bioenergy production. Another fraction of the C is allocated into the stump, which is defined by the cutting height of the harvesting machinery (typically 10 cm above the ground), and the roots. A major part of these plant compartments survives after harvest and stays *in situ*. As a consequence, over a prospective lifetime of 20–30 years these plant compartments form a considerable C sink. In addition, litter and fine root dynamics associated with such systems have the potential to maintain a high level of soil organic matter (SOM), which is beneficial not only because it may help to mitigate climate change due to C sequestration and storage within the soils but also because, in many cases, the soil quality improves with increasing SOM content (Tiessen et al. 2002; Blume et al. 2010). Therefore, an increase in SOM has major positive effects on the resilience of soils, soil fertility, soil water storage capacity, and above all, on the sustainability of agriculture (Blume et al. 2010).

To evaluate C sequestration processes in the soils, it is necessary not only to examine the total C stock but also to determine the decomposability and thus the durability of the soil organic carbon (SOC). In order to estimate the proportion of stabilized organic C in soil, it is a common practice to determine the labile, short to medium term available fractions of soil C. Besides microbial factors such as soil microbial biomass or soil basal respiration, chemical methods such as mild oxidations, weak acid extractions, or physical methods like density fractionation, cold water and hot water extractions are commonly used to quantify the labile soil C pool (Chodak et al. 2003; Ghani et al. 2003; Landgraf et al. 2006). Of these possibilities, the hot water extractable carbon (HWC), as described by Körschens et al. (1990), represents a comparatively easy method to investigate the land use related effects on labile C pools (Böhm et al. 2009a).

In this study, the C sequestration processes in *R. pseudoacacia* plantations of varying age groups were investigated. The focus of the study was on evaluating C storage in aboveground and belowground plant compartments and in the soil.

Materials and Methods

Site Description

Four plantations of *R. pseudoacacia* were established between 1995 and 2007 in the Lower Lusatian lignite mining district in northeast Germany, within the reclamation area of the opencast lignite mining pit of Welzow-Süd (51.6°N, 14.3°E), about 25 km southward of the city of Cottbus (Fig. 1). The area in the district affected by lignite mining activities totaled about ~85,000 ha in 2009. The whole region is comparatively dry with a mean annual precipitation of 556 mm (1971–2000; average values for Germany are around 790 mm year^{-1}) and an average temperature of 9.3°C. Substrates at the sites were derived from overburden tertiary and quaternary sediments dumped during opencast lignite mining. The nutrient poor soils are dominated by loamy sands and sandy loams (Table 1).

The oldest of the four sites is the W95 plantation, which was established in 1995 to investigate the performance of different clones of *Populus* spp. and *Salix* spp. on mining substrates. More detailed information on the scientific survey was published by Bungart and Hüttl (2004). On parts of the experimental site *R. pseudoacacia* was planted, but these plots were not included in the initial experiments. In result, the *R. pseudoacacia* stands were neither harvested nor managed until very recently. The W05, W06, and W07 sites are part of a 170 ha reclamation project, carried out by the mining company Vattenfall Europe Mining. The project started in 2005 with the establishment of *R. pseudoacacia* on a 13.2 ha mining substrate site, around 5 km north of the W95 site. An additional 8.6 ha and 11.7 ha of *R. pseudoacacia* were planted in 2006 and 2007, respectively.

All study sites are located within a small area and W05, W06, and W07 are directly adjoining. The soils at the sites are derived from comparable substrates

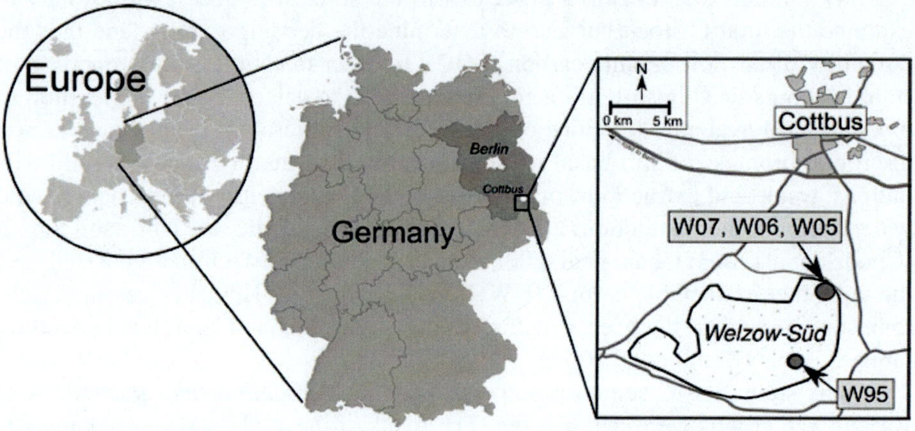

Fig. 1 Outline map of the experimental sites W07, W06, W05, and W95 in the lignite surface mining area "Welzow-Süd" in the region of Lower Lusatia, northeast Germany

Table 1 Description of the experimental sites in Lower Lusatia, northeast Germany

Feature	W07	W06	W05	W95[a]
Area (ha)	11.7	8.6	13.2	ca. 3
Establishment	2007	2006	2005	1995
Elevation (m)	105–112	105–112	105–112	125–130
Slope	Very low	Very low	Very low	Very low
Dominant substrate (USDA classification)	Loamy sand, sandy loam	Loamy sand, sandy loam	Loamy sand, sandy loam	Sandy clay loam
Spacing between trees in one row (m)	0.85	0.85	0.85	0.80
Spacing between tree rows (m)	0.75	0.75	0.75	1.90
Spacing between tree double rows (m)	1.8	1.8	1.8	single row design
Average initial planting density (trees ha^{-1})	9,200	9,200	9,200	6,579
Plantation age at sampling (number of vegetation periods)	2 years	3 years	4 years	14 years

[a]Data from Bungart and Hüttl (2004) and Grünewald et al. (2009)

with similar nutrient status. In contrast to the W95 site, clay contents at the W05, W06, and W07 site were slightly lower and had lower water holding capacities. Nitrogen content and overall soil fertility were generally low for these mining substrates, which are classified as soils in the initial phase of development. All sites were planted with tree saplings directly after amelioration, during which lime and mineral fertilizers were applied to achieve a sustained improvement of soil pH and nutrient supply to aid the subsequent reclamation process (Katzur and Haubold-Rosar 1996). More details on the site preparation and management activities are given by Grünewald et al. (2009).

Sample Collection and Analysis

In each plantation, six sampling plots were established. At the W05, W06, and W07 sites each of these plots comprised of four double rows with 19 trees in each row. At the W95 site, six sampling plots each with an area of about 10 m^2 were randomly selected. Soil samples were taken in the winter of 2008/2009 at five depths (0–3, 3–10, 10–30, 30–45, and 45–60 cm) with an auger (2 cm diameter). In each plot, five cores were taken at random. The samples were pooled (plot-wise and sampling depth-wise) to form a composite sample. After collection, the soil samples were immediately transported to the laboratory and were air dried at room temperature. Samples for soil bulk density determination were taken in 2007/2008 and 2008/2009. Soil sample rings were used to gather 100 cm^3 of soil material at 0–15, 15–30, 30–45, 45–60 cm depth.

Stone content of the air dried soil samples was determined and the samples were homogenized by carefully grinding and sieving through a mesh of 2 mm. Soil carbonate content was measured with a Scheibler device according to DIN-ISO 10693 (DIN 2007). A portion of each sample was air dried (40°C) for determining total organic carbon (TOC) and total nitrogen (TN), measured using a CNS analyzer (Elementar vario EL). For the determination of the HWC, 10 g air dried soil was boiled in 50 ml deionized water for 60 min. After the extracts had cooled, 2 ml of 2 N Mg_2SO_4 solution was added and the extracts were centrifuged at 4,000 rpm for 10 min. TOC concentrations were measured in the decant extracts using a CN analyzer (Shimadzu, Japan). For determination of bulk density, the collected soil cores were dried at 105°C until constant weights. Further evaluation followed the method described in DIN ISO 11272 (DIN 1998). Data from earlier studies carried out in W95, W05, W06, and W07 plantations (Böhm et al. 2009b; Grünewald et al. 2009) were used for evaluating C storage in the living biomass. Total organic C stocks for an individual soil profile with k layers was calculated according to Batjes (1996):

$$TOC_{stock} = \sum_{i=1}^{k} \left(P_i \cdot BD_i \cdot Th_i \cdot \left(1 - \frac{S_i}{100} \right) \right) \cdot 10000$$

where TOC_{stock} is the total amount of organic C (Mg ha^{-1}) over depth, P_i is the proportion of organic carbon (g g^{-1}) in layer i, BD_i is the bulk density (Mg m^{-3}), Th is the thickness (m), and S_i is the volume of the fraction of fragments > 2 mm (%) in this layer. To estimate C accumulation in the soil, the results were compiled as a false chronosequence. Furthermore, information on the lability of SOC was derived from the hot water carbon extractions.

The dataset was tested for significant differences with the nonparametric Mann-Whitney U-test (Mann and Whitney 1947). All statistical calculations were performed using the GNU R software package (Ihaka and Gentleman 1996).

Results

Soil Organic Carbon and Nitrogen

The plantations showed distinct differences in TOC and TN, with higher concentrations in the oldest plantation than in the others (Table 2). The W07 site, which most likely represents the initial situation of reclamation, is characterized by particularly low TOC and TN contents. Furthermore, a general decrease in C and N content with increasing soil depth was observed. The C:N ratios of the investigated soils were relatively high with an average of 25.3 (all plantations, 0–60 cm). The oldest plantation had significantly ($n = 6$, $p < 0.05$) lower C:N values in the uppermost soil layers compared to the lower ones (Fig. 2). Average C:N ratio (all plantations) increased from 19.5 in 0–3 cm to 27.1 in the 30–45 cm soil depth, which declined to 23.3 in the 45–60 cm layer.

Table 2 Soil carbon and nitrogen contents (±standard deviation) at different depths in 2 (W07), 3 (W06), 4 (W05) and 14 (W95) year old plantations of *R. pseudoacacia* (n=6) in Lower Lusatia, northeast Germany

Depth (cm)	W07	W06	W05	W95
TOC (%)				
0–3 cm	0.32±0.026	0.35±0.063	0.66±0.186	2.33±0.368
3–10 cm	0.32±0.128	0.31±0.109	0.40±0.163	1.14±0.165
10–30 cm	0.22±0.141	0.27±0.088	0.37±0.137	1.04±0.112
30–45 cm	0.25±0.146	0.29±0.160	0.38±0.128	0.96±0.155
45–60 cm	0.17±0.116	0.32±0.201	0.37±0.157	0.91±0.131
TN (%)				
0–3 cm	0.02±0.002	0.02±0.007	0.03±0.007	0.15±0.030
3–10 cm	0.01±0.005	0.01±0.002	0.01±0.003	0.05±0.008
10–30 cm	0.01±0.004	0.01±<0.001	0.01±0.003	0.04±0.003
30–45 cm	0.01±0.005	0.01±<0.001	0.01±0.003	0.03±0.003
45–60 cm	0.01±0.005	0.01±<0.001	0.01±0.004	0.03±0.004
Bulk density (g/cm³)	1.40–1.8	1.40–1.8	1.40–1.8	1.6–1.76

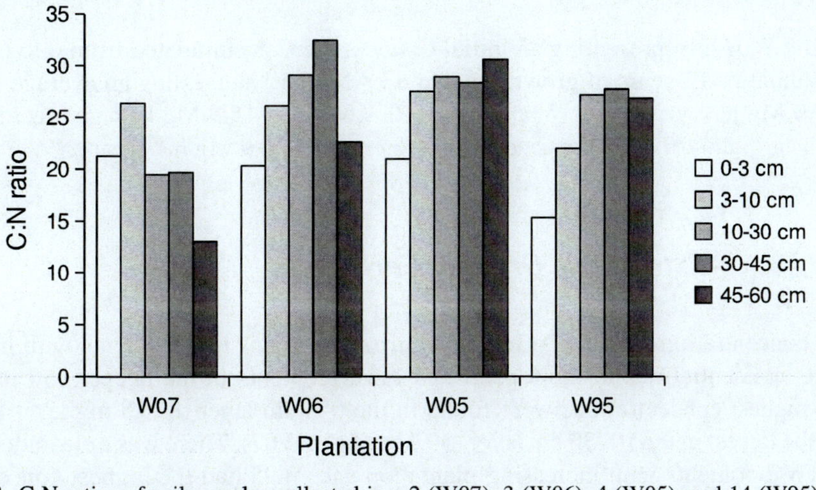

Fig. 2 C:N ratios of soil samples collected in a 2 (W07), 3 (W06), 4 (W05), and 14 (W95) year old plantation of *R. pseudoacacia* at five different depths in Lower Lusatia, northeast Germany

Average bulk densities (0–60 cm) were 1.70, 1.77, 1.73, and 1.72 g cm^{-3} for W95, W05, W06, and W07, respectively. Together with the measured soil C contents, these values were used to calculate soil C stocks (Fig. 3). The values showed a clear increase at all depths with increased plantation age and for all sites a higher C stock was noted at 0–30 cm depth than at 30–60 cm. The TOC stock at 0–60 cm amounted to 22.2±11.3, 29.0±10.7, 38.1±12.5, and 106.0±11.7 Mg ha^{-1}, respectively for W07, W06, W05, and W95 (Fig. 3). The stocks in W95 were significantly higher than the stocks of the other plantations (n=6, $p<0.05$). Likewise, W05 showed significantly (n=6, $p=0.06$) higher values than W07. If the data are used to form a false chronosequence, the average C accumulation rates in the soils can be derived.

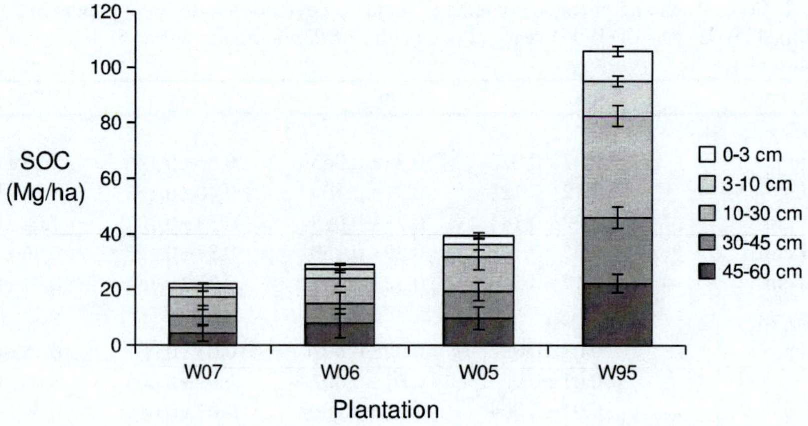

Fig. 3 Soil organic carbon (SOC) stocks in five soil depths for a 2 (W07), 3 (W06), 4 (W05), and 14 (W95) year old plantation of *R. pseudoacacia* in Lower Lusatia, northeast Germany (error bars indicate standard deviation)

Taking W07 as representing an initial C stock, the C accumulated from 0 to 60 cm in W95 after 12 years of growth totaled 83.8 Mg ha^{-1} suggesting an average value of 7.0 Mg ha^{-1} year^{-1}. For W05 the total C stock was 15.9 Mg ha^{-1} giving 7.9 Mg ha^{-1} year^{-1} and for W06 the C accumulation rate was 6.9 Mg ha^{-1} year^{-1}.

Hot Water-Extractable Organic Carbon

The concentrations of HWC (Table 3) mirror the trends in C contents with higher HWC concentrations in the topsoil and lower contents in the deeper soil layers. The highest concentrations were found in the 0–3 cm layer (873.8 mg kg^{-1}: W95) and the lowest in the 10–30 cm layer (50.3 mg kg^{-1}: W07). There was a clear increase of HWC contents with increasing plantation age. W95 had the highest concentration of extractable C with a weighted average of 285.1 mg kg^{-1} (0–30 cm soil depth), compared to the lowest value of 65.4 mg kg^{-1} for W07. Absolute stocks of HWC (Fig. 4) ranged from 0.3 (W07) to 1.4 Mg ha^{-1} (W95). The increase in HWC with increasing plantation age and the distribution of HWC within the soil depth resembled the overall trend in total C stocks (Fig. 3).

The proportion of HWC in TOC (Fig. 5) varied between 1.9 (10–30 cm soil depth for W95) and 3.9% (0–3 cm soil depth for W06). Generally, the HWC fraction declined with increasing soil depth. In the 0–3 cm layer, the proportion of HWC was higher for W06 than for W07, while no differences were observed between these plantations below 3 cm soil depth. At deeper soil depths, slightly higher proportions of HWC were observed for W05. However, this increasing trend was not correlated with age. The proportion of HWC was lower for W95 than that of W05, especially in the 3–10 and 10–30 cm soil layers. To deduce a possible influence of lignite

Table 3 Concentrations of water extractable carbon in the soils (± standard deviation) of 2 (W07), 3 (W06), 4 (W05), and 14 (W95) year old plantations of *R. pseudoacacia* in Lower Lusatia, northeast Germany

HWC (mg kg^{-1})				
Depth (cm)	W07	W06	W05	W95
0–3 cm	114.2 ± 19.86	133.7 ± 55.64	235.3 ± 45.95	873.8 ± 158.41
3–10 cm	87.9 ± 13.03	85.4 ± 12.08	116.3 ± 21.65	275.3 ± 26.08
10–30 cm	50.3 ± 9.24	60.5 ± 10.96	94.4 ± 38.77	200.3 ± 28.76
Weighted average (0–30 cm)	65.4	73.6	113.6	285.1

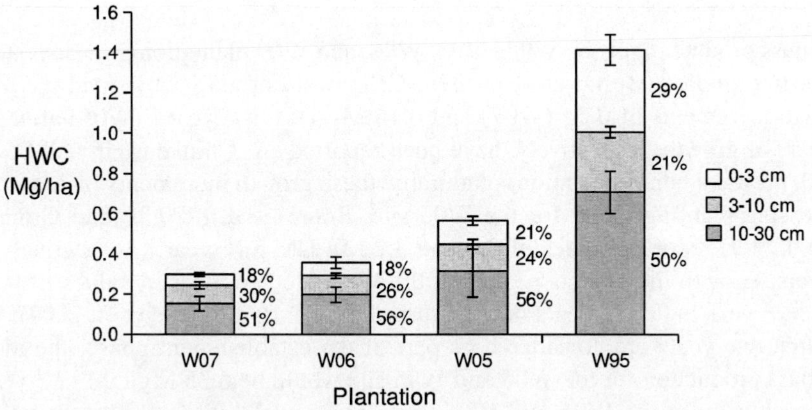

Fig. 4 Stocks of hot water extractable carbon (HWC) in the soils of a 2 (W07), 3 (W06), 4 (W05), and 14 (W95) year old plantation of *R. pseudoacacia* in Lower Lusatia, northeast Germany (error bars indicate standard deviation)

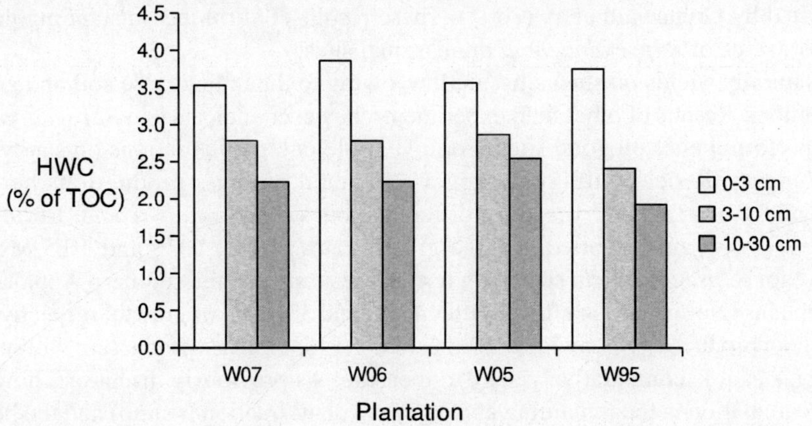

Fig. 5 Proportions of hot water extractable carbon (HWC) in total organic carbon (TOC) for 2 (W07), 3 (W06), 4 (W05), and 14 (W95) year old plantations of *R. pseudoacacia* at soil depths 0–3, 3–10 and 10–30 cm in Lower Lusatia, northeast Germany

traces in the soil on the HWC results, humus poor mineral soil samples from the W07 site were mixed with carbonized wood crumbs collected at the site and with lignite from the strip mine. HWC extraction indicated that only an amount of 0.3–0.8% of the lignite that had been added to the samples could be extracted.

Discussion

Carbon Sequestration in Biomass

Biomass production of the W95, W05, W06, and W07 plantations was investigated in earlier studies (Böhm et al. 2009b; Grünewald et al. 2009). Average annual growth increments of 0.04 (W07) and 0.15 Mg DM ha^{-1} $year^{-1}$ (W06) after 1 and 2 years of growth, respectively, have been reported (A. Quinkenstein and D. Pape, 2010, personal communication). Naturally, these growth increments increased with increasing plant ages and for the W05 site, Böhm et al. (2009b) and Grünewald et al. (2009) reported values of 3.0 and 3.2 Mg DM ha^{-1} $year^{-1}$, respectively after 4 years. Growth increments peaked at the W95 site and reached values of 9.5 Mg DM ha^{-1} $year^{-1}$ after 14 years of undisturbed growth (Grünewald et al. 2009). When the first two years are considered as part of the establishment phase, the average biomass production for the W05 and W95 site would be 6.35 Mg DM ha^{-1} $year^{-1}$.

In addition to the W95 and W05 sites, Grünewald et al. (2009) investigated two other plantations on reclamation sites in the Lusatia region. Depending on the rotation interval, they reported annual aboveground biomass growth rates between 3.1 and 7.4 Mg DM ha^{-1} $year^{-1}$. From a reclamation site in the Helmstedt lignite mining district (central Germany) yields after the first two years of growth ranging between 3.2 Mg DM ha^{-1} (mineral fertilizer) and 4.6 Mg DM ha^{-1} (compost) were reported by Grünewald et al. (2007). These results confirm the outstanding growth performance of *R. pseudoacacia* on marginal sites.

Naturally, yields on such sites are low, owing to the unfavorable soil and growth conditions. Results of other field experiments, however, indicate that *R. pseudoacacia* can perform better on good sites. For 5 year old energy plantations on sandy soils in Hungary, Rédei et al. (2008) reported annual biomass productions between 6.5 and 8.0 Mg DM ha^{-1} $year^{-1}$ for different clones of *R. pseudoacacia*. Boring and Swank (1984) also reported average growth rates of 8.3, 10.2, and 10.5 Mg ha^{-1} $year^{-1}$ for *R. pseudoacacia* stands on fertile, mesic sites in the southern Appalachian Mountains (North Carolina, USA) after 4, 17, and 38 years of growth, respectively.

Shoot productivity is a widely used parameter for biomass production within SRC systems as it is comparatively easy to measure. As previously discussed, however, C accumulation in the remaining aboveground plant fraction (stump) and the below-ground fraction (roots) must be considered as well to get an accurate estimate of the C sequestration potential of such plantations. Investigations of the biomass distribution in shoots, stump, fine roots, and coarse roots for the W95, W05, W06, and

W07 plantations showed that an average of about 60% of the biomass is allocated into the shoots and 40% is distributed into the root system in mature *R. pseudoacacia* plantations (A. Quinkenstein and D. Pape, 2010, personal communication).

Given an average annual growth rate of 6.35 Mg DM ha^{-1} year^{-1} (Grünewald et al. 2009), the total aboveground biomass accumulated in a black locust plantation under the Lusatian conditions after 15 years of growth would approximately be 95.3 Mg DM ha^{-1} or 43.8 Mg C ha^{-1} (assuming an average C content of 46% in the dry matter; D. Pape, 2007, personal communication). An estimated further share of 1.5 Mg C ha^{-1} (3.5% of 43.8 Mg), would be stored within the tree stumps (A. Quinkenstein and D. Pape, 2010, personal communication). Assuming a C allocation pattern of 60% aboveground and 40% belowground, at least 30.2 Mg C ha^{-1} would be stored belowground within the root system. Carbon sequestration in the above- and belowground biomass (excluding leaf biomass) thus gives an accumulation rate of 5.0 Mg C ha^{-1} year^{-1}. This net primary production (NPP) resembles estimates given by Boring and Swank (1984) who reported NPPs of 5.3, 6.85, and 6.90 Mg C ha^{-1} year^{-1} for 4, 17, and 38 year-old stands of *R. pseudoacacia* in North Carolina. Comparable values were reported for poplar plantations on agricultural sites in Brandenburg, Germany, where Quinkenstein et al. (2009) performed a modeling analysis and estimated the annual net C uptake in biomass, litter, and soil for pure short rotation coppices as 6.2 Mg ha^{-1} year^{-1}.

Carbon Sequestration in the Soil

At the time of plantation establishment all sites were in the initial stages of soil formation following mining, with barely any vegetation. Following the establishment of *R. pseudoacacia* organic materials were produced, and leaf litter and root litter were transferred to the soil. Initially, litter production was modest because biomass production of the young trees was low. Field investigations showed that the formation of a closed organic surface layer in these plantations took 3–4 years. In W07 and W06 no organic layer was observed, and only at the W05 and the W95 sites a noteworthy organic layer has been observed. The organic layer within the *R. pseudoacacia* plantations is usually classified as mull humus (with average organic layer thicknesses of 2 cm to 3 cm in W95) and is easily decomposable. Bross et al. (1995), based on a study of the decomposability of *R. pseudoacacia* leaves in Michigan, reported a weight loss of about 80% within 6 weeks for this species. The fine roots and root nodules of *R. pseudoacacia* have a low C:N ratio (N_2 fixing) and are also rapidly decomposed (Boring and Swank 1984), implying that the intermediary C stock in the litter of *R. pseudoacacia* is low compared to the other C stocks in *R. pseudoacacia* plantations.

In the older W95 and W05 plantations, C:N ratios varied between 15 and 21 for the 0–3 cm layer and between 27 and 31 in the subsoil (45–60 cm), implying that the soils are young and comparatively poorly developed (Fig. 2). In the younger plantations of W06 and W07, this trend was not so distinct – C:N ratios of 20 and

21 in the top layer and 23 and 13 in the lowermost layer. Both plantations are young and not well developed (1 and 2 years of age) and therefore the N inputs from above- and belowground litter residues or root exudates are still low and limited to the rooting depth. However, the trend of narrower values at 30–45 cm in W06 and W07 may be interpreted as evidence of N inputs to the deeper soil layers by developing root systems. Annual N fixation of *R. pseudoacacia* also can be high, as reported by Boring and Swank (1984), who measured net annual accretions of 48, 75, and 33 kg N ha^{-1} $year^{-1}$ for 4, 17, and 38 year-old *R. pseudoacacia* stands, respectively.

The TOC stocks (0–60 cm) in the soils investigated in this study increased with increasing plantation age from approximately 22 (W07) to 106 Mg ha^{-1} (W95: Fig. 3). Approximately 50% of this C is stored in the topsoil (0–30 cm) in all four plantations. The stocks of C in the 0–3 cm layer ranged from 1.5 Mg ha^{-1} for the youngest plantation (W07) to 10.8 Mg ha^{-1} for the oldest study site (W95). While evaluating the effects of alley cropping systems on a reclamation site in the mining area of Jänschwalde in Lusatia, Nii-Annang et al. (2009) observed significant increases in soil organic C stocks after 9 years. They compared the agricultural strips with the SRC and found a higher C accumulation under the hedgerows (5.3 and 7.8 Mg C ha^{-1} in the 0–3 cm layer under *R. pseudoacacia* and *Populus* spp. respectively). The corresponding stock in the alleys under rye (*Secale cereale* L.) was only 1.3 Mg ha^{-1}. In the 0–30 cm layer the cumulative stock sums were 16.7, 18.5, and 7.8 Mg ha^{-1} for *R. pseudoacacia*, *Populus* spp., and rye, respectively. These findings are not surprising as woody perennial based land use systems are known to sequester more C within the soils than most row crop agricultural systems due to the higher and long term biomass stock and the more extensive root systems.

When considering the youngest and the oldest sites in this study, the calculated rate of soil C accumulation in the 0–60 cm layer corresponded to an annual C accretion of 7.0 Mg ha^{-1} $year^{-1}$. However, Scholz (2010) investigated 12 year old poplar and willow SRC on agricultural soils in Brandenburg and reported average accumulation rates of 0.8–1.1 Mg ha^{-1} $year^{-1}$ for the 0–30 cm soil layer. The results of this study showed that on average, half of the C is stored at a soil depth of 30–60 cm. By only considering 0–30 cm, as it is usually done for agricultural sites, C sequestration in tree based systems may be underestimated.

The calculated rates of C sequestration in the soil are rather high compared to the below and aboveground biomass supply. Here we did not measure litter-fall biomass, the biomass of the understory, and the biomass of cover crops cultivated in the first two years after planting of trees. These sources need to be considered in further investigations to help explain the increase rates of C stocks in soil. Compared to conventionally managed arable soils, where a dynamic equilibrium of C content is reached over the long term, the C accumulation rates on mining sites are considerably higher. This can be explained by the fact that the initial soils on mining sites have almost no organic C and therefore during the first years of reclamation increase quickly towards an equilibrium C content.

In this study, all of the sites were located in the reclamation area of an active opencast lignite mining site. Although all substrates at the study sites were declared lignite free by the mining company, it is possible that the soils contained traces of lignite, for example, resulting from the deposition of lignite dust coming from the mining operations. It should be borne in mind that having lignite within the soil samples is critical for the TOC results as small amounts of lignite would affect the measured TOC concentrations. Fettweis et al. (2005) investigated the soil C pools of different *Pinus sylvestris* L. plantations on sites with different lignite contents in the Lusatia region. The authors differentiated between lignite derived C and vegetation derived C by measuring the ^{14}C contents in their samples. They found an increase of vegetation derived C within the top soils with increasing plantation age, yet on "lignite rich" soils more than 90% of the C within the deeper soils was lignite derived even after 37 years of growth. Rumpel et al. (2003) reported an accumulation of vegetation derived C for a soil depth of 0–5 cm of 36 ± 9 Mg ha^{-1} after 32 years of growth (approximately 0.8–1.4 Mg ha^{-1} year^{-1}), for comparable study sites in the region and stated that these values are close to the current C stock under ancient forests in the region.

In this study, sampling was restricted to locations free of lignite traces. Accordingly, the results of the HWC measurements showed no signs of lignite contamination of the soil because the proportions of HWC in TOC for the samples did not change substantially between the sites and the total HWC contents reflected the trends in soil C stocks. Furthermore, it was verified with additional measurements that with the HWC method almost no lignite was extracted. Moreover, the authors addressed this issue by compiling the data for the estimation of the C sequestration potential as a false chronosequence ensuring that only the influence of flora (and fauna) was interpreted. To be able to do so, the studied sites must be comparable in every possible way. As can be seen from Table 1, the locations included in this study did not show any remarkable differences for the selected parameters except for a small variation in soil texture between the W95 site and other plantations. Our results confirm that even on marginal mining sites SRC systems of *R. pseudoacacia* sequester high amounts of C within the soil. Such forests like systems under temperate conditions, usually, have higher soil C stocks than conventional agricultural systems.

The humus content of the lignite mining dump soils was very low before starting the reclamation process. Organic matter derived from leaf and root litter resulted in a rapid increase of SOM during the first few years (Table 2). As expected, HWC contents determined at the study sites increased with increasing stand age (Table 3). However, comparing with the values reported by Körschens and Schulz (1999), who compiled HWC classes in terms of the supply of short term available SOM, HWC contents at W07, W06 and W05 were very low. A moderate supply of labile SOC could be detected only at W95. This implies that under *R. pseudoacacia*, ~14 years are probably needed for the buildup of SOM containing a moderate content of short term available organic C compounds. The proportion of HWC in TOC was relatively constant (Fig. 5). After planting *R. pseudoacacia* on lignite mining dump soils, the labile SOC and the total SOC contents and stocks did increase substantially. This indicates a rapid increase of decomposed and stabilized SOC and suggests a high decomposition rate in the soils investigated. The low C:N ratio of black locust litter, and hence high decomposability of the organic material, can surely be considered as one principal reason for this.

The increase of total SOC as well as of HWC occurred not only in the top soil, but also at >30 cm soil depth (Fig. 4). This distribution can be explained by assuming that parts of SOM decomposed or humified near the soil surface were illuviated (deposited in an underlying soil), particularly when intense rain followed distinctly dry periods. Furthermore, the decomposition of dead roots and their rhizodepositions is another potential cause for the accumulation of SOC in deeper soil layers. Carbon rhizodepositions can be significant as documented by ^{14}C experiments by Merbach and Wittenmayer (2004) and are partly depicted by the HWC fraction (Leinweber et al. 1995; Landgraf et al. 2006). Consequently, the increase of HWC in deeper soil depths with increasing stand age could be a result of enhanced rhizodeposition due to the trees having more comprehensive root systems. A translocation of organic matter into deeper soil layers by bioturbation (mixing of soil by organisms) is rather unlikely because in the initial period following spoil dumping, megafauna are still missing.

Conclusions

The total C sequestration rate of SRC systems depends mainly on the biomass productivity of the plantation. For dry marginal sites in East Germany, *R. pseudoacacia* is one of the most productive species because of its tolerance to water stress, especially on sandy soils, and its ability to fix N_2. *R. pseudoacacia* has a high capacity for sequestering C within biomass as well as a high potential to increase soil C stocks on such marginal sites. The increase in the so called stable C fraction is considerable within the first 14 years after beginning the reclamation, however, there is a lack of understanding of the different forms and degree of stabilization of humus. To clarify this, further investigations focused on the quality of the humus composition are necessary.

Acknowledgements This study was part of the ANFOREK project supported by the Vattenfall Europe Mining AG and of the AgroForstEnergie project supported by the German Federal Ministry of Food, Agriculture, and Consumer Protection (project no 22009707).

References

Batjes NH (1996) Total carbon and nitrogen in the soils of the world. Europ J Soil Sci 47:151–163

Blume H-P, Brümmer GW, Horn R, Kandeler E, Kögel-Knabner I, Kretzschmar R, Stahr K, Wilke B-M (2010) Scheffer-Schachtschabel: Lehrbuch der Bodenkunde. Spektrum Akademischer Verlag, Heidelberg, 569 pp

Böhm C, Landgraf D, Makeschin F (2009a) Changes in total and labile carbon and nitrogen contents in a sandy soil after the conversion of a succession fallow to cultivated land. J Plant Nutr Soil Sci 173:46–54

Böhm C, Quinkenstein A, Freese D, Hüttl RF (2009b) Wachstumsverlauf von vierjährigen Robinien. AFZ-DerWald 10:532–533

Boring LR, Swank WT (1984) The role of black locust (*Robinia pseudoacacia*) in forest succession. J Ecol 72:749–766

Bross EL, Gold MA, Nguyen PV (1995) Quality and decomposition of black locust (*Robinia pseudoacacia*) and alfalfa (*Medicago sativa*) mulch for temperate alley cropping systems. Agroforest Syst 29:255–264

Bungart R, Hüttl RF (2004) Growth dynamics and biomass accumulation of 8-year-old hybrid poplar clones in a short-rotation plantation on a clayey-sandy mining substrate with respect to plant nutrition and water budget. Europ J For Res 123:105–115

Chodak M, Khanna P, Beese F (2003) Hot water extractable C and N in relation to microbiological properties of soils under beech forests. Biol Fertil Soils 39:123–130

DIN (1998) Bodenbeschaffenheit: Bestimmung der Trockenrohdichte (ISO 11272:1998). Deutsches Institut für Normung e.V., 10 pp

DIN (2007) Bodenbeschaffenheit: Bestimmung des Carbonatgehaltes-Volumetrisches Verfahren (DIN-ISO 10693). Deutsches Institut für Normung e.V., 6 pp

Fettweis U, Bens O, Hüttl RF (2005) Accumulation and properties of soil organic carbon at reclaimed sites in the Lusatian lignite mining district afforested with *Pinus* sp. Geoderma 129:81–91

Ghani A, Dexter M, Perrott KW (2003) Hot-water extractable carbon in soils: a sensitive measurement for determining impacts of fertilisation, grazing and cultivation. Soil Biol Biochem 35:1231–1243

Grünewald H, Brandt BKV, Schneider BU, Bens O, Kendzia G, Hüttl RF (2007) Agroforestry systems for the production of woody biomass for energy transformation purposes. Ecol Eng 29:319–328

Grünewald H, Böhm C, Quinkenstein A, Grundmann P, Eberts J, von Wühlisch G (2009) *Robinia pseudoacacia* L.: a lesser known tree species for biomass production. Bioenerg Res 2:123–133

Ihaka R, Gentleman R (1996) R: a Language for data analysis and graphics. J Comput Graph Stat 5:299–314

IPCC (2007) Climate change 2007: synthesis report. Contribution of working groups I, II and III to the fourth assessment report of the Intergovernmental Panel on Climate Change. In: Pachauri RK, Reisinger A (eds) IPCC. Cambridge University Press, Geneva, 104 pp

Katzur J, Haubold-Rosar M (1996) Amelioration and reforestation of sulfurous mine soils in Lusatia (Eastern Germany). Water Air Soil Pollut 91(1):17–32

Körschens M, Schulz E (1999) Die organische Bodensubstanz Dynamik - Reproduktion – ökonomisch und ökologisch begründete Richtwerte. Zentrum für Umweltforschung (UFZ), UFZ-Berichte 13, 46 pp

Körschens M, Schulz E, Behm R (1990) Hot water extractable carbon and nitrogen of soils as a criterion for their ability of N-release. Zentralbl Mikrobiol 145:305–311

Landgraf D, Leinweber P, Makeschin F (2006) Cold and hot water extractable organic matter as indicators of litter decomposition in forest soils. J Plant Nutr Soil Sci 169:76–82

Leinweber P, Schulten H-R, Körschens M (1995) Hot water extracted organic matter: chemical composition and temporal variations in a long-term field experiment. Biol Fertil Soils 20:17–23

Mann HB, Whitney DR (1947) On a test of whether one of two random variables is stochastically larger than the other. Ann Math Stat 18(1):50–60

Merbach W, Wittenmayer L (2004) Influence of plant rhizodeposition on C fluxes in soil. Arch Agron Soil Sci 50:99–113

Mitchell CP, Stevens EA, Watters MP (1999) Short-rotation forestry - operations, productivity and costs based on experience gained in the UK. For Ecol Manag 121:123–136

Nii-Annang S, Grünewald H, Freese D, Hüttl R, Dilly O (2009) Microbial activity, organic C accumulation and ^{13}C abundance in soils under alley cropping systems after 9 years of recultivation of quaternary deposits. Biol Fertil Soils 45:531–538

Quinkenstein A, Jochheim H, Schneider BU, Hüttl RF (2009) Modellierung des Kohlenstoffhaushalts von Pappel-Kurzumtriebsplantagen in Brandenburg. In: Reeg T, Bemmann A, Konold W, Murach D, Spiecker H (eds) Anbau und Nutzung von Bäumen auf landwirtschaftlichen Flächen. Wiley-VCH, Weinheim, 379 pp

Rédei K, Osváth-Bujtás Z, Veperdi I (2008) Black locust (*Robinia pseudoacacia* L.) improvement in Hungary: a review. Acta Silv et Lignaria Hung 4:127–132

Rumpel C, Balesdent J, Grootes P, Weber E, Kögel-Knabner I (2003) Quantification of lignite- and vegetation-derived soil carbon using 14 C activity measurements in a forested chronosequence. Geoderma 112:155–166

Scholz V, Ellerbrock R (2002) The growth productivity, and environmental impact of the cultivation of energy crops on sandy soil in Germany. Biomass Bioenergy 23:81–92

Scholz V (2010) Umweltverträglichkeit von Pappeln und Weiden im Vergleich mit anderen Energiepflanzen. Proceedings of the Agrarholz 2010–Symposium held in Berlin, Germany, May 18–19. Agency for Renewable Resources (FNR), 15 pp

Tiessen H, Cuevas E, Chacon P (2002) The role of soil organic matter in sustaining soil fertility. Nature 371:783–785

Does Tree Management Affect Biomass and Soil Carbon Stocks of *Acacia mangium* Willd. Stands in Kerala, India?

T.K. Kunhamu, B. Mohan Kumar, and S. Samuel

Abstract Initial spacing and tree pruning are silvicultural strategies that influence tree growth and productivity, which determine the potential of tree stands to store C in the vegetation and soil. A field experiment was conducted at Thiruvazhamkunnu, Kerala, India in a 6.5 year-old *Acacia mangium* stand to evaluate the changes in vegetation and soil (0–15 cm) C pools as a function of four planting densities (625, 1,250, 2,500, and 5,000 stems ha^{-1}), with and without 50% crown pruning. Both tree planting density and crown pruning significantly ($p<0.01$) influenced the C stocks of *A. mangium* trees. Total vegetation (aboveground + roots) C was highest for the 5,000 trees ha^{-1} (81.82 Mg ha^{-1}) and lowest in the 625 trees ha^{-1} treatment (41.39 Mg ha^{-1}). Soil C stocks also co-varied among the density regimes with the lowest values for treeless control plots. Overall, denser stands promote C storage, but very high stand densities (e.g., 5,000 stems ha^{-1}) may adversely affect tree growth and productivity, reducing vegetation C pools. Likewise, intense pruning may depress the vegetation C pool and would release CO_2 from the pruned biomass, especially if the slash are burnt or decomposed. Pruning effects are, however, dependent on stand density, implying the need for optimizing crown pruning regimes in conjunction with stand density levels. By extension, stand thinning exerts negative or positive feedbacks on biomass accretion depending on stand density, which may also influence the amount of C sequestered by the trees. Information reported in the literature confirms this. Irrigating the trees during water scarcity periods also may promote soil and vegetation C sequestration. But inorganic fertilization may have positive, negative, or neutral effects depending on site fertility, species, fertilizer doses, and the stage of stand development.

T.K. Kunhamu (✉) • B.M. Kumar • S. Samuel
Department of Silviculture and Agroforestry, College of Forestry,
Kerala Agricultural University, KAU P.O., Thrissur, Kerala 680 656, India
e-mail: kunhamutk@yahoo.com; bmkumar.kau@gmail.com; samuel.sijo@gmail.com

B.M. Kumar and P.K.R. Nair (eds.), *Carbon Sequestration Potential of Agroforestry Systems: Opportunities and Challenges*, Advances in Agroforestry 8,
DOI 10.1007/978-94-007-1630-8_12,

Keywords Carbon pools • Crown pruning • Silvicultural practices • Stand density

Introduction

The realization of the role of trees as cheap means to capture and store atmospheric CO_2 in vegetation and soil (Malhi et al. 2008) has generated considerable interest among the climate change mitigation strategists. Economic dimensions of afforestation and reforestation (A/R) based Clean Development Mechanisms (CDM) projects in the developing countries are also increasingly recognized by land managers in disparate parts of the world (e.g., Samek et al. 2011). This, in turn, has prompted screening of fast growing trees with high C storage potential (Paquette et al. 2009). Tropical multipurpose tree species (MPTs) such as *Acacia mangium* Willd. have a vital role in this respect (Awang and Taylor 1993; Shanavas and Kumar 2006). The MPTs' potential for atmospheric CO_2 sequestration, however, is dependent on stand management practices and site characteristics (Nair et al. 2009a, 2010), besides the intrinsic production potential of the species itself. Although silvicultural practices are widely adopted to improve stand productivity, which co-varies with C sequestration potential (CSP), it is still unclear how stand management practices would alter CSP in a given species-climatic and edaphic-time fabric (Dixon 1997). In particular, information on the effects of planting density and thinning on C sequestration by fast growing MPTs is scarce (Jandl et al. 2007). Likewise, crown pruning (live crown) may favor stem development (Beadle et al. 2007), but information on the impact of pruning applied in conjunction with varying population densities on the CSP of woody perennial-based land use systems is not available.

Silvicultural management of MPTs also may augment soil organic matter status triggering overall productivity enhancement, in addition to the climate change moderation effects, implying a win-win situation. Although some information on the aboveground C storage potential of MPTs are available (Nair et al. 2009a, b), very little is known about the changes in soil C storage of MPT stands under differing management regimes. Despite the widespread use of *A. mangium* in the humid tropics (Kunhamu et al. 2009), species-specific information on planting densities and pruning regimes for various agroforestry systems and its impact on C sequestration are lacking. A field trial was, therefore, conducted to evaluate the aboveground, root, and soil C stocks of this species as a function of planting density and tree pruning practices in the humid tropics of peninsular India. In this paper, we report aboveground C stocks of *A. mangium* under differing stand management situations as a surrogate of CSP. Additionally, an attempt was made to synthesize the recent peer-reviewed papers reporting thinning, fertilization, and irrigation effects on C stocks of tropical forest tree species, to provide a comprehensive account on the effects of stand management practices on C sequestration potential of trees.

Materials and Methods

The study was conducted at the Livestock Research Station, Thiruvazhamkunnu, Palakkad in central Kerala, India (11°21'30"N, 76°21' 50"E, ~60 m above mean sea level). The site experiences a warm humid tropical climate with a mean annual rainfall of 2,507 mm (March 2007–March 2008), most of which is received during the southwest monsoon season (June–August) with a secondary peak in September–October. Mean maximum temperature ranged from 27.3°C (July) to 37.7°C (March) and mean minimum temperature from 17.5°C (December) to 24.3°C (May) during the experimental period as per the meteorological records maintained at Thiruvazhamkunnu. Soil of the experimental site is an Ultisol (very deep, clayey, mixed Ustic Palehumults) with an average pH of 5.4 and bulk density of 0.86 g cm^{-3} (0–15 cm; Kunhamu et al. 2010).

Acacia mangium was planted at this site in September 2001 at densities of 1,250, 2,500, and 5,000 stems ha^{-1} (2×4, 2×2, and 2×1 m respectively), with or without crown pruning (see Kunhamu et al. 2010, for details). The trial was laid out in plots of size 15×20 m adopting a factorial randomized block design with three replications. Sixty centimeter wide field risers (30 cm high) on all sides and unplanted (at least 2 m wide) buffer strips separated all plots. Seedling survival was excellent (>95%) and mortality, if any, was replaced immediately – supplemented with pitcher irrigation. Multiple shoots were removed with a sharp knife during the following monsoon period (~9 months age), leaving the leader intact. The plots were manually weeded 2 months after planting and thereafter as and when required. First pruning was carried out in September 2002 to a height of 1.5 m from the ground, when the saplings (20, 35, 70, and 140 per plot corresponding to the stand densities of 625, 1,250, 2,500 and 5,000 trees ha^{-1} respectively) were about a year-old (average height 3.8 m). Subsequently, however, pruning was carried out in August every year (southwest monsoon season). Unlike the first pruning, pruned height was maintained at 50% of total tree height (range: 14.6–16.0 m at 6.5 years of age) in all subsequent operations and all live and some dead branches up to this height were removed with a sharp knife (causing minimal injury to the main shoot). Bordeaux paste was applied on the pruning scar immediately to prevent fungal infections (KAU 2002).

Estimation of Carbon Stocks

Diameter and height of all trees in each plot except the border trees were measured and aboveground biomass of the standing trees computed using allometric equations. The following equation developed for 7 year-old *A. mangium* trees in an adjacent stand at the same location (Kunhamu et al. 2005) was used for this purpose.

$B = 34.63 - 9.89(DBH) + 0.887(DBH)^2$ ($R^2 = 0.97$; SE = 19.51), where B = total aboveground (stemwood + branchwood + foliage) biomass (kg $tree^{-1}$; dry weight basis) and DBH = diameter at breast height (cm).

For estimating the coarse root biomass, the equation developed for 8 year old *A. mangium* trees in West Java (Heriansyah et al. 2007) given below was used.

Root biomass (oven dry) $(W) = -log\ 2.2752\ (D^2H)^{0.9626}$ ($R^2 = 0.85$), where D is tree diameter at breast height (cm) and H is tree height (m).

Carbon stock (reporting age: 6.5 years, which is approximately half the rotation length of 12 years), a proxy of C sequestration potential, was calculated as 50% biomass. Standing stock of C on area basis (per ha) was estimated by multiplying the mean tree values with the corresponding population density because survival was >95%. Owing to a very dense canopy, understory vegetation in the experimental plots was extremely scanty; hence such species were excluded from the C stock computations.

Carbon storage in the soil was estimated as: *C concentration x Bulk density x soil depth* (Anderson and Ingram 1989). Soil C was determined following the Walkley and Black's permanganate oxidation method (Jackson 1958). For this, random soil samples were collected in triplicates from the 0 to 15 cm soil layer in each experimental plot corresponding to the density and pruning treatments. Also, triplicate soil samples were collected from three contiguous treeless plots (81 samples from the *A. mangium* and treeless plots) during February 2008. Although these treeless plots did not form part of the original layout plan of the *A. mangium* trial, they were included as additional treatments (replication-wise) in the soil study. These plots were under miscellaneous herbaceous vegetation since the commencement of the field trial and probably represent the pre-planting soil characteristics. The samples were air dried, passed through 2 mm sieve, powdered, and analyzed. Soil bulk density was determined following the standard soil core procedure (Anderson and Ingram 1989), for which triplicate samples from each plot were randomly collected using a soil core sampler of 5.54 cm diameter (sampling depth: 0–15 cm). The data were subjected to analysis of variance for factorial randomized complete block design (RCBD) using the statistical package MSTATC with stand density and pruning as factors. The treeless plots were included as an additional treatment for analyzing the soil C data. Differences between treatment means for growth and carbon storage parameters were assessed using the LSD test (alpha = 0.05).

Results and Discussion

As can be seen from Table 1, mean diameter, crown width, and mean tree volume of *A. mangium* increased with decreasing planting density. Stand volume and mean annual increment (MAI), however, followed a reverse trend with the high density treatments performing better. Although such density dependent changes in tree growth pattern also have been reported for this stand (2 years of age: Kunhamu et al. 2010), it is probably difficult to predict whether such positive influences would persist for longer periods (e.g., rotation lengths of 12 years). Tree pruning did not significantly ($p < 0.05$) alter most growth parameters. However, radial growth of trees was less in the pruned trees compared to the unpruned ones ($p < 0.05$).

Table 1 Planting density and pruning effects on the growth of 6.5-year-old *Acacia mangium* at Thiruvazamkunnu, Kerala, India

Treatments	Height (m)	DBH (cm)	Crown width (m)	Mean volume (m^3)	Stand volume ($m^3\ ha^{-1}$)	MAI ($m^3\ ha^{-1}\ year^{-1}$)
Planting density (trees ha⁻¹)						
5,000	14.56	10.22^{d}	4.680^{c}	0.08^{b}	433.7^{a}	72.28^{a}
2,500	15.43	11.90^{c}	5.810^{bc}	0.113^{b}	282.1^{b}	47.01^{b}
1,250	15.95	14.93^{b}	7.280^{ab}	0.170^{a}	213.1^{bc}	35.51^{bc}
625	15.08	16.46^{a}	8.500^{a}	0.187^{a}	117.2^{c}	19.54^{c}
Pruning						
50%	14.57	12.72^{b}	6.35	0.115	228.7	38.12
Unpruned	15.94	14.04^{a}	6.78	0.163	294.3	49.04
Spacing × Pruning	**	*	ns	**	ns	ns

Means with the same superscript do not differ significantly
ns not significant
* significant at $p<0.05$; ** significant at $p<0.01$

Table 2 Mean tree carbon stocks in 6.5 year-old *Acacia mangium* stand at Thiruvazhamkunnu, Kerala, India

Treatments	Above ground (kg C tree⁻¹)	Root	Total
Planting density (trees ha⁻¹)			
5,000	13.28^{c}	3.08^{d}	16.37^{c}
2,500	21.47^{c}	4.38^{c}	25.84^{c}
1,250	42.50^{b}	6.95^{b}	49.50^{b}
625	57.54^{a}	8.68^{a}	66.22^{a}
Pruning			
50%	28.55^{b}	4.98^{b}	33.54^{b}
Unpruned	38.84^{a}	6.55^{a}	45.40^{a}
Spacing × pruning	**	**	**

Aboveground carbon = stemwood + branchwood + foliage
Means with the same superscript do not differ significantly
**Significant at $p<0.01$

This is consistent with the findings of Gyenge et al. (2010), who noted a decrease in annual diameter (stem and branch) growth of ponderosa pine (*Pinus ponderosa* Dougl. ex Laws.) following pruning.

Total tree C stocks on a per tree basis (stemwood + branchwood + foliage + roots) ranged from 16.37 to 66.22 kg tree^{-1} (Table 2). Widely spaced stands (625 trees ha^{-1}) recorded the highest total tree C accumulation and it declined with increasing stand density. It is well known that limitations imposed by closely spaced trees on the lateral expansion of crowns and roots may restrain resource acquisition potential of trees leading to reductions in their biomass production. Consistent with this, Kunhamu et al. (2010) using ^{32}P soil injection technique demonstrated that closely spaced *A. mangium* trees had lower lateral spread of roots, implying spatial limitations in resource capture.

Table 3 Above- and belowground carbon stocks on a stand basis as a function of planting density and tree pruning for 6.5-year old *Acacia mangium* at Thiruvazhamkunnu, Kerala, India

Treatments	Above ground (Mg C ha^{-1})	Root	Vegetation	MAI (Mg C ha^{-1} year^{-1})	Soil (Mg C ha^{-1})	Total (vegetation + soil)
Planting density (trees ha^{-1})						
5,000	66.43^{a}	15.39^{a}	81.82^{a}	12.59^{a}	31.79^{ab}	113.62^{a}
2,500	53.67^{b}	10.93^{b}	64.60^{b}	9.94^{b}	34.64^{a}	99.25^{b}
1,250	53.13^{b}	8.68^{c}	61.81^{b}	9.51^{b}	27.02^{bc}	88.84^{b}
625	35.97^{c}	5.42^{d}	41.39^{c}	6.37^{c}	30.01^{abc}	71.40^{c}
Treeless control	–	–	–	–	24.70^{c}	–
Pruning						
50%	45.44^{b}	9.10^{b}	54.55^{b}	8.39^{b}	29.07	84.72^{b}
Unpruned	59.16^{a}	11.10^{a}	70.26^{a}	10.81^{a}	30.19	101.82^{a}
Spacing × pruning	ns	ns	ns	ns	**	*

Aboveground carbon sequestration = stemwood + branchwood + foliage; Means with the same superscript do not differ significantly
MAI mean annual increment in CS, *ns* not significant
*Significant at $p<0.05$; **Significant at $p<0.01$

On a stand basis, however, total C stocks (stemwood + branchwood + foliage + roots) followed a reverse trend and it increased with increasing stand density in the order: 5,000 > 2,500 > 1,250 > 625 stems ha^{-1} (Table 3). Above- and below-ground C stocks and its MAI followed consistent trends in this respect. Almost two-fold increase in stand C stocks has been observed in the dense stand (81.82 Mg ha^{-1}; 5,000 trees ha^{-1}) compared to that of the low density stand (41.39 Mg ha^{-1}; 625 trees ha^{-1}). This is similar to the findings of Shujauddin and Kumar (2003) who reported a near two-fold increase in aboveground C stocks for 8.8 year-old *Ailanthus triphysa* (Dennst.) Alston in Kerala, India for a three-fold increase in stand density, i.e. 59.33 and 26.58 Mg C ha^{-1} at 3,333 and 1,111 trees ha^{-1} respectively.

Continuous annual tree pruning to 50% tree height depressed the C stocks both on individual tree basis (26% less for pruned trees) and on stand basis (22.4% less). Carbon stocks being a function of the overall tree growth (Table 1), this is not surprising. Majid and Paudyal (1992) also noted reductions in tree growth when crown length removal from below exceeded 40% in the *A. mangium* plantation of peninsular Malaysia. In general, removal of higher proportion of sun leaves during pruning operations may adversely affect photosynthetic rates and would depress tree growth (Gyenge et al. 2010). Pruning vs. population density interactions were also significant ($p<0.01$) implying that widely spaced stands showed greater reductions in C sequestration potential consequent to pruning compared to the denser or closely spaced stands (Fig. 1). Widely spaced trees probably lost a considerable part of their functional crowns in such pruning operations. The relative proportion of live crown removal was lower in the denser stands (data not shown). Apart from the direct effects of severe pruning and the consequent lowering of C sequestration potential, burning or decomposition of the pruned materials, if resorted to, may contribute to CO_2 emissions (silvicultural CO_2 emissions; Nair et al. 2010). Nevertheless, in

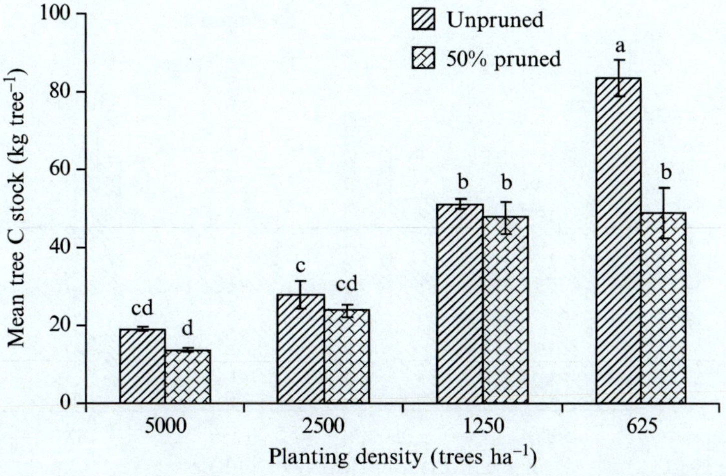

Fig. 1 Mean tree carbon stocks in a 6.5-year-old *Acacia mangium* stand as influenced by planting density and pruning at Thiruvazhamkunnu, Kerala, India. Bars with the same superscript do not differ significantly; errors bars indicate standard errors

situations where the pruned materials substitute fossil fuels, it may help in the climate change mitigation process to the extent that fossil fuels are substituted. Our data do not allow further generalizations on this.

Soil Carbon Stocks

Soil C content in the 0–15 cm soil layer ranged from 24 to 35 Mg ha^{-1} (Table 3). As expected, treeless plots had significantly ($p<0.01$) lower soil C compared to *A. mangium*, implying the role of woody perennials in augmenting the soil C pool. Among stand density treatments, the highest soil C stocks were noted for the 2,500 stems ha^{-1} plot, which was statistically at par with the 5,000 and 625 stems ha^{-1} treatments. Higher litterfall production potential and its faster turnover, being a nitrogen fixing tree, may explain the higher soil organic carbon (SOC) status under *A. mangium* compared to treeless control plots. For example, Kunhamu et al. (2009) reported an average litterfall production of 11.18 Mg ha^{-1} for 9 year-old *A. mangium* trees (stand density, 1,600 trees ha^{-1}), which is comparable to the figures reported for the moist deciduous forest formations in this region (Kumar and Deepu 1992). Variations in the litterfall production of *A. mangium* trees in relation to stand density also has been reported (Kunhamu et al. 2009). Furthermore, among the variable density thinning regimes, heavily thinned stands (533 trees ha^{-1}) showed higher litter decomposition rates (decay coefficient, $k=0.35$) and lower litter half-lives compared with unthinned stands.

The soil C stocks presently reported are also within the range of values for the agroforestry systems in this region. For example, Saha et al. (2010) suggested that

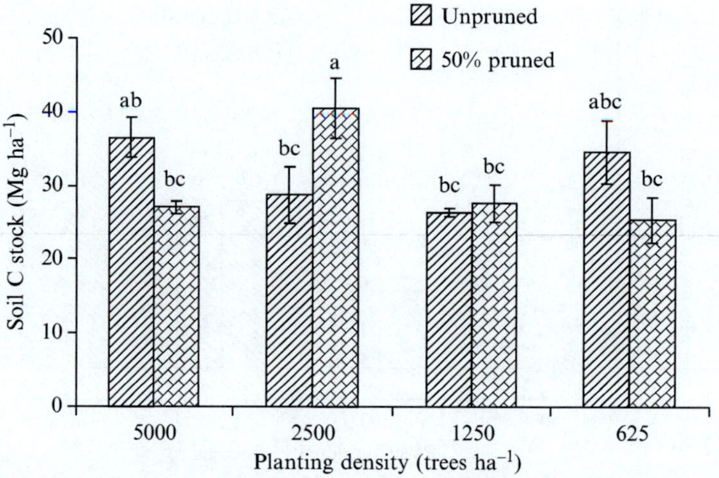

Fig. 2 Soil carbon stocks (0–15 cm soil depth) in a 6.5-year-old *Acacia mangium* stand as influenced by planting density and pruning at Thiruvazhamkunnu, Kerala, India. Bars with the same superscript do not differ significantly; errors bars indicate standard errors

total soil organic carbon (SOC; 0–20 cm) pool of four land use systems in Kerala, viz. coconut (*Cocos nucifera* L.) plantations, homegardens, rubber (*Hevea brasiliensis* H.B.K. M.-Arg.) plantations, and rice (*Oryza sativa* L.) paddy, range from 28 to 37 Mg C ha^{-1}. However, for 6 year old poplar (*Populus deltoides* Bartr)-based agroforestry systems in Punjab, India, Gupta et al. (2009) reported a lower value of 13.3 Mg ha^{-1} (0–15 cm layer). Likewise, for cacao (*Theobroma cacao* L.)-based agroforestry system in West Africa, Isaac et al. (2005) reported 18.2 Mg C ha^{-1} for this soil layer. One major limitation of the present study, however, is that SOC has been estimated for the 0–15 cm soil layer only, which is understandably inadequate to represent the SOC in wooded systems. Trees, by virtue of deep root systems, may influence the SOC levels at deeper soil layers, which call for sampling deeper layers of the soil profile.

Pruning, however, did not substantially alter the surface soil C pool (Table 3) as any reduction in litterfall of severely pruned trees was probably offset by accelerated litter decay rates, owing to greater exposure of the understory and the consequent changes in the microenvironment (George and Kumar 1998). Density vs. pruning interactions, however, suggest high SOC when trees in the intermediate planting density regimes (2,500 trees ha^{-1}) were pruned (Fig. 2). This trend, however, was inverted when planting densities were either 5,000 or 625 trees ha^{-1}; i.e., SOC was higher for these unpruned stands (Fig. 2). While faster decomposition and cycling of nutrients in the intensively pruned stands compared with unpruned stands at moderate stand densities are probable owing to microclimatic modifications, the relatively higher litter inputs in unpruned dense stands may explain the increased SOC in that treatment. Pruning widely spaced trees, as explained earlier, may adversely affect crown expansion of the trees and depress the organic matter inputs.

Management Implications

Carbon stock estimates for 6.5 year-old *A. mangium* in the present study reveal its potential as a promising tree species for C forests. To further enhance the adaptive capacity and mitigation potential of agroforestry systems, tree management offers excellent opportunities. The effects of silvicultural practices are, however, complex but studied only scarcely in the tropics. Table 4 summarizes some of the data on changes in terrestrial C storage of tropical species consequent to adoption of silvicultural practices. Overall, species, site, and stand age are major determinants of optimal population densities, which apart from increasing the quality and quantity of the forest resource, may promote forest CO_2 sequestration and C conservation. Improved nutrition (fertilization) in certain cases enhanced C in the tree biomass (Giardina et al. 2003) and enriched the soil C pool, implying the need for balanced application of nutrient elements. A few workers, however, noted negative effects of N addition (e.g., Jobidon 1993; Luxmoore et al. 2008), and yet others showed neutral effects (Shujauddin and Kumar 2003; Kim 2008). On the whole, inorganic fertilization effects on tree plantations may be positive, negative, or neutral depending on the intrinsic fertility of the site, species, fertilizer doses applied, and the stage of stand development (Nair et al. 2010). Adoption of recommended management practices (RMPs) also may allow steady incorporation of SOC for long periods before reaching equilibrium. Fertigation and irrigation of the stands especially under limited water availability situations would also promote tree growth and C sequestration. However, fossil fuels utilized for silvicultural activities such as fertilization and site preparation, intended to increase C sequestration, may emit CO_2 (Markewitz 2006) and would play a significant negative role in the C balance of forestry and agroforestry systems, which calls for the rational use of silvicultural inputs to augment C pools in soil and vegetation.

Conclusions

This paper portrays the potential of *A. mangium*, a fast growing tropical woody legume, to sequester atmospheric CO_2 and synthesizes the published information (tropics) on how tree and stand management practices influence the vegetation and soil C pools. Higher stand densities (e.g., 2,500 and 5,000 trees ha^{-1}) promoted C storage of *A. mangium*, supporting the adaptive role of stand density regulation in mitigating climate change. Intense pruning (up to 50% of tree height), however, depressed overall tree growth and C stocks of this species. The specific effects of pruning on soil and vegetation, however, were stand density dependent, implying the need for optimizing stand management strategies such as density regulation and crown pruning for improving C sequestration, in conjunction with one another. Likewise, experimental studies on thinning in certain species showed a negative impact of stem removals on C stocks. Irrigation, especially under limited water

Table 4 Some reported values of aboveground carbon stocks of different tree species under varying management conditions

Management practice	Species and location	Experimental system	Remarks	Reference
Thinning	*Tectona grandis* L.f., Kumaon, India	30 year-old stand; regular thinning (30% standing volume removed at 8 year intervals), no thinning, low thinning (10% standing volume removed at 5 and 15 year), and heavy thinning (70% standing volume removed at 15 year with final harvest at 30 year)	Regular thinning reduced C stocks by 42, 38, and 9% compared to no thinning, low thinning, and heavy thinning respectively	Kaul et al. (2010)
Initial density manipulation	*Acacia mangium* Willd., Kerala, India	6.5 year-old stands with 5,000, 2,500, 1,250, and 625 stems ha^{-1}	Biomass C accumulation and soil C storage (0–15 cm) increased with increasing stand density	This study
Pruning	As above	Pruning in conjunction with the four densities mentioned above (up to 50% of tree height vs. unpruned)	Pruning reduced C stock in widely spaced stands more than closely spaced ones	As above
Fertilization	*Eucalyptus saligna* Sm., Hawaii	6-year-old stand; no fertilizer and quarterly application of 65 kg N, 31 kg P, 46 kg K, and one annual addition of 125 kg Ca , 58 kg S, 23 kg Mg and 10 kg Granusol micro-nutrient mix per ha, over 3 years	Aboveground net C storage: 0.5 and 1.0 kg C m^{-2} $year^{-1}$ for control and fertilized treatments respectively	Giardina et al. (2003)
	Ailanthus triphysa (Dennst.) Alston, Kerala, India	4 year-old trees fertilized at 0:0:0, 50:25:25, 100:50:50 and 150:75:75 kg ha^{-1} per year N, P_2O_5 and K_2O	No significant change in C accumulation in stemwood	Shujauddin and Kumar (2003)
Irrigation	*Eucalyptus grandis* Hill ex Maiden × *E. urophylla* S. T. Blake clonal plantation, Bahia, Brazil	3year-old stands micro-sprinkler irrigated (35 mm $week^{-1}$) continuous for 2 years	Aboveground net primary production increased from 0.95 to 2.05 kg m^{-2} $year^{-1}$ (116%)	Stape et al. (2008)
	Dalbergia sissoo, Roxb. ex DC. *Eucalyptus camaldulensis* Dehnh. and *Eucalyptus grandis* Hill ex Maiden. Karnataka, India	3 year-old stands irrigated at 0, 2.5, 5.0 and 7.5 mm per day continuously for 3 years regardless of rainy days	Aboveground C storage (except leaves) increased 150% for *D. sissoo*, 74% for *E. camaldulensis,* and 78% for *E. grandis* at 7.5 mm compared to no irrigation	Hunter (2001)

availability situations, however, promoted it. Inorganic fertilization exerted positive, negative, or neutral influences on vegetation and soil C stocks depending on site fertility, species, fertilizer doses applied, and the stage of stand development. Overall, good silvicultural practices ranging from site preparation to intermediate treatments may favor tree growth and productivity and, by extension, improve above- and below-ground C sequestration, but sometimes may also contribute to C emissions.

References

Anderson JM, Ingram JSI (1989) Tropical soil biology and fertility: a handbook of methods. CAB International, Wallingford, UK, 171 p

Awang K, Taylor D (eds) (1993) *Acacia mangium*: growing and utilization. MPTS Monograph Series No. 3. Winrock International and FAO, Bangkok, 280 p

Beadle C, Barry K, Hardiyanto E, Irianto R, Junarto Mohammed C, Rimbawanto A (2007) Effect of pruning *Acacia mangium* on growth, form and heart rot. For Ecol Manage 238:261–267

Dixon RK (1997) Silvicultural options to conserve and sequester carbon in forest systems: Preliminary economic assessment. Crit Rev Environ Sci Technol 27(Special):139–149

George SJ, Kumar BM (1998) Litter dynamics and cumulative soil fertility changes in silvopastoral systems of a humid tropical region in central Kerala. Int Tree Crops J 9:267–282

Giardina CP, Ryan MG, Binkley D, Fownes JH (2003) Primary production and carbon allocation in relation to nutrient supply in a tropical experimental forest. Glob Change Biol 9:1438–1450

Gupta N, Kukal SS, Bawa SS, Dhaliwal GS (2009) Soil organic carbon and aggregation under poplar based agroforestry system in relation to tree age and soil type. Agroforest Syst 76:27–35

Gyenge JE, Fernańdez ME, Schlichter TM (2010) Effect of stand density and pruning on growth of ponderosa pines in NW Patagonia, Argentina. Agroforest Syst 78:233–241

Heriansyah I, Miyakuni K, Kato T, Kiyono Y, Kanazawa Y (2007) Growth characteristics and biomass accumulations of *Acacia mangium* under different management practices in Indonesia. J Trop For Sci 19(4):226–235

Hunter I (2001) Above ground biomass and nutrient uptake of three tree species (*Eucalyptus camaldulensis, Eucalyptus grandis* and *Dalbergia sissoo*) as affected by irrigation and fertilizer, at 3 years of age, in southern India. For Ecol Manage 144:189–199

Isaac ME, Gordon AM, Thevathasan N, Oppong SK, Quashie-Sam J (2005) Temporal changes in soil carbon and nitrogen in West African multistrata agroforestry systems: a chronosequence of pools and fluxes. Agroforest Syst 65:23–31

Jackson ML (1958) Soil chemical analysis. Asia Publishing House, New Delhi, 498 p

Jandl R, Lindner M, Vesterdal L, Bauwens B, Baritze R, Hagedorn F, Johnsong DW, Minkkinen K, Byrne KA (2007) How strongly can forest management influence soil carbon sequestration? Geoderma 137(3–4):253–268

Jobidon R (1993) Nitrate fertilisation stimulates emergence of red raspberry (*Rubus idaeus* L.) under forest canopy. Fert Res 36:91–94

KAU (2002) Package of practices recommendations: crops, 12th edn. Directorate of Extension, Kerala Agricultural University, Thrissur, 278 p

Kaul M, Mohren GMJ, Dadhwal VK (2010) Carbon storage and sequestration potential of selected tree species in India. Mitig Adapt Strateg Glob Change 15:489–510

Kim C (2008) Soil carbon storage, litterfall and CO_2 efflux in fertilized and unfertilized larch (*Larix leptolepis*) plantations. Ecol Res 23:757–763

Kumar BM, Deepu JK (1992) Litter production and decomposition dynamics in moist deciduous forest of the Western Ghats in Peninsular India. For Ecol Manage 50:181–201

Kunhamu TK, Kumar BM, Vishwanath S (2005) Tree allometry, volume and aboveground biomass yield in a seven year-old *Acacia mangium* Willd. stand at Thirvazhamkunnu, India. In: Tewari VP, Srivastava RL (eds) Multipurpose trees in the Tropics: Management and Improvement Strategies. Proceedings of the international symposium. Multipurpose trees in the tropics: assessment, growth and management. Scientific Publishers, Jodhpur, pp 415–421

Kunhamu TK, Kumar BM, Viswanath S (2009) Does thinning affect litterfall, litter decomposition, and associated nutrient release in *Acacia mangium* stands of Kerala in peninsular India? Can J For Res 39:792–801

Kunhamu TK, Kumar BM, Viswanath S, Sureshkumar P (2010) Root activity of young *Acacia mangium* Willd trees: influence of stand density and pruning as studied by ^{32}P soil injection technique. Agroforest Syst 78:27–38

Luxmoore RJ, Tharp ML, Post WM (2008) Simulated biomass and soil carbon of loblolly pine and cottonwood plantations across a thermal gradient in southeastern United States. For Ecol Manage 254(2):291–299

Majid NM, Paudyal BK (1992) Pruning trial for *Acacia mangium* Wild. plantation in Peninsular Malaysia. For Ecol Manage 47:285–293

Malhi Y, Roberts JT, Betts RA, Killeen TJ, Li W, Nobre CA (2008) Climate change, deforestation, and the fate of the Amazon. Science 319:169–172

Markewitz D (2006) Fossil fuel carbon emissions from silviculture: Impacts on net carbon sequestration in forests. For Ecol Manage 236:153–161

Nair PKR, Kumar BM, Nair VD (2009a) Agroforestry as a strategy for carbon sequestration. J Plant Nutr Soil Sci 172:10–23

Nair PKR, Nair VD, Kumar BM, Haile SG (2009b) Soil carbon sequestration in tropical agroforestry systems: a feasibility appraisal. Environ Sci Policy 12(8):1099–1111

Nair PKR, Nair VD, Kumar BM, Showalter JM (2010) Carbon sequestration in agroforestry systems. Adv Agron 108:237–307

Paquette AJ, Hawryshyn A, Senikas V, Potvin C (2009) Enrichment planting in secondary forests: a promising clean development mechanism to increase terrestrial carbon sinks. Ecol Soc 14(1):31

Saha SK, Nair PKR, Nair VD, Kumar BM (2010) Carbon storage in relation to soil size-fractions under tropical tree-based land use systems. Plant Soil 328:433–446

Samek JH, Skole DL, Klinhom U, Butthep C, Navanugraha C, Uttaruk P, Laosuwan T, Dumrongsukit S, Sangkanukij P, Kulwong T (2011) Inpang community network agroforestry carbon bank in northeast Thailand. In: Kumar BM, Nair PKR (eds) Carbon sequestration potential of agroforestry systems: opportunities and challenges. Springer, Dordrecht, pp 263–280

Shanavas A, Kumar BM (2006) Physical and mechanical properties of three agroforestry tree species from Kerala, India. J Trop Agric 44:23–30

Shujauddin N, Kumar BM (2003) *Ailanthus triphysa* at different densities and fertilizer regimes in Kerala, India: growth, yield, nutrient use efficiency and nutrient export through harvest. For Ecol Manage 180:135–151

Stape JL, Binkley D, Ryan MG (2008) Production and carbon allocation in a clonal Eucalyptus plantation with water and nutrient manipulations. For Ecol Manage 255(3–4):920–930

Part III
Policy and Socioeconomic Aspects

Can Forest Carbon Finance Influence Land Tenure Security in Project Areas? Preliminary Lessons from Projects in Niger and Kenya

André Rodrigues de Aquino, André Aasrud, and Leticia Guimarães

Abstract Land tenure security is often considered a requirement for forest carbon (C) finance investments, as a way to ensure that C emission reductions can be legally delivered to buyers and to promote permanence of emission reductions from forest activities. Areas under unsecure land tenure are regarded as less attractive for C finance investments. Nevertheless, there is limited research on how C finance may conversely affect land tenure. This paper aims to contribute to this debate by exploring how land tenure security of local landholders and communities in reforestation areas may be affected through forest C projects. By clarifying C ownership, a C transaction creates a new stick to the bundle of rights associated with land (emission reductions from sequestered C), which normally is created with clear tenure. This can in turn directly affect the overall land tenure for those landholders involved in the transaction. Carbon ownership clarification can be achieved through institutional arrangements, such as community-level benefit-sharing agreements or legal private contracts. Adequate design and implementation of C finance transactions can also influence land tenure security, mainly by supporting the organizations of local landholders around rural institutions and encouraging the involvement of national land agencies in the implementation of the project. These institutional arrangements prompted by C transactions can result in overall increased land tenure security for landholders and communities in the reforestation project area, as evidenced by two case studies from agroforestry projects in Niger and Kenya.

Keywords Benefit-sharing • Carbon ownership • Community forestry

---

A.R. de Aquino (✉) • A. Aasrud
Carbon Finance Unit, The World Bank, 1818 H St, NW, Washington, DC 20433, USA
e-mail: adeaquino@worldbank.org; andre.aasrud@iea.org

L. Guimarães
School of Public Policy, University of Maryland, 2101 Van Munching Hall,
College Park, MD 20742, USA
e-mail: leticiagsguimaraes@gmail.com

B.M. Kumar and P.K.R. Nair (eds.), *Carbon Sequestration Potential of Agroforestry Systems: Opportunities and Challenges*, Advances in Agroforestry 8,
DOI 10.1007/978-94-007-1630-8_13,

Introduction

Carbon (C) finance is a mechanism for climate change mitigation. Land use activities can generate emission reductions by sequestering C from the atmosphere (afforestation and reforestation) or avoiding emissions (reduced emissions from deforestation and forest degradation: REDD). Recently, there has been increased debate on the impact of forest C on land tenure and the implications for the distribution of revenues from C transactions (Luttrell et al. 2007; Eliasch 2008; Pirard 2008; Cotula and Mayers 2009). Land tenure and its relation to property rights to forest C are also highlighted as key issues in the climate change mitigation debate (Barnes and Quail 2009).

Sub-Saharan Africa is responsible for 15% of global emissions from land use change and forestry. Land-based activities, such as reforestation, agroforestry, and agriculture offer the largest potential for climate change mitigation in sub-Saharan Africa. Nevertheless, up to now, the region accounts for only 1% of all registered Clean Development Mechanism (CDM) forest projects. A number of barriers for scaling up C finance in the continent have been identified, including those related to insecure property rights to land (Desanker 2005). Issues around land tenure, such as those related to property rights to emission reductions and the relationship between land tenure and the adoption of sustainable land use practices, are key to the climate change agenda in Africa.

Although much has been written about the importance of clear land tenure for the development of forest C projects (Unruh 2008; Quan and Dyer 2008), the impact that C finance transactions may potentially have on land tenure is less explored. Rather than focusing on secure land tenure as a prerequisite for C finance investments, this paper investigates the potential influence that forest C finance projects may have on land tenure security by assessing two ongoing CDM agroforestry projects in Niger and Kenya.

The theoretical framework that supports this hypothesis is developed in the three main sections of this paper. The first section focuses on land tenure security definitions and its connection to sustainable land management (SLM). The second part examines the use of C finance to promote SLM activities (including reforestation) emphasizing the role of clear carbon ownership to attract carbon investments. The paper then explores the connection between forest C finance and land tenure security, outlining the theoretical framework used in this paper. The core hypothesis advanced – namely that forest C finance may positively affect the land tenure security of landholders that participate in forest C projects under certain conditions – is tested in two reforestation projects currently under implementation in Niger (*Acacia senegalensis* Plantation Project) and Kenya (Community Forestry around Mount Kenya) with the support of the BioCarbon Fund (details in Table 1).

This study draws on both secondary and primary data sources. The theoretical framework draws on a literature review of land tenure and agroforestry studies, e.g., forest C projects and their connections with land tenure, investment incentives, and SLM. Secondary data for the case studies come from the literature on land tenure in Niger and Kenya as well as national land and forest legislations, in addition to information from the BioCarbon Fund (BioCF) project documents

Table 1 Details on BioCF projects in Niger and Kenya

Project	Location	Species planted	Project entity	Other parties	Expected project area	Expected ERs	Other co-benefits
Acacia senegalensis plantation	Niger	*Acacia senegalensis* (drought- tolerant native species)	*Achats Services International* (ASI)	• Ministry of Rural Development and Ministry of Agriculture • 30 *grappes* (community associations)	12,000 ha	0.78 $MtCO_2e$ by 2017	• Increased land tenure security • Reduced erosion • Combat desertification • Improved soil fertility • Additional revenue from the sale of Arabic gum, grains and forage • Economic empowerment • Strengthen local organizations • Increased social cohesion
Aberdare Range/ Mt. Kenya small scale reforestation	Kenya	Mix of native species	Green Belt Movement (GBM)	• Kenya Forest Service • Communities organized under Community Forest Associations	1,643 ha	0.34 $MtCO_2$ by 2017	• Increased land tenure security • Reduced erosion • Regulation of water flows • Increased local biodiversity • Access to forest products (fuel wood, charcoal, medicinal plants, bee-keeping) • Employment opportunities • Increased technical capacity on sustainable land management • Stronger local institutions and social cohesion

Source: BioCF project reports. Available at http://wbCfinance.org/Router.cfm?Page=BioCF. Accessed 28 Aug 2010 (e.g., project design, contracts, minutes of meetings with communities, etc.). The case studies were also informed by primary data, collected through semi-structured interviews with project developers and other local actors in 2009.

Land Tenure, Land Tenure Security, and Sustainable Land Management

Land tenure is the bundle of rights over natural resources that characterize the relationship among individuals and groups with respect to land (FAO 2002). As social conventions protected by the government (statutory rights) or the community (customary rights), land rights allow individuals and/or groups to gain from different benefit streams related to the land. For the purpose of this study, tenure security is defined as the individual's confidence that her/his rights will be recognized by others and protected when challenged, as well as the ability of the individual to reap the benefits of labor and capital invested in that land (Bruce and Migot-Adholla 1994; FAO 2002).

Increased land tenure security – improved tenure clarity and certainty – may be achieved through different means, among them the recognition of one's rights by the community (social recognition of one's rights), government (political recognition), and/or formal legal systems (legal title, contracts, etc.). At the local or community level, the process of increasing land tenure security may be triggered by projects that use tools of tenure change to mitigate investors' risks. Among these tools are the community legislation, contracts, and projects of economic leverage (benefit-sharing schemes: Bruce 1986).

Given the long gestation period dedicated to tree planting and the absence of additional guarantees and incentives for investing in forestry, forest C projects are likely to be developed in lower risk areas with clearer land tenure. Different tenure systems can ensure access to land and stimulate investments in land improvement (Migot-Adholla et al. 1991; Perez et al. 2007). With the right institutions to ensure the compensation for labor and other long-term investments on land, customary tenure systems may provide the right set of incentives needed to foster the development of forestry activities (Migot-Adholla et al. 1991; Omura 2008).

Carbon Finance and Sustainable Land Management

Carbon finance is a market-based mechanism for climate change mitigation consisting basically of emission reduction emission reductions (ERs) transactions in both compliance and voluntary markets (Samek et al. 2011). These ERs are generated through projects that reduce greenhouse gas (GHG) emissions or sequester C from the atmosphere. The most important market for C projects in developing countries at present is the Clean Development Mechanism (CDM), created as part of the Kyoto Protocol.

Carbon finance can serve as an incentive for the promotion of SLM by compensating landholders[1] for the adoption of certain land-use practices leading to C sequestration or avoidance of greenhouse gas emissions. Carbon finance for afforestation and reforestation (the only forest activities accepted under the CDM) rewards landholders for the adoption of land-use practices leading to the accumulation of C in the vegetation biomass (above- and belowground), in dead wood and litter, and in the soil, beyond what would have happened in the baseline scenario.

Carbon finance transactions can be regarded as a type of payment for environmental service scheme whereby a well-defined ecosystem service (C storage or avoiding emissions) is purchased by at least one buyer from a minimum of one provider and the payment is contingent upon the delivery of the service. The degree to which C finance provides a real incentive towards the adoption of SLM practices depends on a number of variables, including the opportunity cost incurred by landholders and the expected revenues from C finance and other products. In turn, the expected revenue from C finance for reforestation is a function of two main variables: the price of C and the amount of C sequestered by unit of land. The price of C responds to many forces in the C market, including the usual demand and supply forces. The second variable, the amount of C sequestered by unit of land, is a function of the type of species being planted and the manner in which the plantation is managed (Kunhamu et al. 2011).

Forest Carbon Finance and Land Tenure Security

Unclear land tenure is often seen as an impediment for the successful implementation of afforestation and reforestation projects (Unruh 2008). The main argument is that C investors would not be willing to promote projects in areas under unclear land tenure, due to their high delivery risks and questions about transferability of assets. Forest C investors tend to favor projects in areas under clear tenure mainly for two reasons. First, clear land tenure is often associated with clear C ownership, reducing the risks that the asset purchased may be legally disputed. Second, project implementation in areas with higher levels of tenure security is more likely to be successful (in terms of higher trees growth and lower mortality rate), leading to lower delivery risks, since clear tenure creates incentives for adoption of SLM activities.

The BioCF experience, however, shows a more nuanced reality. The BioCF has signed C purchase contracts (called Emission Reductions Purchase Agreement – ERPA) with project developers carrying out activities in areas with a broad range of land tenure conditions, including fully titled private land, titled community land, state-owned land, untitled community land, untitled private land, among others

[1]Landholder here is broadly understood as the individual or community participating in a C finance transaction regardless of the ownership of the land. Landholders encompass fully titled private landowners, titled community lands, untitled private and community landowners, governments, etc.

(The BioCarbon Fund Experience: Insights from Afforestation/Reforestation Clean Development Mechanism Projects, *forthcoming*). The key has been the design of institutional and contractual arrangements, to reassure investors that, despite the often unclear tenure condition in the project area, mechanisms would be put in place to: clarify C ownership (and the legality of the C transaction) and ensure adequate project implementation. These institutional arrangements took into consideration both customary and statutory land rights, giving the BioCF flexibility to operate under different tenure conditions.

Clarifying Carbon Ownership

Clear C ownership is a key element of any C finance transaction. Investors in forest C projects need the assurance that the emission reductions they are purchasing can be legally transferred to them without restrictions. Carbon ownership agreements are created to allow projects to trade C as a commodity. As mentioned by Saunders et al. (2002), a C entitlement will create new property, a new stick to be added to the bundle of rights already associated with forests.

The possibility that the revenue from C finance may affect the local land tenure condition has been explored in the literature. Some argue that C payments may increase the value of common property land which, in turn, can lead to conflicts over the tenure of resources and land. In situations where land resources are subject to multiple uses and claims, revenues from C payments may intensify tension among resource users (Perez et al. 2007). Others also highlight the potential risk of capture of C revenues by powerful actors (Cotula and Mayers 2009).

Resolving the uncertainties surrounding legal title to the sequestered C is critical to securing its market value in a forest C transaction (Barnes and Quail 2009). Since most countries presently do not have specific legislation on ownership of ecosystem services, C ownership needs to be determined through other mechanisms. Among these mechanisms are: (1) private contracts signed between the parties before the implementation of forest C projects; and (2) benefit-sharing mechanisms among landholders and community members. Additionally, C ownership is also partially clarified through the Letter of Approval issued by the host country's Designated National Authority, a requirement for CDM projects under the Kyoto Protocol. The Letter of Approval can be seen as the State authorization of the actual transfer of ownership of the emission reductions (Streck 2009).

Private Contracts

Private contracts are the main mechanism used in C finance transactions to clarify C ownership and create a commodity that can be traded. Contracts that clarify C ownership can potentially separate this right from broader rights to the forest

Table 2 Carbon aggregator and contract agreements in the BioCF projects in Niger and Kenya

Project	Carbon aggregator	Legal/Contractual agreement
Acacia senegalensis plantation, Niger	*Achats Services International* (ASI)	Contract between ASI and local community cooperatives (*grappes*). Each community decided individually how to distribute the income from the project (*clé de repartitión de revenus*)
Aberdare range/Mt. Kenya small scale reforestation, Kenya	Green Belt Movement (GBM)	Tri-party agreement between GBM, Kenya Forest Service and local Community Forest Associations

Source: BioCF project reports. Available at http://wbCfinance.org/Router.cfm?Page=BioCF. Accessed 28 Aug 2010

and land while defining responsibilities and liabilities (Cotula and Mayers 2009). According to Bruce (1986) "contracts can be used as a tool for regulating tenure arrangements between groups or individuals or even with the government". Contracts can be flexible enough to reflect the concrete tenure conditions on the ground when establishing the rights and responsibilities of the parties. In the BioCF, two types of contracts are used to clarify C ownership in a project: ERPAs signed between the emission reductions seller and the buyer; and C transfer Subsidiary Agreements, signed between the participating landholders and the C aggregator.

The ERPA is a legally-binding contract between buyers and sellers of emission reductions that sets out the rights and responsibilities of parties to a C transaction, such as the volume of emission reductions being transacted, the price, and the delivery schedule of emission reductions and payments. In the case of reforestation projects, the contract normally includes terms related to the permanence of the planted trees, such as restriction on tree cutting for the duration of the contract. The ERPA addresses C ownership directly, seeking to assert the legal owner of the emission reductions and the conditions under which those can be transferred to the buyer. It aims to reduce the risks that could arise from the emission reduction transfer, including those related to ownership.

Subsidiary agreements, in the context of C transactions, refer to private contracts signed between participating landholders and a C aggregator who transacts emission reductions, signing the ERPA on their behalf. Hence, a subsidiary agreement is a key element in a C transaction, as it transfers the legal right to transact C emission reductions to an aggregator, while specifying the roles and responsibilities of each party in this transfer (including price, payment methods, emission reductions delivery schedule, among others). Aggregators can be different type of organizations, from private companies, to NGOs to the government itself. Table 2 provides the details in this respect on the BioCF projects of Niger and Kenya.

In the *Acacia senegalensis* Plantation Project (ASPP) in Niger, the local private company *Achats Services International* (ASI) acts as the C aggregator and as project developer. In the ERPA, ASI asserts its exclusive ownership of the emission reductions to be generated by the project, as well as its control over the project land area. Exclusive ownership of ERs, in turn, is achieved by ASI's signing a sub-agreement with each of the 30 *grappes* representing the communities and landholders participating in the project.

These subsidiary agreements, declared as a condition of effectiveness of the ERPA at the request of the World Bank, attest and recognize that the *grappes* have the tenure over the land where C reforestation activities are to take place and the ownership of the ERs to be generated. The subsidiary agreements also determine that the ASI will trade the Arabic gum – to be explored once the acacias are mature – in international markets after paying the *grappes* a fair price for this product.

Subsidiary agreements vary substantially in form and content, but they are an important instrument to reduce the delivery risks for buyers by clarifying the rights and responsibilities in relation to the land and resources, and hence C ownership. Subsidiary agreements are particularly important when many landholders are involved in a C transaction, such as in reforestation projects with multiple small landholders or on community lands.

In some cases, additional legal mechanisms (such community by-laws) are necessary for the establishment and operation of community groups or cooperatives. The Aberdare Range/ Mt. Kenya Small Scale Reforestation project exemplifies this. In terms of land tenure, the key document is the forest license provided by Kenya Forest Service (KFS) to the Community Forest Associations (CFAs), granting these communities the user rights to the project area. Under the Kenya Forest Act of 2005 the establishment of the CFAs is a requirement for this concession. The Forest Act also requires CFAs to establish a CFA constitution (by-law) and a site management plan (Community Forest Management Agreements). The Community Forest Management Agreements are legally binding for both parties, and can be revoked by KFS if the CFAs do not fulfill the requirements.

Once tenure rights in the project areas have been clarified by contracts defining C rights, landholders have a greater certainty that their rights are respected by their peers and protected by the government when challenged, giving them a greater sense of tenure security (FAO 2002). In some of the areas where C finance transactions take place, the subsidiary agreement for the C transaction may be the first contract signed by landholders and could itself constitute a type of legal proof of their rights to the land. It should be noted, however, that the strength of this contract will depend on its recognition by peers (including customary authorities) and the State.

Benefit-Sharing Schemes

Another important mechanism used to clarify C ownership in forest C projects is benefit-sharing schemes. Benefit-sharing schemes aim at clarifying the distribution of the monetary and non-monetary benefits flowing from the emission reduction transactions among the providers of the service. Some of the issues to be agreed upon by the service providers (in this case, the landholders undertaking reforestation activities) in such a scheme are how much each participant will receive, at what frequency, at what level (community, individual), and through what payment method (in-kind, cash). Benefit-sharing schemes may be consolidated through social agreements and/or formal contracts.

In order to contribute to effective implementation of forest C projects, benefit-sharing schemes must be well understood by the local landholders/communities and be socially accepted. Clear benefit-sharing schemes contribute to C ownership clarification by determining who is to benefit from C payments and how these are to be delivered, avoiding, or minimizing future conflicts when resources start flowing. Benefit-sharing schemes can also directly affect the land tenure security of the participating landholders by ensuring that they receive the returns from the investments they make on the land. The ability of reaping the returns on the investment on land is a major component of land tenure security (Bruce and Migot-Adholla 1994).

Evidence from the case studies attests to the importance of the design of benefit-sharing schemes. In the case of Niger, the very process of defining the benefit-sharing scheme (*clé de repartitión de revenu*) has been crucial in increasing social cohesion within the community, since it gave them the chance to reflect collectively on how to use the resources coming from C sales (and from other revenue streams, such as the sale of Arabic gum). It also allowed for increased understanding that C payments would only be made if the asset (C sequestration) was delivered, which meant the need for communal work in tree planting and maintenance.

These "*clé de repartitión de revenus*" are community level agreements on how revenues from C and Arabic gum trading are to be distributed among the members of the *grappe*, These arrangements are encompassed by the subsidiary agreement signed with the C aggregator and result from a consensus within the community, which should avoid conflicts once the revenues start flowing. They have been fully bargained between the *grappes* and ASI, and within the community, and have led to a common understanding about roles and responsibilities in the deal. The obligations of each party are observable through the project monitoring for the planting and through the disbursement of the funds to the *grappes*.

These agreements however also face challenges. The C payments, for instance, are taking longer to be delivered than initially foreseen due to delays in the CDM regulatory process (mainly the process to get the project validated by a third-party) and the slow growth of the plants. Carbon is also a very new and complex concept. Therefore, communities take time not only to fully internalize the concept, but also to trust such an abstract deal. Despite a lot of efforts of communication from the management unit of the project, there is a continued need for further communication with the participants in the project.

In Kenya, the main benefit for the communities from the project in the short term is the direct compensation for planting trees from the Green Belt Movement (GBM). This short-term benefit is similar to GBM's regular (non-CDM) reforestation projects where a small monetary reward is provided upon planting of seedlings and later on for maintenance of the trees for the first couple of years. In this project, this will provide income of approximately 10 Ksh per tree (around US$130/ha) directly to the groups involved in the tree planting. Additional revenue will be generated from the sale of C. This amount will depend on the actual growth rates of the trees and based on estimated growth rates would amount to about $1 million by 2017.

In terms of benefit-sharing, there is a risk that long-term development benefits may not fairly distributed among the community as a whole, and/or those who put most effort into the initial tree-planting phase. Often benefits are captured by the most powerful and wealthy members of the community.[2] To mitigate this risk the formula for determining the sharing of C revenues should be based on fully participatory and transparent decision-making and subsequent management of any projects that are selected by the community as a whole. It is also important that C revenues are utilized for community projects that potentially benefit all members of the community.

The contracts between GBM and the CFAs include a requirement to ensure fair and equitable sharing of the benefits accrued from the project. However, the key document and process in this context is the development and implementation of the Community Forest Management Agreements, which along with the social structure and capacity of the CFAs, will be the key elements in securing an equitable benefit-sharing under the project.

Ensuring Adequate Project Implementation

In reforestation projects, adequate implementation entails measures to minimize tree mortality and maximize growth rates. The measures put in place to achieve these efficiency goals – strengthening of rural institutions and the involvement of land agencies in project implementation – also have some impact on the level of land tenure security of local landholders.

Strengthening of Rural Institutions

The implementation of forest projects in areas where communities manage natural resources is a challenging task. If there are no incentives for individuals to participate in local activities, and no institutions to locally enforce the rules agreed upon in the design of the project, the implementation is likely to fail. As stated by Perez et al. (2007):

> Community based C sequestration cannot operate and be sustained in a vacuum. Such activities must be backed up by strong rural organizations, legitimate and representative leadership, client-driven extension, local capacity building, and informed and enabling policies. Coordinated interventions to strengthen this institutional scaffolding and to advance favorable policy reforms will need to complement the efforts to stimulate widespread adoption of technical solutions.

[2] Kamweti and Acworth (2006).

Rural institutions, such as local cooperatives, can mobilize landholders around a common goal, increase their negotiating power vis-à-vis outside actors, and foster shared interests. Strong local institutions are necessary to coordinate joint efforts as community reforestation projects. The importance of strong rural institutions for adequate project implementation is exemplified in the BioCF projects. Local cooperatives such as the *grappes* in Niger and the CFAs in Kenya are key in organizing local landholders around tree planting efforts. By facilitating the decision-making process among various landholders, these institutions foster agreement among community members and contribute to the clarification of land use rights. In turn, clarification of user rights has a positive effect on increasing land tenure security for local landholders.

In Kenya, social mobilization and capacity building for community groups is at the very core GBM operations. The basis of GBM is the mobilization of thousands of women's groups who establish tree nurseries and plant indigenous trees on their farms and on public lands. This focus is also reflected in this project and in the agreement between GBM and the CFAs where the enhancement of women's livelihood is stated as a vision of the project.

Involvement of National Land Agencies in Project Implementation

The involvement of national land agencies such as the Local Land Tenure Commissions in Niger in the process of project design and implementation can also contribute to adequate project implementation, while having a direct effect on land tenure at the local level. As the BioCF experience shows, national land agencies (and their regional or local branches) are mainly called upon during the project design phase, when the land tenure situation is assessed in the project area, and during project implementation, when benefit-sharing schemes are crafted. Carbon investors see their involvement as crucial to ensure that the national land legislation will be respected. Carbon investors may even fund part of the activities of these agencies (such as the case in Niger) to ensure their adequate participation in project design.

Through their involvement, project participants ensure that the activities on the ground have the explicit recognition of the national government, including the user rights of landholders involved in a C transaction. By participating in discussions on the institutional arrangements created within the project, and recognizing their legitimacy, these agencies contribute to strengthening local landholders' security of tenure. They can also go as far as titling the lands where projects are being developed, as seen in the Niger case. Even though land titling is not synonymous with land tenure security, it may function as one more instrument to define rights over the land.

Carbon Finance Influence on Land Tenure Security in Project Areas – Evidence from the Case-Studies

Forest C finance projects can contribute to increase land tenure security of those landholders and communities participating in C finance projects through project design, preparation, and implementation. Figure 1 provides a graphic representation of the theoretical framework outlined above. Evidence from BioCF projects in Niger and Kenya support that.

In both cases, the efforts in the project areas to clarify C ownership and ensure adequate project implementation triggered the process of local land tenure securitization. To define who has the right to C, rights to other resources had to be asserted. In both projects, the participating landholders had their user rights to the land recognized by the government. Even though laws from both countries already encompassed the possibility of statutory recognition of customary user rights, forest C projects' institutional arrangements and investments prompted the organization of these individuals and communities and the recognition of their claims by the government. Details on the land tenure situation in Niger and Kenya before and after the project are summarized in Table 3.

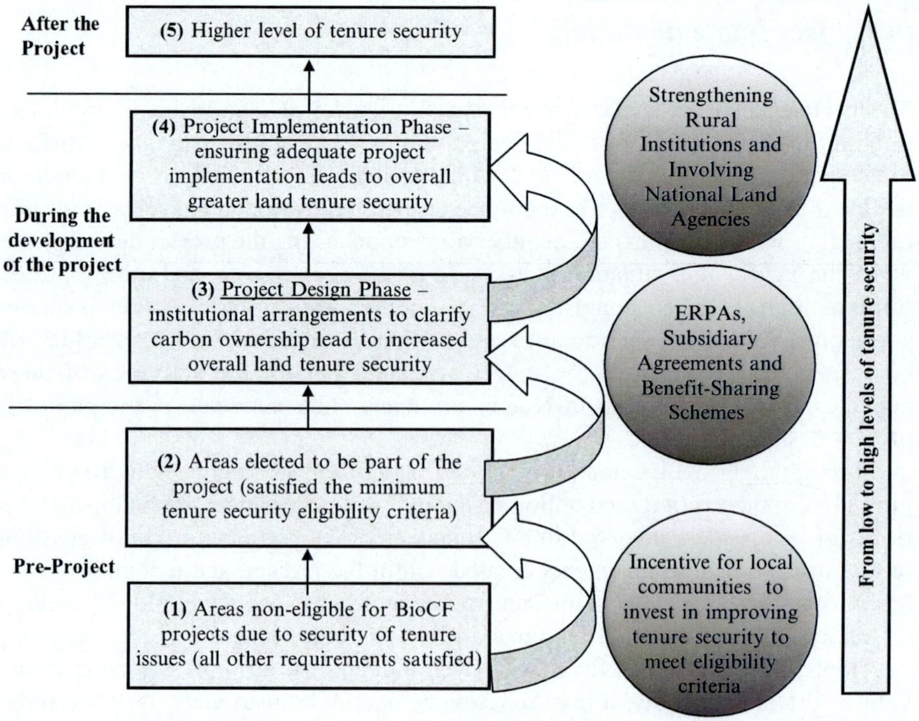

Fig. 1 Conceptual model of forest carbon project's influence on the level of land tenure security at the project area (*ERPA* emission reduction purchase agreement)

Table 3 Land tenure situation in project areas in Niger and Kenya

	Land tenure situation	
Project location	Before the project	After the project
Niger	• Untitled private land • Vacant land • Classified forests	• Rural concessions – customary user rights recognized by the government and private owners • Private lands – titled delivered by the government
Kenya	• Gazetted public land under the control of Kenya Forest Service (KFS)	• Community Forestry Associations were granted forest licenses by the KFS through which the government recognizes their user rights

Source: BioCF project reports. Available at http://wbCfinance.org/Router.cfm?Page=BioCF. Accessed 28 Aug 2010

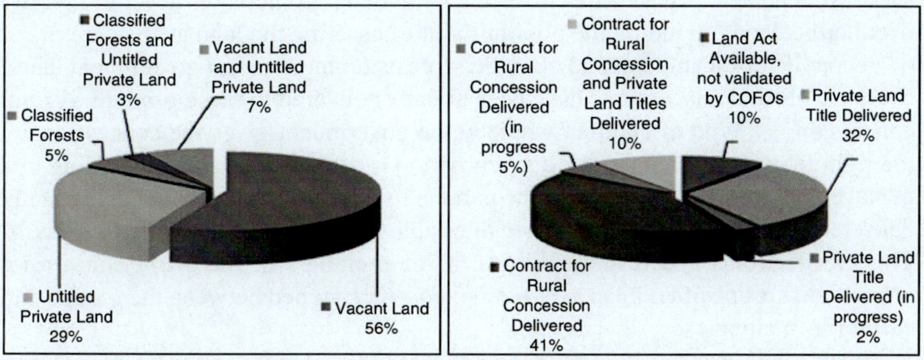

Fig. 2 Land tenure condition before and after project implementation in Niger (as of June 2009). The pre-project data is based on Moussa, Y., Bagnou, A., Lecko, M., 2006 (Républuque du Niger, Ministère du Développement Agricole. Appui aux commissions foncières départementales, aux services techniques et aux grappes pour l'élaboration des actes de sécurisation foncière dans le cadre de la plantation de gommiers. Report prepared for the BioCarbon Fund, personal communication) whereas the land tenure condition as of June 2009 is based on a field visit to each site carried out by the Management Unit of the project

Land Tenure Security Changes in the Niger Acacia senegalensis Plantation Project (NASPP)

In Niger, the project sites were selected based on a feasibility study undertaken by ASI and the Ministry of Rural Development after an assessment of the potential project areas. The areas chosen for the project were severely degraded to avoid competition with other land uses such as agriculture and grazing and displacements and/or access restrictions. The project also avoided sites that would require expropriation and areas where tenure was disputed. The pre- and post-project land tenure conditions in the project areas vary from site to site and are summarized in Fig. 2.

In the pre-project phase, there were three main types of land tenure situations in the project area:

1. Untitled private land: a type of individual property recognized by the customary leaders and not formally titled according to statutory law;
2. Vacant land: a type of communal area, where no proof of property right can be established but that formally belongs to the State;
3. Classified forests: lands formally titled to the State and managed by the state body in charge of forest management.

Every site in the project has gone through some sort of change in its tenure status since the inception of the project. Despite the complex differences in each specific case, two main generalizations concerning this change can be made. Firstly, on untitled private lands, private property rights assigned by customary leaders to individual landholders in the past and are now getting statutory recognition. The assignment of private titles gives the landholder an increased level of security over her/his land, including the possibility of transacting the land as they see fit.

Secondly, in vacant land and classified forests (commons and government land), contracts of rural concessions have been already delivered to some *grappes*. A rural concession is a type of contract whereby the government gives the concessionaire the right to explore the land for a given period of time (renewable) according to a management plan agreed between the parties. In this project, this contract is usually delivered to the *grappe* for the purpose of establishing a plantation of *A. senegalensis*. The specific roles and responsibilities of the members of the *grappe* in a rural concession are specified in an *exploitation contract*, signed between the *grappe* and each of its members.

Vacant lands have traditionally been a source of land tenure conflict in Niger as no clear property rights can be asserted over it. As discussed by Roncoli et al. (2007), for the case of Mali, "unused land was declared state land, (...) leading to the destruction of the resources". Through a Rural Concession, the state is devolving land rights to communities and allowing for secured community management of the land recognized by the state. A similar situation is found in the classified forests, which are transferred from the private domain of the state to communities through a Rural Concession.

As it is widely known, changes in land tenure are a long-term process. In this case, the changes are still ongoing. Few private land titles and rural concession contracts have been delivered to the landholders and *grappes* respectively. Nevertheless, the project has been instrumental in triggering this process.

Land Tenure Security Changes in the Aberdare Range/Mt. Kenya Small Scale Reforestation Initiative

Kenya's Forest Act 2005 is the major legal instrument that regulates land and resources tenure in the country. This is the first legislation in Kenya to acknowledge the importance of sustainable forest management for C sequestration and other environmental services, but still falls short of defining C reductions ownership (although it does classify it as a non-timber product). In the Act, customary rights are recognized (Forest Act 2005, section 40, f)[3] and community participation in managing and improving the forestry sector is seen as important.

Although the project sites are fairly dispersed geographically, the land tenure condition is homogeneous across the different sites: gazetted public land under the control and administration of the Kenya Forest Service (KFS). This tenure classification is fixed, and would be very difficult to change. The control and timber rights of KFS have been solidified through the project and associated contractual process. At the same time, local communities' land tenure security was strengthened by the project activity and the C specific contractual agreements. As a result, communities now have more certainty of reaping the benefits from investments they make on land.

Conclusions

This study provides evidence that forest C projects can positively affect the level of land tenure security of participating landholders. Realizing this potential, however, entails proper project design and implementation, leading to high transaction costs (negotiating private contracts, establishing benefit-sharing agreements, negotiations). In view of the current low prices for forest C credits in international markets, private C investors may not be willing to cover these costs. This could discourage investment in agroforestry projects (especially in sub-Saharan projects), despite the potential positive social benefits from these projects.

The lessons from this study are also relevant to the discussions on REDD. The use of institutional arrangements to clarify C ownership and ensure C permanence could be an alternative for advancing the REDD agenda in the face of unclear land tenure over large forest areas. In defining their national frameworks for the implementation of REDD, countries should draw on experiences from CDM projects.

The paper also highlights the importance of equitable benefit-sharing mechanisms designed in a participatory way. Nevertheless, since C payments have not yet started in the two projects analyzed, not all the hypotheses presented here could be tested on the ground. Once revenues start to flow, the extent to which benefit-sharing schemes actually avoid conflicts and benefit less powerful individuals within these communities will be an important topic for research.

References

Barnes G, Quail S (2009) Property rights to carbon in the context of climate change. School of Forest Resources and Conservation, University of Florida. Available at siteresources.worldbank.org/INTIE/Resources/Barnes.doc. Accessed 8 Nov 2010

[3]Kenya National Land Policy – Ministry of Lands, National Land Policy Secretariat, May 2007; Kenya Forest Act – Ministry of Lands, National Land Policy Secretariat, 2005.

Bruce JW (1986) Land tenure issues in project design and strategies for agricultural development in Sub-Saharan Africa. Land Tenure Center, University of Wisconsin, Madison, 190 p

Bruce JW, Migot-Adholla SE (eds.) (1994) Searching for land tenure security in Africa. The World Bank, Kendall/Hunt Publishing Company, Dubuque, 282 p

Cotula L, Mayers J (2009) Tenure in REDD – start-point or afterthought? Natural resource issues. No. 15. International Institute for Environment and Development. London. 56 p. Available at www.cbd.int/doc/meetings/for/wscb-fbdcc…/wscb-fbdcc-01-oth-07-en.pdf. Accessed Dec 2010

Desanker PV (2005) The Kyoto Protocol and the CDM in Africa: a good idea but…. Unasylva 222(56):24–26

Eliasch J (2008) Climate change: financing global forests: the Eliasch review. Office of Climate Change, UK, 273 p

FAO (2002) Law and sustainable development since Rio: legal trends in agriculture and natural resources management. FAO legislative study no. 73. FAO Legal Office, Rome. For more information: http://www.fao.org/DOCREP/005/Y3872E/y3872e00.htm#Contents. Accessed 12 Nov 2010

Kamweti D, Acworth J (2006) Green Belt Movement BioCarbon Project, Environmental and Social Impact Assessment Study. Prepared by KAMFOR Ltd. for the BioCarbon Fund

Kunhamu TK, Kumar BM, Samuel S (2011) Does tree management affect biomass and soil carbon stocks of *Acacia mangium* Willd. Stands in Kerala, India ? In: Kumar BM, Nair PKR (eds.) Carbon sequestration potential of agroforestry systems: opportunities and challenges. Springer, Dordrecht, pp 217–228

Luttrell C, Schreckenberg K, Peskett L (2007) The implications of carbon financing for pro-poor community forestry. Forest policy and environment programme, Overseas Development Institute, Forestry Briefing, London, 145 p

Migot-Adholla S, Hazell PB, Blarel B, Place F (1991) Indigenous land rights systems in Sub-Saharan Africa: a constraint on productivity? World Bank Econ Rev 5:155–175

Omura M (2008) Property rights and natural resource management incentives: do transferability and formality matter? Am J Agric Econ 90(4):1143–1155

Perez C, Roncoli C, Neely C, Steiner JL (2007) Can carbon sequestration markets benefit low-income producers in semi-arid Africa? Potentials and challenges. Agric Syst 94(1):2–12

Pirard R (2008) The fight against deforestation (REDD): economic implications of market-based funding, vol 20, Idées Pour le Debat. Institut du Développement Durable et des Relations Internationales, Paris

Quan J, Dyer N (2008) Climate change and land tenure: the implications of climate change for land tenure and land policy. FAO, Land tenure working paper 2. IIED/University of Greenwich/FAO. 2008. Available at ftp://ftp.fao.org/docrep/fao/011/aj332e/aj332e00.pdf. Accessed 17 July 2010

Roncoli C, Jost C, Perez C, Moore K, Ballo A, Cissé S, Ouattara K (2007) Carbon sequestration from common property resources: lessons from community-based sustainable pasture management in north-central Mali. Agric Syst 94:97–109

Samek JH, Skole DL, Klinhom U, Butthep C, Navanugraha C, Uttaruk P, Laosuwan T (2011) Inpang carbon bank in northeast Thailand: a community effort in carbon trading from agroforestry projects. In: Kumar BM, Nair PKR (eds.) Carbon sequestration potential of agroforestry systems: opportunities and challenges. Springer, Dordrecht, pp 263–280

Saunders L, Hanbury-Tenison R, Swingland IR (2002) Social capital from carbon property: creating equity for indigenous people. Phil Trans R Soc Lond A 360:1763–1775

Streck C (2009) Rights and REDD+: legal and regulatory considerations. In: Angelsen A, Brockhaus M, Kanninen M, Sills E, Sunderlin WD, Wertz-Kanounnikoff S (eds.) Realising REDD+: national strategy and policy options. CIFOR, Bogor, pp 151–162

Unruh JD (2008) Carbon sequestration in Africa: the land tenure problem. Glob Environ Change 18(4):700–707

Constructing Public Policy in a Participatory Manner: From Local Carbon Sequestration Projects to Network Governance in Chiapas, Mexico

Celia Ruiz-De-Oña-Plaza, Lorena Soto-Pinto, Stephanie Paladino, Federico Morales, and Elsa Esquivel

Abstract The *Scolel Té* project is a long standing experiment in carbon (C) sequestration through agroforestry and forestry systems. Developed in Chiapas, México, this project has evolved since 1996 into a solid model to manage C stocks in indigenous small farmer (*campesino*) landholdings, to be sold in a voluntary C market and to use the C credits for financing conservation and restoration activities. The experience of *Scolel Té* has matured into a well structured system for C transactions, the Plan Vivo System, which is now being applied in other countries of Latin America and Africa. This model of C marketing has been so successful that decision makers and other stakeholders from the environmental policy arena in Chiapas have decided to adopt and modify it with the aim of transforming it into a state wide program of ecosystem services: the Chiapas Program for Ecosystem Services Compensation (PECSE). The final design and implementation of PECSE is done by a policy network called Group of Ecosystem Services for Chiapas (GESE) — a group of public and private stakeholders. The challenges for GESE will be to overcome the internal problems of coordination and to develop a political lobby that would implement the PECSE. This effort, however, is triggering an

C. Ruiz-De-Oña-Plaza • F. Morales
PROIMMSE, Universidad Nacional Autónoma de México, Calle Cuauhtemoc, 29200, San Cristobal de las Casas, Chiapas México
e-mail: celia.ecosur@gmail.com; fmorales@servidor.unam.mx

L. Soto-Pinto (✉) • S. Paladino
El Colegio de la Frontera Sur, ECOSUR, Carretera Panamericana y Periférico Sur s/n, San Cristóbal de las Casas, Chiapas 29290, México
e-mail: lsoto@ecosur.mx; macypal@gmail.com

E. Esquivel
Cooperativa Ambio, Cuitlahuac 30, San Cristóbal de las Casas, Chiapas 29290, México
e-mail: elsaesquivelb@hotmail.com

B.M. Kumar and P.K.R. Nair (eds.), *Carbon Sequestration Potential of Agroforestry Systems: Opportunities and Challenges*, Advances in Agroforestry 8,
DOI 10.1007/978-94-007-1630-8_14,

ongoing environmental governance process with implications at local, national, and international levels that could reconfigure existing strategies to tackle the problem of climate change.

Keywords Climate change • Carbon credits • Environmental governance • Policy networks

Introduction: Environmental Governance and Carbon Sequestration Projects

Climate change has been recognized as the main environmental problem today. Although numerous studies have been conducted on the processes of mitigation and adaptation, those addressing the issue of public policies and governance in carbon (C) projects are still scarce. Governance here refers to the alliance of public and private actors to build up public policies in an interactive way (Rhodes 1996; Koiman 2004). Novel arrangements for environmental governance have emerged in the form of policy networks. This means that environmental governance is increasingly the result of diverse interests, activities, and capacities of a variety of stakeholders, including governments, civil society organizations, academic institutions, and international organizations (Lemos and Agrawal 2006). Participatory and collaborative forms of governance are expected to lead to more effective improvements in environmental quality (Newig 2007).

Climate change is typically a matter of network governance. The multiplicity of stakeholders and interests involved in it call for solutions based on consensus rather than on market transactions exclusively. Perspectives on market as the main regulator for the delivery of natural resources and their commoditization have been modified as a result of the market limitations to conserve ecosystems (Hodgson 2008). A good example of this is the Payments for Environmental Services (PES) projects currently being implemented that consider social and ethical factors (Bracer et al. 2007; Jacka et al. 2008).

Here it is argued that a critical step in developing a successful PES strategy is its effective linkage with public policies in an integrated and multi-sectoral approach. This document shows how a successful, locally generated C sequestration project called *Scolel Té*, itself borne out of the alliance and interactions of different stakeholders (indigenous farmers, scientists, and nongovernmental organizations), stimulated the emergence of a regional network of civil society and government institutions focused on creating consistent guidelines for a state wide program of PES in Chiapas, México (Gibbs et al. 2002).

The chapter describes the *Scolel Té* project in terms of its current status, institutional structure, main achievements/impacts, and the key factors that work towards its permanence and stability. It also describes how the project's model of C transaction was used to develop a public policy program – the Program for Ecosystem Services Compensation for Chiapas (PECSE) – through an environmental governance process involving a policy network of private and public sectors called the Group of Ecosystem Services for Chiapas (GESE).

Scolel Té: A Project to Sell Carbon from Agroforestry Systems in Chiapas, Mexico

In the southern state of Chiapas, Mexico, a pilot project that uses forest and agroforestry (AF) systems to sequester C was initiated in 1996 through the collaboration of indigenous farmers' organizations, research institutions, and groups from the civil society. Its main objective was to improve the living standards of participating communities, using voluntary C credit payments to help conserve and restore forestry resources (Soto-Pinto et al. 2005). This effort was later turned into a permanent project called *Scolel Té*. Since 1997, the *Scolel Té* producers have been selling C sequestered in their AF plots to national and international organizations through the voluntary C market. Initially, the C sequestered was sold to FIA (Federation Internationale de l'Automovile), which agreed to buy 5,500 Mg C per year at US $12 per Mg. Since 2001, there has been a 45% increase in the amount of C sold. As of 2006, the project has sold a total of 98,754 Mg C to different buyers, such as Future Forest, Lloyd, Key Travel, The Nature Conservancy, Workers of The World Bank, and the UK Department for International Development (DFID). Ambio, a locally based cooperative in Chiapas, operates the project and deals with the administration of payments and its distribution to farmers. Ambio is also in charge of monitoring tasks, training local technicians, and fostering relations with participants in the project; its institutional structure is outlined in Fig. 1. To date, 62 communities (677 producers) have participated through a variety of forestry and AF systems. These involve up to 500 individual plots, consisting of 2,000 ha in C sequestration activities, 2,660 ha in avoided emissions activities, and more than 7,500 ha in conservation and restoration activities.[1]

The participants are smallholders, 50% of whom belong to five different Maya language groups (*Tzeltal, Tzotzil, Ch'ol, Tojolabal* and *Lacandon*). In spite of their cultural and ecological differences, the participating communities experience certain common socioeconomic problems related to land use such as strong pressures on land and other natural resources, high rates of deforestation, high levels of social marginalization, and the disruption of social and economic structures (for instance, through migration, loss of traditional knowledge, and lack of economic alternatives). While the majority of farmers participate in the project as individuals, using their own family managed landholdings for the C projects, some communities also participate on a collective basis, enrolling communally owned forest lands in the project. Individual plot sizes range from 1 to 10 ha.[2]

---

[1] Vargas-Guillen et al. (2009).

[2] Ambio (2006).

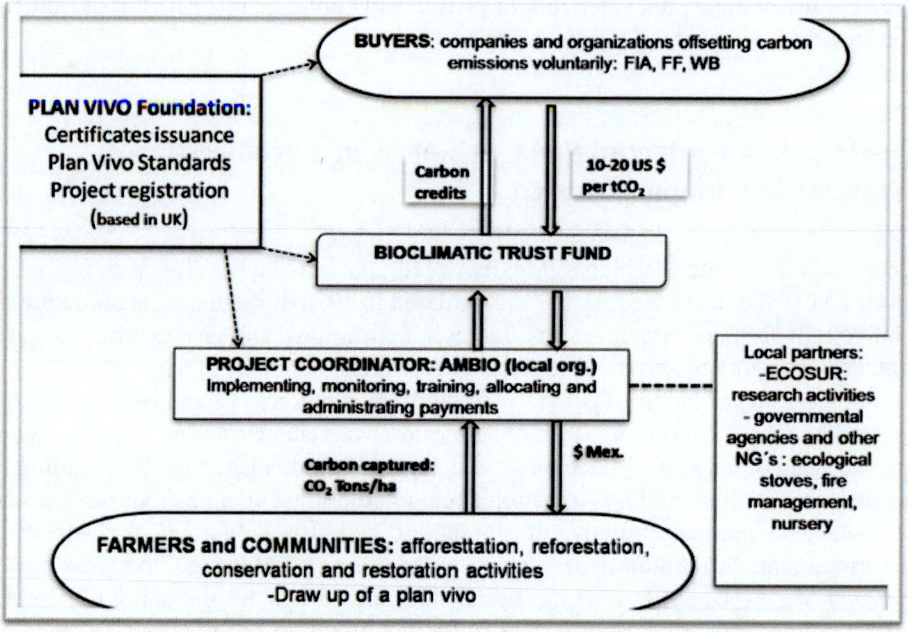

Fig. 1 Institutional structure of Scolel Té. *Solid arrows* show flows of carbon and money equivalent and *dotted arrows* show administrative procedures and knowledge exchange amongst the actors (Source: Adapted from Ambio's presentation of the project)

Farmers are involved with the project decision making process through their working groups. Each working group will be represented by a representative, who will attend all the six monthly meetings to bring suggestions to the Ambio headquarters and pass the information on to members of the working group. Since 2008, regional and community technicians' teams were formed in order to promote *Scolel Té* within new communities. Most of these technicians were previously producers themselves, which facilitates communication and gives a better insight to the new producers (for example, they can talk from a producer's point of view, which avoids mistrust, and also talk in the local language).

After more than 10 years of operation, the *Scolel Té* has become a well known and established project that has developed its own methodology and a set of standards, centered on the *Plan Vivo system.*[3] This system has been developed recently based on *Scolel Té* experience for setting up C sequestration projects under a registered C standard, the *Plan Vivo Standard*. The Plan Vivo System is governed by a Scottish charity, The *Plan Vivo* Foundation, which publicizes the projects to potential buyers and has also developed sister projects in other countries, including Uganda, Mozambique, Malawi, Tanzania and Nicaragua (http://www.planvivo.org, accessed October 2010).

[3] BDRT (2008).

Institutional Mechanisms of *Scolel Té*

Scolel Té organizers designed a bottom up approach for C transactions consistent with those initiatives that considered C as an added benefit and not as the main incentive behind the project. Therefore, *Scolel Té*'s main thrust was on identifying the best land use practices for communities in a participatory mode; and only afterward, the organizers considered how to derive a C product that could be sold in the voluntary market (Tipper 2003). The procedure that emerged from this included three main components: first, the planning process for establishing AF and forestry systems for C sequestration (see Schroth et al. 2011); second, the process of registering the potential C gains in order to sell them in a voluntary market; and third, selling the C and issuing the certificates of C credits to the owners.

Carbon Sequestration Through Forestry and Agroforestry Systems

The individual farmers or communities decide to participate in the project after attending an educational workshop on AF systems (AFS), climate change, and C sequestration services. The participants then start a planning and design process for AFS that includes an action plan called "*Plan Vivo*", which uses participatory maps, work schedules, estimation of costs, and other tools (Beniest 1994). This planning method helps the farmers to design AFS, make decisions, and identify the technical or social constraints (Soto-Pinto et al. 2008). It became a standard element of the project methodology as *Scolel Té* expanded its geographic scope over the years.

A standard *Plan Vivo* is developed in three steps: first, a simple map of the farmer's land, indicating the distribution of existing land uses (crops, fallow land, forests, rivers or streams, pastures, etc.) is drawn. Secondly, the areas for establishing AF and the choice of systems are decided, wherein the farmers specify the AF arrangements, species to be introduced, and in what densities, whether to include any associated crops, and the details on planting and maintenance activities. Finally, they estimate the costs of labor and the materials needed and also decide on a calendar of operations (i.e., when to carry out the activities; Fig. 2). A screening process for the potential participants is also built into the project, in that it helps farmers to opt out if they do not have sufficient land or other resources to preserve livelihood activities. This implies, however, that participation in the project is not feasible for farmers without a certain minimum level of resources (E. Corbera , N. Kosoy and M. Martínez-Tuna , 2006, personal communication). After drawing up the *Plan Vivo*, it is registered in a database held at the Ambio's headquarters, to serve as the baseline for monitoring tree plantings.

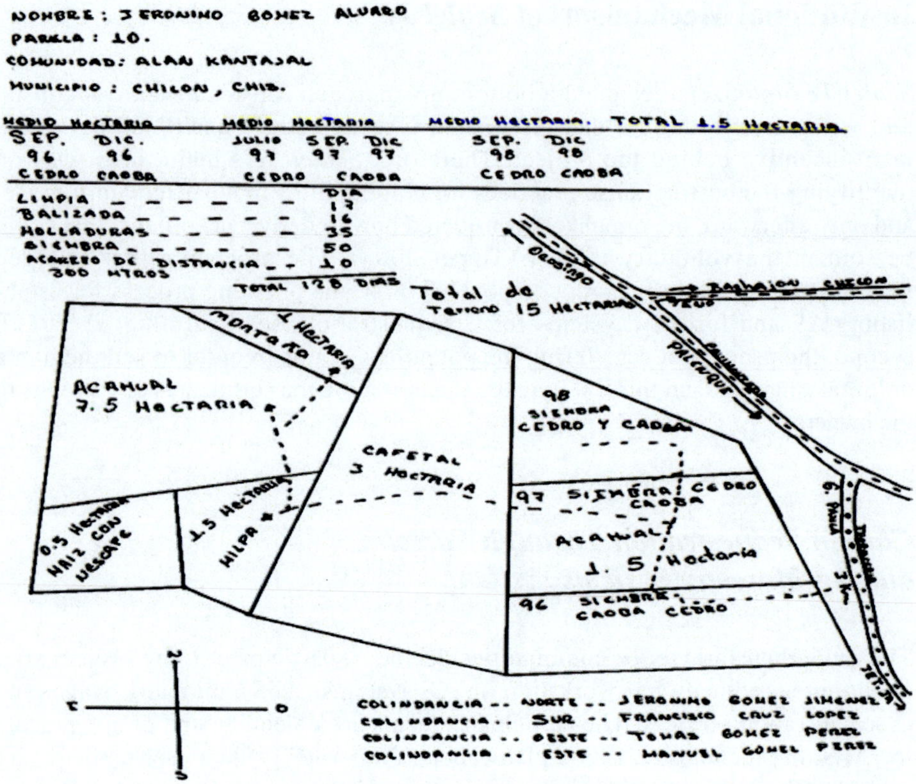

Fig. 2 Example of Plan Vivo drawn by farmers in Scolel Té (Source: Ambio's headquarters archives)

Monitoring C Sequestration and the System of Payments to the Farmers

Trees planted are the key to generate trust on the demand side (those who are going to buy the C captured by the trees planted). In order to achieve this, a fairly strict system of monitoring has been evolved by the project. A team of local technicians monitors 100% of the registered plots and between 10% and 20% of the project area is monitored by the Ambio's professional team that organizes, supervises, and supports the entire procedure. Monitoring consists of filling out a form annually with information on the performance of the plantations. It includes parameters such as: the degree to which *Plan Vivo* goals are achieved, tree mortality, growth measurements, tree species richness, health conditions, and the requirements of pruning, shade management, or clearing, with final remarks from the local technician. The monitoring system is reviewed by an independent, third party verifier, Smartwood (http://www.rainforest-alliance.org/forestry.cfm?id=smartwood_program; accessed March 2010), which guarantees transparency and refines the procedures (Fig. 2).

Under *Plan Vivo*, the farmers commit to maintain the AFS for a period of 15 years (to avoid land use changes that could result in C loss). They also receive a kind of "bankbook" for the C account, in which the total quantity of C to be sequestered is shown along with the equivalent amount of money (Tipper 2003). Ambio and their partners have standardized the estimates of C sequestered by each system over time. Carbon sequestered above a certain baseline forms the basis for payments received by the farmers (de Jong 2001; de Jong et al. 2000). The fee for C sequestered is paid to the farmer ex-ante (i.e., before the C is actually stored in the system), as described below, but the payments are withheld if targets are not reached. Moreover, only after the Ambio's technicians have verified that the trees are actually planted and that other associated tasks related to maintenance of the land have been accomplished, payments are released.

Since most of the labor and other investments take place during the establishment phase of the AFS, the main portion of the money equivalent to the C sold is distributed during the first few years itself. The payments are distributed in four installments of 18% each paid during the first three consecutive years and in the fifth year, and a final installment in the eighth year. A minimum 10% buffer is deducted from each sale agreement with a community or producer in order to raise a contingent fund to cover up the risks and uncertainties in the delivery of C credits, e.g., non-compliance by producers or any other risks that can threaten tree planting, such as natural disasters. Through this system of risk buffering, permanence is guaranteed (Sandie Fournier, Plan Vivo Foundation, June 2010, personal communication).

Carbon Credits and Their Sale in the Voluntary Market

The information gathered by technicians during the monitoring process are captured in the data base of *Plan Vivo* maps, which enables the Foundation to assess the progress of the project towards expected emission reductions. Once this assessment is done, *Plan Vivo* Foundation issues the C certificates, which the buyers will be able to acquire in the voluntary markets. These certificates have a unique serial number representing the C credits bought by a particular buyer, thus the project ensures that the same quantity of C is only sold once. The money from the C sales goes to a trust fund, called the *Fondo Bioclimático* that is managed by *Plan Vivo* Foundation who acts as an escrow agent. The payments received by farmers come from this trust fund. Because the payments are made upfront, risks of failure and overestimation of C benefits exist. If this happens, corrective actions can be instituted or compensation made from the buffer fund, referred to before.

Outcomes of the Project

The organizational structure of the project favours mainly ecological benefits. It also ensures that the payments are made to the farmers in accordance with their contributions. However, economical impacts and the farmer organizations' involvement in project decision making fall short of expectations. The following sections elucidate these issues in greater detail.

Environmental Impacts: The Ecological Benefits of Agroforestry Systems

By incorporating AFS, *Scolel Té* allowed substantial C sequestration benefits to be integrated into the regional production systems, along with other gains such as ecosystem restoration and conservation of natural resources. The implicit theme here is that rural landscapes actively managed or modified by humans are very important loci for environmental services (Harvey et al. 2006). For instance, organic shade grown coffee (*Coffea* spp.), improved fallows, and silvopastoral systems have demonstrated the value of providing environmental services, due to their complex structure and species diversity (Perfecto et al. 2003; Harvey et al. 2006). Moreover, AFS such as *taungya* (maize, *Zea mays* L., in association with trees) and improved fallows also have proven C additionality in aboveground biomass compared to traditional maize systems. Improved fallows and coffee systems are also good options for carbon conservation or sequestration, and for avoided deforestation projects (Soto-Pinto et al. 2010), since large areas were transformed from forest to secondary forest during the past few decades in Mexico (Masera et al. 1997).

To achieve the above mentioned benefits, participating farmers engage in a planning and design process (Raintree 1987), where they select the AF prototypes, species to be planted, and appropriate spatial and temporal arrangements for planting, as mentioned earlier. Shaded coffee with timber trees, *taungya*, improved fallows, pine plantations, and conservation and restoration are the most frequently selected designs (Soto-Pinto et al. 2010). Such designs usually reflect the biophysical, technical, economical, and social conditions and livelihood systems of the locality, as well as the personal interests of the farmers concerned (Vanclay et al. 2006).

Along with C sequestration, these systems are designed to help address other problems such as low productivity of swidden farming systems, inefficient land utilization, land scarcity and degradation, non-availability of forest products (timber and firewood), and low income levels (Nelson and de Jong 2003). For instance, the combination of commercial timber species with agricultural crops has contributed to the re-evaluation of the maize, coffee, and livestock farming systems. Aside from C sequestration, AFS have also shown great potential for increasing the products and services from limited space, intensifying land use while incorporating ecosystem conserving measures, and biodiversity conservation (Soto-Pinto et al. 2010). Table 1 summarizes the most frequently chosen AFS by *Scolel Té* participants and the relative amounts of C credit payments and C sequestration.

Social Impacts: Agroforestry to Avoid Conflicting Land Uses

Scolel Té's origin as a project designed to have a positive impact on the indigenous livelihoods and landscapes, with participation of farmer organizations, is particularly important in providing social benefits to the local community. The project experience during the past more than a decade shows that the focus on AFS has allowed farmers

Table 1 Agroforestry systems implemented in the Scolel Té project, Chiapas, Mexico according to carbon sequestration capacity, carbon payments by system, and farmer participation

Agroforestry systems	Area (ha)	Carbon sequestration (Mg C ha^{-1})	Number of producers/ communities per system	Unitary payment per ton of carbon per system (US$)
Taungya	107.5	99.0	134	8
Improved fallows in tropical area	398.0	96.0	304	8–13
Improved fallows in sub-tropical area	256.0	45.7	91	8
Coffee diversification with timber trees	163.1	39.0	75	8
Conservation in tropical area	6493.0	325.0	5 (communities)	4–6
Restoration in sub-tropical area	157.0	44.7	6 (communities)	8
Living fences and pastures in tropical area	256.8	43.0	182	8–10
Living fences and pastures in sub-tropical area	109.0	27.9	62	8–10

Source: Technical specifications of *Scolel Té* data base

to integrate trees into existing production systems without disrupting subsistence activities.[4] Participants strategically adopted and modified the systems chosen in order to find synergies and avoid negative impacts on labor, land, and livelihoods. For instance, many explicitly chose to plant trees in association with crops (e.g., *taungya*, coffee, fruit trees, pasture), in view of the spatial and temporal complementarities in resource use and the potential for diversified production (e.g., maize and trees). Additional synergies making the AF approach attractive to participants are that the C payments in many cases may help subsidize the production costs (e.g., labor, other inputs) of the co-planted crops too; i.e., the care and management of the trees indirectly benefit associated perennial crops such as coffee, especially during off years when prices are too low to warrant labor investments on them. Along with this, farmers anticipate non-economic benefits from the project such as learning of new skills, better familiarity with, and appreciation for the possibilities of silviculture, and leaving behind a legacy of tree planting.

Adverse Impacts of the Project: Economic Impact on Livelihoods, Future Uncertainties, and Carbon Complexities

In economic terms, however, the project has had little substantive impact on the participants' overall economic status. The data suggest that C payments could

[4]Paladino (2008).

range from 1% to 25% of overall household income.[4] However, the absolute numbers involved are small, with 2008 payments for 1 ha in a high carbon capture region of the state reaching only 20–25% of the net income that could be earned by putting that same hectare in maize.[4] These payments have not been sufficient to capitalize changes in livelihood strategies or techniques that could substantially boost the household income. Nevertheless, depending on individual circumstances and the AFS chosen, the C payments are typically more than sufficient to cover the costs incurred for establishing the trees.[4] Revenues from sales of the trees for timber, and ultimately the possibility of developing sustainably managed, smallholder-based, forestry practices could become a significant contribution to the rural economy, but this is yet to be realized and subject to many uncertainties. The realization of this potential may, in fact, surpass the lifetime of many of the older participants.

The relatively long timeline for realizing these economic benefits is a potential vulnerability of this approach to C sequestration, since farmers or their heirs could be tempted to convert the plots to other uses before the trees are saleable.

The low C sequestration prices, the uncertainty associated to obtaining timber benefits, and the technical and administrative complexities of C trading have been major disincentives of farmer participation.

The strategic involvement of farmer organizations in the project was high in the early years of the project but it got diminished due to internal political fights and economic backlashes, amongst other factors (Nelson and de Jong 2003). Implicit in this is that the greatest investment in strategic skill building and institutional capacity has been centered on Ambio itself. In recent years, however, there has been more emphasis on hiring and training participant representatives to work on technical and recruitment roles, as well as on addressing wider questions of farmer participation (Sotero Quechulpa, 2008, personal communication).

Key Determinants of the Stability of *Scolel Té* Model for Carbon Sequestration in Indigenous Communities

In spite of the obstacles encountered, *Scolel Té* strategy of conservation and restoration via C marketing keeps growing and evolving. After more than 10 years of existence, its consolidation and expansion would not have been possible without a continuing process of learning, based on self reflection, evaluation, and continuous adaptation to new challenges (Sotero Quechulpa, 2010, personal communication). Despite this ongoing evolution, the original objectives of the project have been preserved. Together with the ability to learn and evolve, these objectives have made this project one of the most trusted and prestigious C initiatives, recognized as an example of best practices in forestry (Chappel 2008).

A number of strategic factors that work towards the stabilization and permanence of the project have been identified. These factors grouped in four dimensions are

Table 2 Characterization of Scolel Té project Chiapas, Mexico according to factors promoting permanence

Dimensions and guiding principles	Key factors
Strategy of coordination and cohesion of actors and their coalitions	Successful articulation with actors from the International level (University of Edinburgh and Plan Vivo Foundation)
	Positive alliances and coalitions with governmental agencies and other NGO´s
	Strong interaction with local leaders in the communities
	Coordination with research institutions that generates new knowledge, contributes to the diffusion of the project and strengthen methodology to assess carbon stocks
	Faithfulness associated to interpersonal relations
Flexibility and simplicity of operating rules	Solid and well defined mechanisms of carbon transaction that generates confidence amongst buyers
	Strong monitoring system at the local level
	A system legitimated through international certification
	Training of local technicians
	"Plan Vivo" planning as a course of action for producers
Efficient and transparent use of resources	Self-sustaining project via carbon credits
	Divers sources of resources via other projects and alliances
	Human resources: volunteers, students, independent researchers
Discourse	Payments for environmental services are internationally promoted schemes for conservation and for climate change mitigation, specifically voluntary carbon markets have potential to trigger environmental and social benefits for the local communities involved (Chappel 2008)

Source: Authors' elaboration according to dimensions of a policy arrangement defined by Arts and Leroy (2006)

summarized in Table 2. It is worth noting that other PES strategies have also been launched in Chiapas and at the national level. Examples include the program designed and implemented by the National Forestry Commission (CONAFOR) and subsidized by The World Bank.[5] This program applies to highly biodiverse communal forest and is mainly focused on watershed and biodiversity conservation, and C fixation by forest and AF practices. A summary of the impacts of CONAFOR PES program can be found in Corbera et al. (2009).

[5]CONAFOR (2007).

Table 3 Salient attributes of the Group of Ecosystem Services for Chiapas (GESE)

Stakeholders	Origin	Objectives	Resources	Structure
Governmental agencies at federal and regional levels; national and international NGOs; research institutes and universities; and non-profit associations	Voluntary participation starting with a consultancy on PES, executed by Ambio in response to a call by the Chairman, State Commission of Sustainable Forestry	• To design and implement a state wide program of PES (PECSE) • To lobby for the inclusion of PES in the environmental agenda government • To generate and exchange knowledge on PES	Voluntarily granted by partners occasionally	Inter-institutional group with tripartite representation - elected amongst members Open memberships Five working commissions

PES Payments for Environmental Services; *PECSE* Program for Ecosystem Services Compensation

Scaling Up Scolel Té to a Public Policy Program: Steps Towards a Process of Network Governance

Given these characteristics, the *Scolel Té* experience has been seen as a model for an expanded PES program that could be developed with the participation of a broader set of stakeholders. In 2007, Ambio, the government agencies dealing with forest conservation and protected areas management, and other national and international NGOs have joined to form a policy network to lobby for the inclusion of the PES strategy in the environmental agenda of the state government. This network is called the Group for Ecosystem Services of Chiapas (GESE) and it foresees the possibility of conserving more natural resources and reaching out to more rural communities. The main objective of GESE is to design a Program for Ecosystem Services Compensation for Chiapas (PECSE) and to find ways to implement it as a networked strategy. This initiative triggered a process of strategic coordination among the stakeholder, but poses huge challenges in terms of task allocation, resource distribution, information management, articulation of competing interests, and the construction of a common view to which all parties must agree and commit (Table 3).

Despite such challenges, there are important advantages to implementing programs in a network fashion (Slaughter and Zaring 2006). These include the possibility of integrating a range of opinions and perspectives that, in turn, may enrich and grant legitimacy to the program; an exchange of information and its diffusion at all levels that eventually will strengthen links between the public and private sectors; and the coordination of policies in order to achieve a more efficient use of scarce resources and a better correspondence between the society's expectations and needs, and the government programs.

After 2 years of monthly meetings, the GESE network has achieved some of its goals in terms of putting the issue of PES into the government agenda in Chiapas. For example, the GESE network is in charge of the environmental services section for the future Action Plan of Climate Change for Chiapas, launched by the Ministry of Environment and Urbanism of the state government. In addition, a network of networks initiative is just emerging with the aim of developing a REDD pilot project in Chiapas, in cooperation with the Mexican Carbon Program (PMC). This initiative links national, state, and local efforts to develop a robust methodology for local level monitoring of the C stocks in forests under the REDD scheme (F. Paz, 2010, personal communication). It will require the involvement of local organizations at community level to conduct monitoring activities of land use changes. The role of GESE organizations, in coordinating the grass root organizations and building a network of local and community technicians that can generate data for national C stock accounting, will be critical to the success of this initiative. Ambio has been playing a central role in the above initiatives and in providing key information based on *Scolel Té* experience to implement the technical aspects of these strategies. Although the existing PECSE proposal adopts the technical and organizational facets of the *Scolel Té* C transactions model (Fig. 3), it falls behind in the establishment of institutional mechanisms that promote an integrated regional approach.

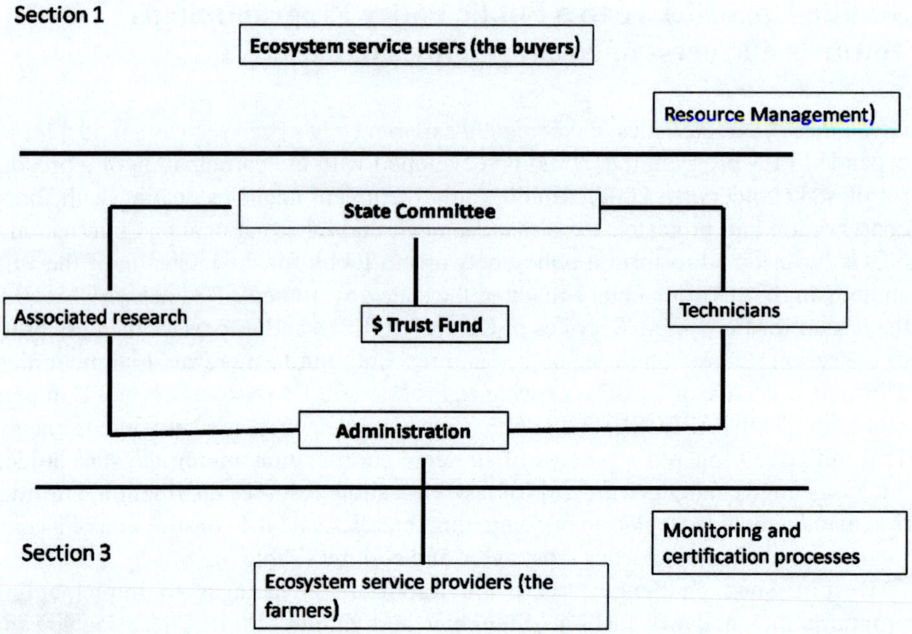

Fig. 3 Projected structure of PECSE program. Note the similarities with Scolel Té institutional architecture (Source: Vargas-Guillen et al. 2009)

Some Final Remarks

Some aspects of the model of PES created by *Scolel Té* have been adapted into a proposal for a public policy program, as a process of environmental governance. The strategy on PES, as it is being implemented in the *Scolel Té* project coincides to some degree with the perspectives promoted by the Chiapas state government. There are new developments in the state legislation in relation to PES strategy that provide a legal framework for PES implementation at the state level.[6] The Strategic Development Plan for Chiapas (Plan de Desarrollo Chiapas Solidario 2007–2012; www.chiapas.gob.mx/plan/; accessed March 2010), the vital document for planning state policies, also includes PES as one of the key strategies for conservation. This has resulted in the creation of specific departments to deal with ecosystem services in Chiapas.

This is not the place to expound on the characteristics of political dynamics in Chiapas and Mexico in the field of the environmental public policies. It should however be noted that, Mexico and Chiapas have emerged as world leaders through their innovative experiences in dealing with climate change. The past Conference of the Parties, COP16, of UNFCCC (the United Nations Framework Convention on

[6]Zorrilla-Ramos (2006).

Climate Change) was held in Cancún, and Mexico City will host the full meeting of UNFCCC in 2012, where the Kyoto Protocol will be renegotiated.

Should the GESE network and its efforts succeed, this would prove to be a worldwide example and a strong argument in favor of the final approval of REDD strategies in a post-Kyoto environment (F. Paz, 2010, personal communication). This has enormous implications for developing countries and emerging economies like Mexico in terms of obtaining funding to conserve forests and combat climate change. At the international level, the value of the GESE network rests upon these considerations. Networking around a PES political strategy at the regional level has proven to be neither easy nor quick, but it could open up democratic structures for managing natural resources, with a potentially win-win scenario for all stakeholders.

Acknowledgements Authors would like to thank Consejo Nacional de Ciencia y Tecnología for support (EP-2004-C01-46244, FORDECYT-ECOSUR 116306) and to Ulises Ramírez Mijangos for data management, Sandie Fournier and Alexa Morrison from *Plan Vivo* Foundation for their comments on the final draft. We are also grateful to the two anonymous referees for the valuable comments, suggestions, and corrections made on a previous version of the paper.

References

Ambio C (2006) Scolel Te. Reporte Anual 2005. Cooperativa Ambio, San Cristóbal de las Casas, 12 p

Arts B, Leroy P (eds.) (2006) Institutional dynamics in environmental governance. Springer, Dordrecht, 283 p

BDRT (2008) *Plan Vivo* system and standards. Carbon management and rural livelihoods. BioClimate Research and Development, Edinburgh

Beniest J (1994) Diagnosis and design exercise book. International Centre for Research in Agroforestry, Embu, 59 p

Bracer C, Scherr S, Molnar A, Sekher M, Ochieng B, Sriskanthan G (2007) Organization and governance for fostering pro-poor compensation for environmental services ES scoping study issue paper no 4 ICRAF Working Paper no 39. World Agroforestry Centre, Nairobi, 55 p. Available at http://www.worldagroforestry.org/downloads/publications/PDFs/wp14961.pdf. Accessed Oct 2010

Chappel A (2008) Making the voluntary carbon market work for the poor. Current and future roles. Forum for the Future. DFID. Available at www.forumforthefuture.org.uk. Accessed Jan 2010

CONAFOR (2007) En marcha proyecto especial de servicios ambientales del bosque. México Forestal. Revista electrónica de la Comisión Nacional Forestal

Corbera E, González Soberanis C, Brown K (2009) Institutional dimensions of payments for ecosystem services. An analysis of Mexico's carbon forestry programme. Ecol Econ 68:743–761

de Jong BHJ (2001) Uncertainties in estimating the potential for carbon mitigation of forest management. For Ecol Manag 154:85–104

de Jong BHJ, Ochoa-Gaona S, Castillo-Santiago M, Ramírez-Marcial N, Cairns M (2000) Carbon flux and patterns of land use/land cover change in the Selva Lacandona, Mexico. Ambio 29:504–511

Gibbs D, Jonas A, While YA (2002) Changing Governance Structures and the Environment: Economy–Environment Relations at the Local and Regional Scales. Journal of Environmental Policy & Planning 4:123–138

Harvey CA, Sánchez AM, Vilchez S, Hernández B, Sáenz JC, Maes JM, Casanoves F, Sinclair FL (2006) Patterns of animal diversity associated with different forms of tree cover retained in agricultural landscapes. Ecol Appl 16:1986–1999

Hodgson G (2008) What is wrong with mainstream economics? And how could economics be improved? Available at http://www.feed-charity.org/our-differences-with-mainstream-economics.htm. Accessed March 2009

Jacka BK, Kouskya C, Simsa KR (2008) Designing payments for ecosystem services: lessons from previous experience with incentive-based mechanisms. Proc Natl Acad Sci USA 105:9465–9470

Koiman J (2004) Gobernar en gobernanza. Revista Instituciones y Desarrollo Institut Innternacional de Governabilitat de Catalunya 16:171–194

Lemos MC, Agrawal A (2006) Environmental governance. Annu Rev Environ Resour 31:297–325

Masera O, Ordonez J, Dirzo R (1997) Carbon emissions from Mexican forests: the current situation and long-term scengarios. Clim Ch 35:265–295

Nelson KC, de Jong BH (2003) Making global initiatives local realities: carbon mitigation projects in Chiapas, Mexico. Glob Environ Change 13:19–30

Newig J (2007) Does public participation in environmental decisions lead to improved environmental quality? Towards an analytical framework. Comm Coop Particip (International Journal of Sustainability Communication) 1(1):51–71. Available at www.ccp-online.org/docs/artikel/01/03_Forschung_Newig_final.pdf. Accessed Oct 2010

Paladino S (2008) The social and economic impacts of campesino adoption of Agroforestry systems for carbon capture – case study of participants in Scolel Te', Chiapas, Mexico El Colegio de la Frontera Sur, San Cristóbal de las Casas, Chiapas, Mexico and Consejo Nacional de Ciencia y Tecnología, Mexico City. Internal Project Report.

Perfecto I, Mas A, Dietsch TV, Vandermeer J (2003) Conservation of biodiversity in coffee agroecosystems: a tri-taxa comparison in southern Mexico. Biodivers Conserv 12:1239–1252

Raintree JB (1987) The state-of-art of agroforestry diagnosis and design. Agroforest Syst 5:219–250

Rhodes R (1996) The New Governance: governing without government. Polit Stud 44:652–667

Schroth G, Laderach P, Dempewolf J, Philpott S, Haggar J, Eakin H, Castillejos T, Garcia J, Soto Pinto ML, Hernandez R (2009) Towards a climate change adaptation strategy for coffee communities and ecosystems in the Sierra Madre de Chiapas, Mexico. Mitigation and Adaptation of Strategies for Global Change 14:605–625

Schroth G, da Mota MSS, Hills T, Soto-Pinto L, Zepeda Y, Wijayanto I (2011) Linking carbon, biodiversity and livelihoods near forest margins: the role of agroforestry. In: Kumar BM, Nair PKR (eds.) Carbon sequestration potential of agroforestry systems: opportunities and challenges. Springer, Dordrecht, pp 179–200

Slaughter A-M, Zaring D (2006) Networking goes international: an update. Annu Rev Law Soc Sci 2:211–229

Soto-Pinto L, Jiménez-Ferrrer G, Vargas A, de Jong B, Bergsma B, Esquivel-Bazán E (2005) Experiencia agroforestal para la captura de carbono en comunidades indígenas de México. Revista Forestal Iberoamericana 1:44–50

Soto-Pinto L, Jiménez-Ferrer G, Lerner-Martínez T (2008) Diseño de sistemas agroforestales para la producción y la conservación. Experiencia y tradición en Chiapas. ECOSUR, México, 70 p

Soto-Pinto L, Anzueto-Martinez M, Mendoza V, Jimenez-Ferrer G, de Jong B (2010) Carbon sequestration through agroforestry in indigenous communities of Chiapas, Mexico. Agroforest Syst 78:39–51

Tipper R (2003) El apoyo a la participación de campesinos indígenas en el mercado internacional de servicios de carbono: el caso de Scolel Té. In: Pagiola S, Bishop J, Landel-Mills N (eds.) La venta de servicios ambientales del bosque. SEMARNAT. INE, Mexico, pp 355–370

Vanclay J, Prabhu R, Sinclair F (2006) Realizing community futures. A practical guide to harnessing natural resources. Earthscan, London, 162 p

Vargas-Guillen A, Aguilar-Martínez S, Castillo-Santiago MA, Esquivel-Bazán E, Hernández-Vázquez MA, López-Gómez AM, Quechulpa-Montalvo S (2009) Programa estatal para la compensación por servicios ecosistémicos. Una propuesta para Chiapas. Series Acciones/No5. Corredor Biológico Mesoamericano, CONABIO, SEMARNAT and Global Environmental Facility, México, D.F., 54 p

Zorrilla-Ramos M (2006) Elaboración de la estrategia regional de compensación por prestación de servicios ambientales en el área del Corredor Biológico Mesoamericano-México. Fase 1. Diagnóstico de instrumentos legales y de política pública. CBM, Informe final, 120 p

Inpang Carbon Bank in Northeast Thailand: A Community Effort in Carbon Trading from Agroforestry Projects

Jay H. Samek, David L. Skole, Usa Klinhom, Chetphong Butthep, Charlie Navanugraha, Pornchai Uttaruk, and Teerawong Laosuwan

Abstract Carbon (C) is a new commodity that is now traded in financial markets and there is potential for farmers adopting agroforestry to sell C in addition to traditional timber and non-timber agroforestry commodities. Implementing agroforestry C offset projects is a challenging task, however, and it requires new, market-approved, C accounting methods that reduce transaction costs. This paper describes the Inpang Carbon Bank project in Northeast Thailand developed through collaboration between the Inpang Community Network, scientists at the Department of Forestry, Michigan State University (USA), Mahasarakham University (Thailand), and colleagues at the National Research Council of Thailand. Under this project a new protocol has been developed, which is in review by the Chicago Climate Exchange, besides an on-line C offset monitoring, verification, and reporting management system, called Carbon2Markets. A cost recovery analysis for the Inpang Carbon Bank smallholder teak (*Tectona grandis* L.f.) offset project shows that C would have to be sold at a value not less than US$1.66 per Mg CO_2.

Keywords Carbon bank • Carbon financial markets • Carbon sequestration • Climate mitigation

---

J.H. Samek (✉) • D.L. Skole
Department of Forestry, Michigan State University, East Lansing, MI 48823, USA
e-mail: samekjay@msu.edu; skole@msu.edu

U. Klinhom • C. Navanugraha • P. Uttaruk • T. Laosuwan
Mahasarakham University, Tambon Kamriang, Kantarawichai District, Mahasarakham 44150, Thailand
e-mail: usa_klinhom@yahoo.com; encnv@hotmail.com; pornchai.u@msu.ac.th; teerawong@msu.ac.th

C. Butthep
Office of International Affairs, National Research Council of Thailand, 196 Phaholyothin Road, Chatuchak, Bangkok 10900, Thailand
e-mail: butthep@msu.edu

B.M. Kumar and P.K.R. Nair (eds.), *Carbon Sequestration Potential of Agroforestry Systems: Opportunities and Challenges*, Advances in Agroforestry 8,
DOI 10.1007/978-94-007-1630-8_15,

Introduction

Greenhouse gas emissions from deforestation and forest degradation and the climate change mitigation potential of forested landscapes are well documented (IPCC 2007). Pressures on tropical forest resources by local people may be alleviated through adoption of agroforestry (Angelsen and Kaimowitz 2004; Montagnini and Nair 2004). Agroforestry, the use of trees on farm, including the domestication of indigenous trees, provides a variety of potential income streams from both timber and non-timber products (Michon and de Foresta 1996; Simons and Leakey 2004). It can also play an important role in sustaining a variety of ecosystem services (Jose 2009), including climate mitigation through carbon (C) sequestration (Nair et al. 2009). Furthermore, C itself is now a commodity trading on a number of greenhouse gas or "carbon" financial markets, both regulatory and voluntary (Kossoy and Ambrosi 2010). Agroforestry, therefore, has the potential to both mitigate climate change and provide an additional income stream to farmers, beyond the income generated from traditional timber and non-timber products.

Recognizing that disperse small scale agroforestry farms in developing countries are sequestering C in biomass and therefore mitigating climate change, there are a number of challenges linking agroforestry farmers to buyers willing to pay for C offsets. Transaction costs to implement any forestry C offset project are non-trivial. These costs include identifying and demarcating project boundaries, collecting field based biometric data, C measuring and monitoring tools and tasks, third party verification, and project reporting. Furthermore, the heterogeneous nature of agroforestry (spatial planting configurations and species diversity) adds greater complexity to biotic C accounting. Markets expect C offsets to be real, verifiable, and permanent, and C offset methods and protocols are designed to ensure these requirements. While there are a growing number of newly proposed forest C accounting methods and protocols, there are currently only a few that are market accepted, and these do not include the broad range of agroforestry systems practiced in developing countries. For example, under the Kyoto driven regulatory market, the United Nations Framework Convention on Climate Change (UNFCCC) Clean Development Mechanism (CDM) approved methodologies only provide for Afforestation/Reforestation C offset projects (UNFCCC 2010a, b). The largest voluntary C market, the Chicago Climate Exchange (CCX), currently only includes Afforestation/Reforestation and Sustainably Managed Forest Projects (CCX 2010).

This paper documents the collaborative international and institutional efforts to develop an agroforestry C sequestration offset project in Thailand that uses advanced Internet based geospatial tools in a C management application which functions as a monitoring, reporting, and verification system for C offset projects. One outcome of this project is a new protocol for "Biotic Carbon Sequestration in Small Scale Agroforestry Systems in Developing Countries" and a smallholder teak offset project submitted to the CCX. Researchers from the National Resource Council of Thailand (NRCT), Mahasarakham University in Thailand, and Michigan State University in the United States have partnered in this project with a farmer's association in Northeast Thailand, called the Inpang Community Network (Inpang). Members of the Inpang network call this their "Carbon Bank" project. Their concept of a "Carbon Bank" is one in which living trees in the agroforestry system have a marketable value beyond the traditional market commodities associated with timber or non-timber forest products. The value is in the tree's C. These trees provide multiple benefits to the farmers including sequestering C in biomass, much like a bank secures gold or currency.

Methods

Genesis of the Inpang Network Carbon Bank Project

The Inpang network began in 1987 with a group of village leaders in Ban Bua Village, Tambon Kut Bak, Kut Bak District, Sakon Nakhon Province (17° 5′ 14″N, 103° 49′ 21″E). In order to break the cycle of debt from cash cropping, the farmers began to transform their farm landscapes from costly high input, chemical dependent monocultures to diverse agroforestry systems that included rice (*Oryza sativa* L.) for consumption and a wide variety of woody perennials. From a small group of 12 members, the Inpang network has now grown to over 4,000 members in five provinces in northeast Thailand (Fig. 1), with linkages to many other farmer groups

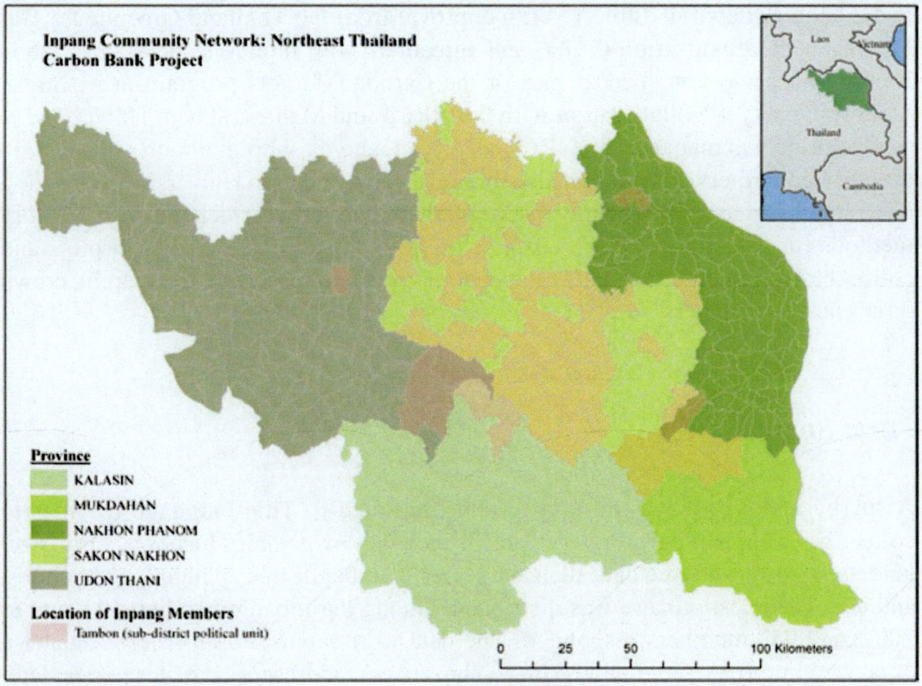

Fig. 1 Map of Inpang member's locations by *tambon* (sub-district) in five provinces in northeast Thailand

throughout Thailand. Inpang members grow hundreds of native woody perennials aimed at promoting the use of forest products from on-farm resources, rather than harvesting and collecting from the natural, protected forests in the nearby Phuphan National Park. The network currently markets a number of secondary products, including *makmao* (*Antidesma acidum* Retz.) fruit juice, wine, and various herbal medicines.

Capacity Building and Networking

The concept of selling C, or actually, selling *sequestered atmospheric carbon dioxide*, stored in trees, as a farm level commodity is not a simple notion easy to comprehend. Clear understanding of the C cycle, climate change, and the role of forests and trees in the context of climate change are also not easily attained. To build capacity in understanding these concepts and to develop a collaborative project concept, a series of meetings and workshops were held, starting in 2005, with national level agencies [National Research Council of Thailand (NRCT), Office of Natural Resources and Environmental Policy and Planning (ONEP), Land Development Department (LDD), and Royal Forest Department of Thailand (RFD)], regional collaborators of this project (Mahasarakham University and Mahidol University), and Inpang members. Details of the nine meetings and workshops which have taken place between 2005 and 2009 are shown in Table 1. With approval from the Thailand Greenhouse Gas Management Organization (TGO) and agreement with Inpang leaders, the Carbon Bank project was launched as part of the Carbon2Markets program at Michigan State University in collaboration with the NRCT and Mahasarakham University.

The project is managed by NRCT as a pilot activity who maintain contact with the enrolled farmers directly and also through Mahasarakham University colleagues. NRCT and Mahasarakham University collaborate to train farmers in tree inventory methods for their agroforestry farms, e.g., how to establish permanent plots and collect basic biometric tree data: diameter at breast height (DBH), tree height, crown dimensions, and others.

Farm Surveys

A survey instrument was developed and translated to Thai language in order to collect basic Inpang agroforestry data. It included ownership, location, farm size, land use history, and tree data such as species, number of trees planted, age of trees, and use. The questionnaire was distributed among the Inpang members in October 2007, and 957 members responded. The data were input into an Access database organized in three related tables: ownership (name, address, size of farm area, land use history data), list of trees species (Latin and local names), and an agroforestry table (owner id, species planted, age of trees by species, number of trees planted

Table 1 Meeting, trainings, and workshops supporting the Inpang Carbon Bank project

Date	Location	Participant groups	Number of participants
Dec. 2005	Bangkok, Thailand	ONEP, NRCT, LDD, RFD, Mahasarakham University, Kasetsart University, Mahidol University	26
Apr. 2007	Inpang Learning Center, Kut Bak, Sakon Nakhon	NRCT, Mahasarakham University, Inpang Community Network	48
Apr. 2007	LDD, Bangkok, Thailand	LDD, ONEP, NRCT, RFD, Mahasarakham University, Suranaree University, Mahidol University	14
Aug. 2007	Inpang Learning Center, Kut Bak, Sakon Nakhon	NRCT, Mahasarakham University, Inpang Community Network	55
Sep. 2007	Inpang Learning Center, Kut Bak, Sakon Nakhon	NRCT, Mahasarakham University, Inpang Community Network	31
Oct. 2007	Inpang Learning Center, Kut Bak, Sakon Nakhon	NRCT, Mahasarakham University, Inpang Community Network	38
Jan. 2009	Bangkok, Thailand	Mahidol University	75
Jan. 2009	Inpang Learning Center, Kut Bak, Sakon Nakhon	NASA Science Mtg participants, NRCT, Mahasarakham University, Inpang Community Network	28
Jun. 2009	Bangkok, Thailand	TGO	14

by species, and use of tree species). The data from this initial survey provided preliminary information upon which an Inpang Carbon Bank information system was developed. These preliminary data were not sufficient for C accounting or registering agroforestry C offset projects. However, the data allowed us to stratify the diversity of Inpang Carbon Bank agroforestry practices in multiple ways: geographically, by species composition, and by the extent of area under agroforestry. Using the location information in the database, we developed an Inpang Carbon Bank Geographical Information System. Figure 2 shows how these data were used to stratify the Inpang Carbon Bank agroforestry areas and prioritize C offset project planning activities.

Internet-Enabled Geographical Information System (GIS) Content Management System

The requirements of C financial markets demand scientifically robust, accurate methods to measure and monitor biotic C in order to show that sequestered C in offset projects is *real*, *verifiable,* and *permanent*. To meet these requirements we developed an Internet-enabled GIS content management system (www.carbon2markets.org). The system serves multiple functions: archives C offset biometric and geospatial data (GIS and satellite remote sensing), uses C accounting models and algorithms

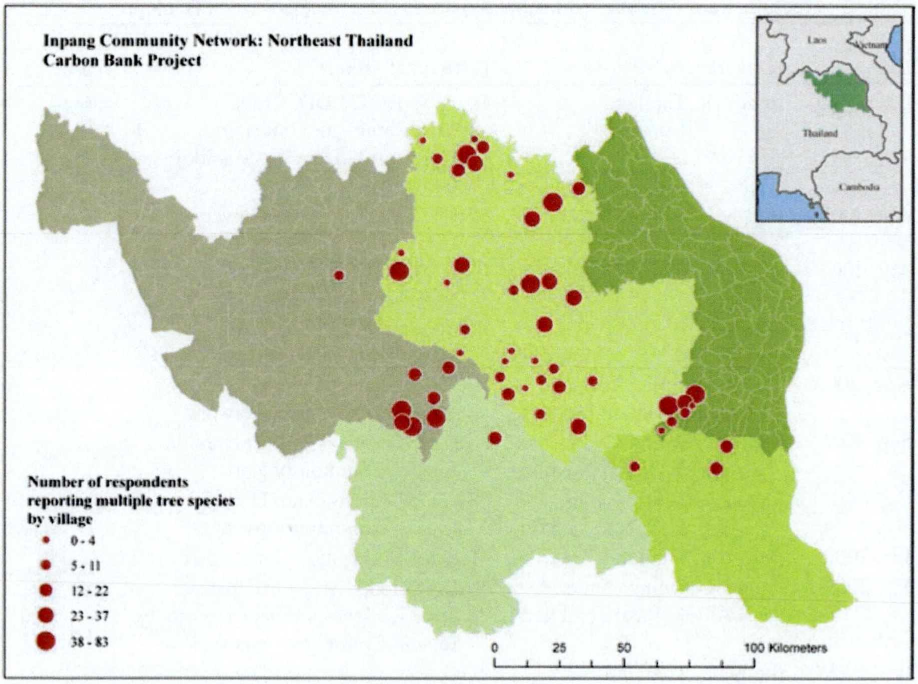

Fig. 2 Geographical Information System (GIS) for the Inpang Carbon Bank preliminary data: village-wise distribution of respondents reporting planting multiple agroforestry species

to calculate C stock baselines and future projections of C sequestration in registered areas, provides monitoring functions using annual or bi-annual hyper-resolution remote sensing satellite data (e.g. IKONOS, GeoEYE, QuickBird), allows for verification and validation of C offsets with access to all data (down to the tree level biometric data) and C accounting protocols (e.g., what specific allometric equations have been used to calculate biomass and C), and serves as a link between smallholder farmers in developing countries and the C financial markets.

Carbon Accounting Protocol for Smallholder Teak

The first offset activity of the Inpang Carbon Bank project enrolled 98 farmers and 114 smallholder teak (*Tectona grandis* L.f.) woodlots. The smallholder teak C offset project had 44 teak areas owned by Inpang members, and 54 additional areas owned by non-Inpang farmers (10 in Uttaradit Province, 20 in Nakhon Sawan Province, and 24 in Nong Bua Lumphu Province). The average size of the teak areas was less than 3 ha and the total area enrolled in Carbon2Markets was 283.27 ha. One hundred and seventy permanent plots with a minimum size of 20 × 25 m were established and 13,021 teak trees tagged and their dbh and height measurements recorded.

We calculated the baseline C stock (reporting period: 2009) of the total enrolled area using an allometric equation developed for aboveground teak biomass in Thailand (Eq. 1: Petmark and Sahunalu 1980). The IPCC (2003) model (Eq. 2) was used for computing the belowground biomass. The projected rate of C sequestration was derived based on the 2009 C stock with teak age ranging from 5 to 18 years (the oldest stand planted in 1992). To estimate the baseline C stock of the total enrolled area we calculated the CO_2 stock (Mg CO_2) for each tree in each sample plot (Eqs. 1–4), stock per hectare for each enrolled teak woodlot (Eq. 5), and the total stock for each enrolled teak woodlot (Eq. 6). The baseline C stock for the project is the sum for all teak areas. For a number of the larger teak areas (>2 ha) we established more than one sample plot to capture the variability in growing conditions (e.g., soil nutrient and moisture differences, besides topographic relief). We used the average stock of all plots within a teak area (if more than one), multiplied by area to estimate the C stock of each woodlot (Eq. 6).

$$AGB\left(\text{aboveground biomass, kg}\right) = Ws + Wb \tag{1}$$

Where Ws = stem biomass (kg): $log Ws = 0.9797 * log(D\text{^}2H) - 1.6902$, Wb = branch biomass (kg): $log Wb = 1.0605 * log(D\text{^}2H) - 2.6326$, D = tree diameter at breast height (cm), H = tree height (m)

$$BGB\left(\text{belowground biomass, kg}\right) = exp\left(-1.0587 + 0.8838 * ln\left(AGB\right)\right) \tag{2}$$

$$C\left(tree\right) = \left(\left(AGB + BGB\right) / 1000\right) * 0.5 \tag{3}$$

$C\ (tree)$ = amount of carbon in each tree within a plot (Mg)

$$CO_2\left(tree\right) = C * \left(44/12\right) \tag{4}$$

$CO_2\ (tree)$ = amount of carbon dioxide in each tree within a plot (Mg)

$$CO_2 plot = \sum CO_2\left(tree\right) * \left(10000 / plot\ area\right) \tag{5}$$

$CO_2 plot$ = the amount of CO_2 per hectare for each plot sampled (Mg)

$$TotalCO_2 for\ each\ teak\ areas\left(Mg\right) = AveCO_2 plot * teak\ site\ area \tag{6}$$

$AveCO_2\ plot = \Sigma$ CO_2 plot within a single teak area/total number of plot within a single teak area (Mg), teak site area = total area of teak site (ha)

Economic Viability of the Smallholder Teak Project

A cost recovery analysis was performed for the Inpang Carbon Bank smallholder teak offset project using our project preparation and development costs to date, and assuming standard verification and certification costs. The project preparation and development costs included: workshops and training, field data collection, data analysis, project management and administration, and the drafting of the CCX Project Implementation Document (PID). This analysis is simply a sum of the total costs divided by the expected amount of C sequestered by the project over a 15 year period. It was used to derive the rate at which the C would need to be sold in order to recover the costs of implementing the project and to determine the economic viability of the project (greater the difference of this value below the market value of C, greater is the viability of the project).

Results

Participatory Agroforestry Mitigation

The early results of the Inpang Carbon Bank project are promising. The Inpang community network, already an environmentally conscious farmers' organization, embraced the practice of mitigating climate change through the ecosystem services provided by trees and woody perennials. After some training (Table 1), the farmers also appeared to understand the basics of the C cycle and how local level land use and land cover (e.g., agroforestry) can positively impact the global phenomenon of climate change resulting from anthropogenic emissions of greenhouse gases. At the national level in Thailand, within the agencies responsible for climate related policy and project oversight (such as NRCT, ONEP, and TGO), there has been an increasingly greater acceptance of the idea that forestry and agriculture should be part of the solution set for combating climate change. They also started recognizing the role for the voluntary markets especially with respect to agroforestry and smallholder farmers whose on-farm activities help sequester C.

The smallholder teak C offset project initiated by the Inpang Carbon Bank activity enrolled nearly 300 ha of teak woodlots. The farmers who signed project enrollment agreements have been encouraged not to harvest the trees for a period of at least 15 years and to replant with other long-lived woody perennials, wherever harvesting was unavoidable. Carbon losses from tree cutting have been monitored as part of the project. We calculated the baseline C for the enrolled lands as 44,801 Mg CO_2 and estimated a conservative annual rate of sequestration at 10.62 Mg CO_2 ha^{-1} based on the plot level C stock data and age of teak stands. This rate is consistent with the values reported in the literature (Mittelman 2000; Pandey and Brown 2000; Reid and Stephen 2001). At 10.62 Mg CO_2 ha^{-1} $year^{-1}$, the project is expected to sequester an estimated 45,125 Mg CO_2 over a 15 year period.

The cost of developing the Thailand smallholder teak project was approximately US$30,000.00 (Table 2). The total project area of 283.27 ha was estimated to sequester 45,125 Mg CO_2 over a 15 year period. To recover the costs of establishing this project, the C would have to be sold for about US$0.67 per Mg CO_2. If we include an estimated verification and certification cost of US$15,000

Table 2 Costs associated with Thailand Teak Agroforestry Carbon Project

Activity	Cost (US$)
Workshops and community training	6,000
Field Data Collection – GPS / Biomass data in permanent plots	10,000
Data analysis & satellite imagery	4,000
Project management	8,000
Project Implementation Documentation (PID)	2,000
Sub-total	30,000
Verification and certification (15,000 × 3)	45,000
Total	75,000

All costs except verification/certification are estimates from NRCT program expenses and MSU funded projects' expenses. The verification and certification costs are based on Merger (2008) which compares a number of standards operating in the voluntary C market

every 5 years (total project cost of US$75,000), then the break even cost would rise to US$1.66. This would be the cost recovery fees prior to any financial benefits going to the farmers, and assumes that the full amount of C would be sold over the 15 year period (no reserve lost to leakage). With forestry C being sold in the voluntary, over-the-counter (OTC), markets at US $8.44 per Mg CO_2 (Hamilton et al. 2010), this project could potentially realize a profit of US$305,855.00 for payment to the enrolled 98 farmers ((US$8.44 × 45,125 Mg CO_2) minus US$75,000).

All 114 enrolled teak areas have been uploaded to the Carbon2Markets on-line management system. The system reports data and information for the project at the total aggregate level down to the individual tree measurements within a plot. The system shows both tabular and geographic data. Figure 3 is an example of an enrolled teak woodlot boundary with high resolution satellite data and the list of tree data within one of the inventory plots.

Based on this smallholder teak project, we developed a new protocol for "Biotic Carbon Sequestration in Small Scale Agroforestry Systems in Developing Countries", which is currently under review by the CCX forestry offsets committee. We have also submitted to the CCX a Project Implementation Document (PID), "Small Scale Agroforestry Development in Thailand" for trading C offsets generated by the smallholder teak project.

Inpang Agroforestry Practices and Potential for C Offsets

The data from 957 Inpang respondents show that these farmers have planted and are managing a great diversity of tree species on their farms. Inpang respondents identified 254 different woody perennial species on their agroforestry farms. Timber is not the only reason for Inpang farmers to plant and manage trees on their farms.

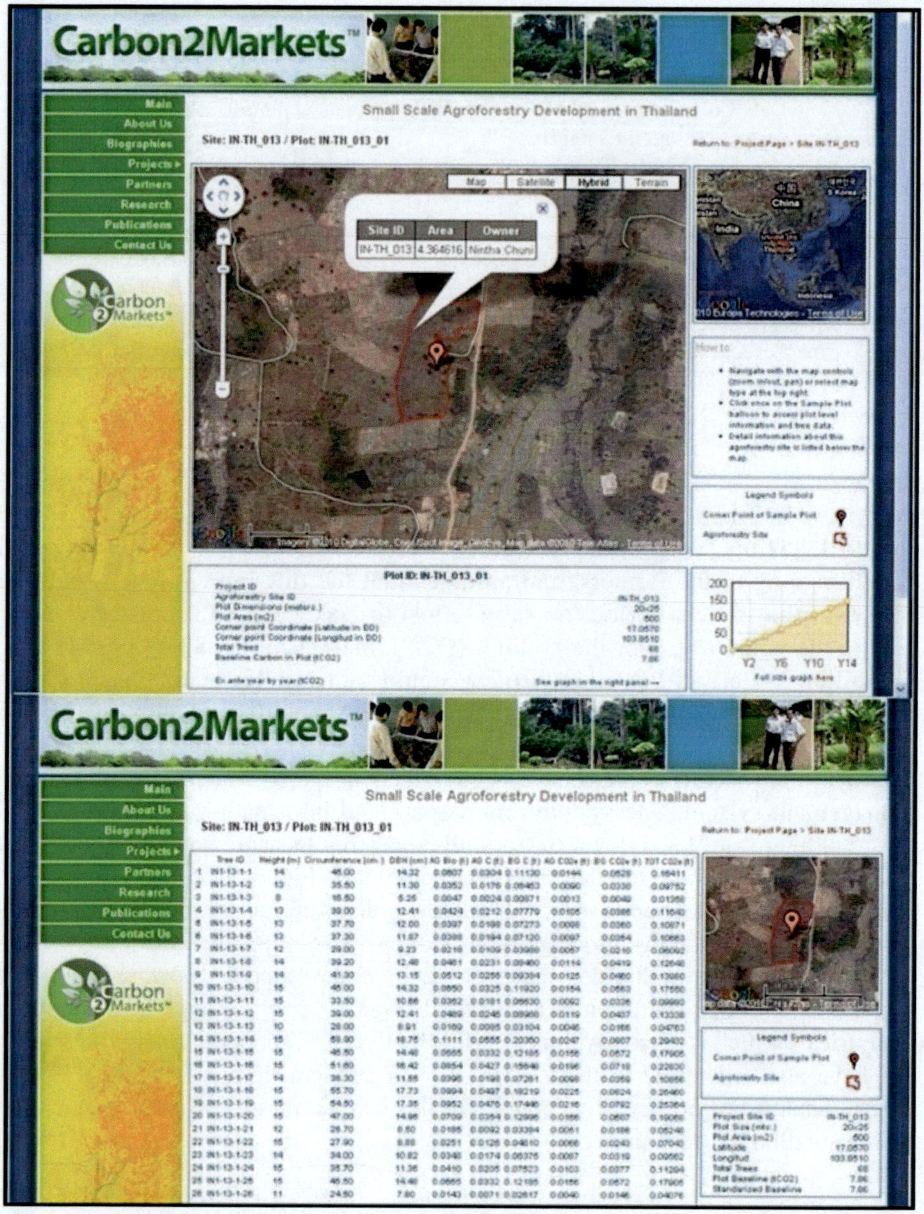

Fig. 3 Carbon2Markets offset registry showing a registered smallholder teak area that is part of the Inpang Carbon Bank project

Of the 254 species identified, 142 (55.91%) species bear useful fruits, 216 (85.04%) are used for woodfuel, resins, latex (sap) is collected from 49 (19.29%) species, 166 (65.35%) are used for construction, 183 (72.05%) in cooking, and 186 (73.23%) are used for their herbal medicinal properties. *Shorea obtusa* Wall. ex Blume is the

Table 3 List of top 15 species by total trees planted reported by Inpang members (n = 957)

Species	Total trees planted	Number of households planting
Shorea obtusa Wall. Ex Blum	146,564	509
Dipterocarpus tuberculatus Roxb.	129,306	426
Hevea brasiliensis H.B.K. M.-Arg.	99,720	114
Xylia xylocarpa (Roxb.)Taub.	93,210	510
Pterocarpus macrocarpus Kurz	78,456	491
Eucalyptus sp.	59,268	72
Sindora siamensis var. *maritima* (Pierre) K.Larsen & S.S.Larsen	50,741	246
Mangifera indica L.	38,133	360
Tectona grandis L.f.	30,769	116
Cratoxylum formosum (Jack) Dyer	20,034	135
Dimocarpus longan Leenh.	17,643	165
Lagerstroemia floribunda Jack	15,172	86
Tamarindus indica L.	13,843	287
Terminalia alata Heyne ex Roth.	13,327	140
Dipterocarpus obtusifolius Teijsm. ex Miq.	12,762	123
Croton argyratus Blume	12,359	65
Afzelia xylocarpa (Kurz)Craib	12,293	73
Aporusa villosa (Wall. ex Lindl.) Baill	11,645	73
Irvingia malayana Oliv. ex A.W.Benn.	8,198	83

dominant species planted by the respondents (Table 3). Known in Thailand by the common name "Teng", *S. obtusa* is a member of the family Dipterocarpaceae, and is a valuable hardwood. Other dominant tree species in terms of numbers of trees planted (Table 3) include *Hevea brasiliensis* Willd. ex A.Juss. for latex, *Dipterocarpus tuberculatus* Roxb. for fuelwood and medicinal herbs, and *Mangifera indica* L. for fruit.

More than half (55.11%) of the Inpang agroforestry trees have multiple uses as reported by Inpang members: timber, fuelwood, sap (including resins and latex), fruit, medicinal herbs, animal fodder, cooking spices, and others. For example, *Irvingia malayana* Oliv. ex A. Benn., known locally as *kabok* bears edible fruit and is also used for fuelwood and construction timber. *Phyllanthus emblica* L. produces an edible, acidic berry like, fruit with herbal medicinal properties. Various parts of the tree are also be used to make yellow dyes, and the tree is reported by some to be used for fuelwood. Certain woody perennials are also planted because they create the ecological conditions which bring additional co-benefits. One example of this is *Xylia xylocarpa* (Roxb.) Taub. (Daeng or Iron wood). Weaver ants (*Oecophylla smaragdina* Fabricius) are known to build nests on its leaves. The weaver ant eggs are harvested by local people for consumption or sold in the market. *Xylia xylocarpa* and other species (*Dipterocarpus alatus* Roxb. ex G.Don, *Pterocarpus macrocarpus* Kurz., *I. malayana*, *Adenanthera pavonina* L, and *Hopea odorata* Roxb.) can also create favorable habitat for edible mushrooms (Fig. 4).

Fig. 4 Inpang member's agroforestry farm: habitat favorable to non-tree products – edible mushrooms and weaver ant (*Oecophylla smaragdina* Fabricius) eggs

Discussion

The diversity of trees grown and managed by Inpang members and agroforestry replacing annual agricultural crops is an opportunity to develop more C offset projects with the Inpang network. This is clearly a positive outcome of the capacity building programmes initiated under this project. The land use change from annual crops to long-lived woody perennials means that C sequestration and mitigation are real. Implicit in this is also a greater awareness among the community members about the potential role of agroforestry in climate change mitigation. The diverse use of the trees grown and managed, beyond timber, bodes well for permanence at least in the 15–30 year period.

Valuing the ecosystem services of agroforestry systems is consistent with the tenets of the "Sufficiency Economy" formulated by His Majesty King Bhumibol Adulyadej of Thailand since 1970s (Krongkaew 2003; Chalapati 2008) and is embraced by the Inpang Community Network. In 1994 the King outlined the model farm designed to achieve self sufficiency under his "New Theory" initiative (Mongsawad 2010). At the level of an average smallholder farm in Thailand (2.4 ha), the model farm is expected to promote self reliance and risk aversion through establishing diverse farming landscapes that include water reservoirs for fish ponds and dry season cultivation (30%), a portion for rice cultivation (30%), areas for fruit and other crops (30%), and a smaller area for housing and animal husbandry (10%). The approach encourages natural methods for soil, pest, and weed management (Bhumibol 2007; UNDP 2007; Mongsawad 2010). Sufficiency economy and the New Theory model farm directly contrasts the monoculture of maize (*Zea mays* L.), cassava (*Manihot esculenta* Crantz) and sugarcane (*Saccharum officinarum* L.), which dominated Northeast Thailand since the 1970s (Ekasingh et al. 2007). Such farming systems have depleted soil nutrients, increased farmer debt, left households vulnerable (Cho and Zoebisch 2003; Rigg and Salamanca 2009), and degraded the ecosystems (Howeler 1991). The Inpang Carbon Bank centered on smallholder agroforestry landscapes twins the potential of selling C as a commodity with other timber and non-timber market opportunities and the ecosystem services of agroforestry.

There are a number of other existing C mitigation projects that include smallholders and agroforestry in developing countries, e.g., the Plan Vivo C mitigation projects in Tanzania, Mexico, Mozambique, and Uganda (Plan 2010). These projects, however, include other forest C mitigation components such as afforestation/reforestation, forest conservation, and avoided deforestation. The Plan Vivo is a "standard" rather than a protocol. A protocol or method such as the CCX Forestry Carbon Sequestration protocol or the UNFCCC CDM approved "Afforestation/Reforestation with Trees Supported by Shrubs on Degraded Land", on the other hand, are market approved methods. They define how to develop Certified Emission Reductions (CERs), Verified Emission Reductions (VERs), or Certified Financial Instruments (CFIs), which are tradable commodities on C market trading platforms. Plan Vivo approved projects receive Plan Vivo certificates, which represent "long term carbon benefits (VERs)" and are traded only in the over-the-counter voluntary markets (Plan 2010).

The Plan Vivo standard is a set of best practices to ensure that a forest C mitigation project provides equitable distribution of benefits, ensures livelihood needs are met, includes local people in the development and management of the project, and supports biodiversity and environmental services (Ruiz-De-Oña-Plaza et al. 2011). Plan Vivo projects emphasize capacity-building, long term C benefits, diversifying livelihoods, and protecting biodiversity (Plan 2010). The Inpang Carbon Bank activities also follow the same principles. Two main differences between our approach to agroforestry C offsets and the Plan Vivo standard are the use of an Internet-enabled content management application, which uses GIS and remote sensing data analysis, and our efforts to develop a new CCX market approved agroforestry C offset protocol.

In order for agroforestry C to be developed as a commodity it must, however, be economically viable. Dixon et al. (1994) estimated the financial cost of C sequestration in agronomic, agroforestry, and forest systems as US$1–69/Mg C. Cacho (2009) observed that smallholders may be constrained from participating in C markets due to the high transaction costs, but that economies of scale, particularly in the number and size of farms enrolled in a C offset project, can make a project economically feasible. Not all agroforestry C offset projects are, however, cost prohibitive. Sathaye et al. (2001) assessed C mitigation potential and costs in seven countries including agroforestry in China and Mexico and concluded "that about half the mitigation potential of 6.2 Pg C between 2000 and 2030 in the seven countries could be achieved at a negative cost, about 5 Pg C at a cost less than US$20 per Mg C, and much of the rest at costs ranging up to US$100 per Mg C." The range of costs for C sequestration in the tropics as shown from a survey by de Jong et al (2004) is in the range of US$1.00 to US$35.10 per Mg of C. Our smallholder teak project falls within this range.

Certainly there are costs associated with "bringing" the C to market. The "markets" are still evolving and the activities associated with implementing projects are still being developed and tested. There are market, institutional, and social barriers that all bear costs. Our activity has focused on addressing the social barriers through capacity building and training local people in the Inpang Network, the institutional barriers through partnering with NRCT and TGO, and the market barriers through technological innovations in managing C offset projects. The Internet-based C offset management system is truly in a software R&D phase and the costs associated with this aspect will be greatly reduced when the system matures.

The advantage of the Carbon2Markets system for smallholder C offsets is in lowering the transaction costs associated with field level measurements and verification. The fully developed on-line C management system will allow a farmer's agroforestry C offset field to be enrolled through on-line tools that will register the field boundaries, calculate the baseline C, estimate leakage from on-farm management practices and report future amounts of C sequestration (ex ante projections). The on-line tools will require input data uploaded to the database. Field boundary will either be delineated using a hand held GPS receiver, uploaded via the Internet, and entered as coordinates on-line, or drawn on-screen using hyper-resolution satellite imagery (1 m or less) and Internet-GIS tools.

Can smallholder agroforestry C offset projects of this type be scaled up? With an on-line Internet-GIS management application, such as Carbon2Markets, and the potential to use satellite remote sensing data to directly measure and monitor C sequestration in biomass, it is feasible. Aggregating individual farm parcels is already a function of the Carbon2Markets software application. The smallholder teak activity in Thailand is a demonstration itself in how farms spread across five provinces can be managed as a single project and scaled up to national levels. Ground-based measurements will likely be a requirement for project validation and verification for all projects in the foreseeable future. However, the use of satellite remote sensing measurements of C in biomass that have been validated and calibrated with ground based measurements means larger areas can be assessed with fewer field data requirements. This bodes well for project scalability. Costs of hyper-resolution satellite data are still quite high (e.g. US$35.00 per km^2 for precision-level geo-referenced IKONOS data), but are expected to come down as more satellites are launched and additional providers enter the market.

In addition to developing and implementing cost effective agroforestry C accounting methods for C financial markets, projects require contractual obligations to ensure commitment to permanence, usually for a minimum of 15 years. Furthermore, in order to realize payments for agroforestry C offset agreements, linkages to responsible agencies must be forged. These will not always be the same from country to country or even between different projects within the same country. Developing partnerships at the national level with agencies that are responsible for natural resources management, forest resources, agricultural lands, and climate policy and action is, perhaps, a pre-requisite for successful projects implementation. The near future, especially as C markets mature and stabilize, as expected, we foresee the growth of an industry of "Carbon Offset Providers" or "Aggregators" in developing countries that will facilitate contractual, legal obligations and payment transactions for offset projects.

International agreements and national legislation also contribute to the advancement of agroforestry C offsets projects as well as to the stability and growth of C markets. Within the UNFCCC there is still no firm or clear path for agroforestry C offset projects within the CDM. In the United States, Congress has yet to pass a climate bill that would legislate a cap and trade system for regulating greenhouse gas emissions. It would be naïve to underestimate the impact any future U.S. legislation and whatever form the post-Kyoto commitment takes in both the regulatory and the voluntary markets with respect to agroforestry C sequestration offsets. With that in mind, it would be a folly not to recognize that opportunities currently do exist to sell agroforestry C sequestration offsets in the voluntary market but that much work is still to be completed.

Mitigation of climate change from offsets, including forestry offsets, provide a common good for a global problem. However, not all offsets are equal. Agroforestry and plantations both sequester C, of course, and mitigate climate change. Permanence concerns aside, there are potential environmental and social co-benefits associated with agroforestry that are often absent in plantations. It is not unimaginable, to consider differential pricing for agroforestry C sequestration offsets. Payments for such activities support not only the positive impacts on climate change, but also reward agricultural practices that encourage diversification, biodiversity, soil health, and entrepreneurialism. Inpang farmers demonstrate such business-mindedness in their marketing of secondary products (wines, juice, herbal medicines, liquor, cosmetics, organic fertilizers, and others) made from the diversity of plants being grown in their agroforestry farms. Constraints to the adoption to agroforestry that are financial or economic may be overcome by paying farmers, in advance, for C offsets that will be achieved through planting woody perennials on farms.

Future work under the Inpang Carbon Bank project is to move from simple single-species smallholder agroforestry systems (such as the teak woodlots) to more complex multi-species agroforestry systems. These species-rich Inpang agroforestry areas are also on spatially dispersed farms and require the aggregation services of the Carbon2Markets system. More complex agroforestry systems will necessitate additional C accounting models, beyond those used for single species. New protocols are currently under development for C accounting, measurement, and monitoring in these more complex agroforestry systems. The challenge is to develop methods to ensure robust estimates of baseline C stock and the rates of sequestration that are also able to be implemented in timely, cost effective ways. The Inpang Carbon Bank project is developing methods that couple *in situ* biomass data with analysis of hyper resolution (1 m or less) satellite remote sensing data and C bookkeeping models to estimate landscape level biomass and C.

Conclusions

While progress has been made with the Inpang Carbon Bank project we are cautiously optimistic about realizing agroforestry C offset transactions and the potential economic benefits to rural farmers in the near term. Carbon financial markets that allow biotic C sequestration offsets are nascent, complex, and rapidly evolving. Offset allowances, market pricing, and market regulations are intertwined with national and international policies and legislations. Over-the-counter voluntary markets are perceived to be soft, and yet regulatory markets do not allow agroforestry C offsets.

Widespread adoption of agroforestry C projects is, of course, not solely dependent on any post-Copenhagen agreement or on US climate change legislation. The most basic infrastructure for the selling of biotic C as a commodity is still being developed. This infrastructure includes the tools and techniques to ensure C sequestration measurements are real and verifiable. Market accepted rules and protocols that support such projects are still being developed. The economic barriers to implementing smallholder C offset projects are high today, but will not stay that way as technological advances in Internet computing are realized. Challenges in the ways in which payments are distributed are many and the legal frameworks for risk assumption are still difficult to overcome and work through. Best practices will emerge as project activities move forward in a path-finding manner.

The goals of promoting and developing agroforestry C offset projects, like the Inpang Carbon Bank are worthy of further efforts. Recognizing that the rural poor in developing countries are perhaps the most vulnerable to climate change impacts, it is possible to help tackle two important problems through a single intervention. Agroforestry activities that benefit farmers with a second, additional, income stream from selling sequestered C can help fight rural poverty and mitigate climate change. Agroforestry practices that transform landscapes from annual crops, which often require high fertilizer inputs, to C rich areas with trees and other woody perennials offer multiple benefits to farmers at their local level. In addition, it also helps the global community in fighting to stem the tide of greenhouse gas emissions and climate change.

Acknowledgments We thank the following for their support of the Inpang Carbon Bank project: Dr. Anond Bunyaratvej, Secretary General, National Research Council of Thailand; Ms. Choosri Keedumrongkul, Director, Office of International Affairs, National Research Council of Thailand, Dr. Prasertsuk Chamornmarn, Deputy Executive Director, Thailand Greenhouse Gas Management Organization; Dr. Jeeraphan Suksringarm, Dean of the Faculty of Science, Mahasarakham University; Ms. Siritorn Dumrongsukit and Ms. Patoo Sangkanukij, support staff at the National Research Council of Thailand; the Inpang Community Network leaders and youth group (especially Mr. Tawatchai Kulwong Mr. Serm Uduomna, Mr. Charatsaiphon Udomsab, Mr. Krongsak Rachrongchai,

Mr. Saman Butwang, Mr. Set Mekseangsi, Mr. Sangwian, Mr. Pongpat Rattananiyakun); Mr. Oscar Castaneda and Dr. Eric Kasten, Global Observatory for Ecosystem Services, Carbon2Markets, Department of Forestry, Michigan State University. Financial and in-kind support for this work provided by: the Asia-Pacific Network for Global Change Research (APN ARCP2007-09NSY; PI David L. Skole, Michigan State University); the Office of International Affairs, National Research Council of Thailand; Faculty of Science, Mahasarakham University; Inpang Community Network; Michigan State University (grants from the MSU Vice-President for Research, Research Excellence Fund, for the Carbon2Markets project and from the MSU Environmental Stewardship Campaign, Reducing Greenhouse Gas Emissions on the MSU Campus through commitments to the Chicago Climate Exchange) and the United States National Aeronautics and Space Administration, Land Cover / Land Use Change Program (NNH07ZDA001N-LCLUC "Enhancing Global Scale Observations And Information On Tropical Forest Change Using Landsat Global Data Remote Sensing").

References

Angelsen A, Kaimowitz D (2004) Is agroforestry likely to reduce deforestation? In: Schroth G, da Fonseca GAB, Harvey CA, Vasconcellos HL, Gascon C, Izac AMN (eds.) Agroforestry and biodiversity conservation in tropical landscapes. Island Press, Washington, DC, pp 87–106, CAB International

Bhumibol A (2007) King Bhumibol Adulyadej: the great king of communication. Foreign Office, Public Relations Department, Office of the Prime Minister, Bangkok, 129 p

Cacho OJ (2009) Economics of carbon sequestration projects involving smallholders. In: Lipper L, Sakuyama T, Stringer R, Zilberman D (eds.) Payment for environmental services in agricultural landscapes economic policies and poverty reduction in developing countries. Springer and FAO, Rome, pp 77–102

Chalapati S (2008) Sufficiency economy as a response to the problem of poverty in Thailand. Asian Soc Sci 4(7):3–6

Chicago Climate Exchange (CCX) (2010) Forestry carbon sequestration offsets. Available at http://www.chicagoclimatex.com/content.jsf?id=242. Accessed 19 Feb 2010

Cho KM, Zoebisch MA (2003) Land-use changes in the upper Lam Phra watershed northeastern Thailand: characteristics and driving forces. J Agric Rural Dev Trop Subtrop 104(1):15–29

de Jong BHJ, Ochoa Gaona S, Montalvo SQ, Bazán EE, Hernández NP (2004) Economics of agroforestry carbon sequestration. A case study from southern Mexico. In: Alavalapati J, Mercer E (eds.) Valuing agroforestry systems: methods and applications, Advances in Agroforestry 2. Kluwer Academic, Dordrecht, pp 123–138

Dixon R, Winjum JK, Andrasko KJ, Lee JJ, Schroeder PE (1994) Integrated land use systems: assessment of promising agroforestry and alternative land use practices to enhance carbon conservation and sequestration. Clim Chn 27:71–92

Ekasingh B, Sungkapitux C, Kitchaicharoen J, Suebpongsang P (2007) Competitive commercial agriculture in the Northeast of Thailand. The World Bank, Washington, DC, 165 p

Hamilton K, Chokkalingam U, Bendana M (2010) State of the forest carbon markets 2009: taking root & branching out. Ecosystem MarketPlace. World Bank, Washington, DC, 8 p. Available at http://www.ecosystemmarketplace.com/pages/dynamic/resources.library.page.php?page_id=7525&section=our_publications&eod=1. Accessed Nov 2010

Howeler RH (1991) Long-term effect of cassava cultivation on soil productivity. Field Crop Res 26(1):1–18

IPCC (2003) Good practice guidance for land use, land-use change and forestry. In: Penman J, Gytarsky M, Hiraishi T, Krug T, Kruger D, Pipatti R, Buen-dia L, Miwa K, Ngara T, Tanabe T, Wagner F (eds.) IPCC National Greenhouse Gas Inventories Programme and Institute for Global Environmental Strategies (IGES) Hayama, Kanagawa, Japan. Available at: http://www.ipcc-nggip.iges.or.jp/public/gpglulucf/gpglulucf_ ~ contents.htm . Accessed 18 Nov 2010

IPCC (2007) Climate change 2007: IPCC fourth assessment report (AR4). In: Solomon S, Qin D, Manning M, Chen Z, Marquis M, Averyt KB, Tigora M, Miller HL (eds.) The physical science basis. Cambridge University Press, Cambridge, UK/New York, 996 p

Jose S (2009) Agroforestry for ecosystem services and environmental benefits: an overview. Agroforest Syst 76:1–10

Kossoy A, Ambrosi P (2010) State and trends of the carbon market 2010. The World Bank, Washington, DC, 78 p. Available at http://www.economicsclimatechange.com/2010/06/state-and-trends-of-carbon-market-2010.html. Accessed 1 Nov 2010

Krongkaew M (2003) The philosophy of sufficiency economy. Kyoto review. Available at http://kyotoreview.cseas.kyoto-u.ac.jp/issue/issue3/article_292.html. Accessed 1 Nov 2010

Merger E (2008) Forestry carbon standards 2008: a comparison of the leading standards in the voluntary carbon market and The State of Climate Forestation Projects. Carbon Positive, 71 p. Available at www.carbonpositive.net/fetchfile.aspx?fileID=133. Accessed Nov 2010

Michon G, de Foresta H (1996) Agroforests as an alternative to pure plantations for the domestication and commercialization of NTFPs. In: Leakey RRB, Temu AB, Melnyk M, Vantomme P (eds.) Domestication and commercialization of non-timber forest products in agroforestry systems, Non-wood forest products No. 9. FAO, Rome, pp 160–175

Mittelman A (2000) Teak planting by smallholders in Nakhon Sawan, Thailand. Unasylva 51(2):62

Mongsawad P (2010) The philosophy of the sufficiency economy: a contribution to the theory of development. Asia-Pacific Dev J 17(1):123–143

Montagnini F, Nair PKR (2004) Carbon sequestration: an underexploited environmental benefit of agroforestry systems. Agroforest Syst 61:281–295

Nair PKR, Kumar BM, Nair VD (2009) Agroforestry as a strategy for carbon sequestration. J Plant Nutr Soil Sci 172:10–23

Pandey D, Brown C (2000) Teak: a global overview. Unasylva 51(201):3–13

Petmark P, Sahunalu P (1980) Primary production of teak plantations. I. Net primary production of thinned and unthinned plantations at Ngao, Lampang. Forest Research Bulletin, Faculty of Forestry, Kasetsart University, Bangkok, Thailand. Available at http://www.en.mahidol.ac.th/journal/20081/1supawan_p_f.pdf. Accessed 25 Nov 2010

Plan Vivo (2010) Web site: http://planvivo.org.34spreview.com/#. Accessed 23 June 2010

Reid R, Stephen P (2001) The farmer's forest: multipurpose forests for Australian farmers. RIRDC publication no. R01/33. RIRDC, Canberra, 167 p

Rigg J, Salamanca A (2009) Managing risk and vulnerability is Asia: a (re)study from Thailand, 1982–83 and 2008. Asia Pac Viewp 50(3):255–270

Ruiz-De-Oña-Plaza C, Soto-Pinto L, Paladino S, Morales F, Esquivel E (2011) Constructing public policy in a participatory manner: from local carbon sequestration projects to network governance in Chiapas, Mexico. In: Kumar BM, Nair PKR (eds.) Carbon sequestration potential of agroforestry systems: opportunities and challenges. Springer, Dordrecht, pp 247–262

Sathaye JA, Makundi WR, Andrasko K, Boer R, Ravindranath NH, Sudha P (2001) Carbon mitigation potential and costs of forestry options in Brazil, China, India, Indonesia, Mexico, the Philippines, and Tanzania. Mitig Adapt Strat Glob Chang 6:185–211

Simons AJ, Leakey RRB (2004) Tree domestication in agroforestry. Agroforest Syst 61:167–181

UNFCCC (2010a) Clean development mechanism: methodologies: afforestation / reforestation methodologies. Approved A/R Methodologies. Available at http://cdm.unfccc.int/methodologies/ARmethodologies/approved_ar.html. Accessed 19 Feb 2010

UNFCCC (2010b) Clean development mechanism: methodologies: small scale afforestation/reforestation CDM methodologies. Methodologies for small scale A/R CDM project activities. Available at http://cdm.unfccc.int/methodologies/SSCAR/approved.html. Accessed 19 Feb 2010

United Nations Development Programme (UNDP) (2007) Thailand Human Development Report 2007: Sufficiency Economy and Human Development. UNDP, Bangkok, 130 p

The Socioeconomic Context of Carbon Sequestration in Agroforestry: A Case Study from Homegardens of Kerala, India

Subhrajit K. Saha, Taylor V. Stein, P.K. Ramachandran Nair, and Michael G. Andreu

Abstract While the biological and ecological role of agroforestry (AF) on climate change mitigation has received considerable research attention lately, the role of socio-psychological factors in this context has been left largely unexplored. Socio-psychological variables such as culture, demography, economy, and social values play important roles in farmers' decision making with the land management, which in turn influence the ability of AF systems to sequester carbon (C). This chapter presents a case study from Thrissur, Kerala, India, which examined how different socio-psychological factors influence soil C sequestration through land management decisions in tropical homegardens (HGs), a popular agroforestry system in the tropics. This study used the Theory of Planned Behavior (TPB) as the theoretical framework to understand homegarden owners' perceptions on the adoption of five land management practices (i.e., tillage, tree planting, plant residue incorporation, manure usage, and fertilizer applications), which are known to impact C sequestration. Data collected using focus group and household interviews were analyzed by regression statistics. Results indicated that farmers' decision making processes were most influenced by factors such as ancestors and education, followed by peers, financial condition, and economic importance of the AF land holding. The results of this case study will not only benefit researchers and extension practitioners, but can also contribute to the policy platform to recognize the role of socio-psychological factors in agricultural decision making.

S.K. Saha
Ashoka Trust for Research in Ecology and the Environment (ATREE), Bangalore, India
e-mail: subhrajit_s@yahoo.com

T.V. Stein (✉) • P.K.R. Nair • M.G. Andreu
Center for Subtropical Agroforestry, School of Forest Resources and Conservation, University of Florida, P.O. Box 110410, Gainesville, FL 32611–0410, USA
e-mail: tstein@ufl.edu; pknair@ufl.edu; mandreu@ufl.edu

B.M. Kumar and P.K.R. Nair (eds.), *Carbon Sequestration Potential of Agroforestry Systems: Opportunities and Challenges*, Advances in Agroforestry 8,
DOI 10.1007/978-94-007-1630-8_16,

Keywords Agricultural practices • Decision making • Human dimensions • Socio-psychological factors • Theory of planned behavior

Introduction

The world's knowledge of agroforestry (AF) has dramatically increased over the last several decades mainly through innovative ecological and biological research. This research has led to a better understanding of AF's importance in carbon (C) sequestration and global climate change. However, AF's interdisciplinary research should continue to expand into the social sciences in order to better understand the mechanism of landowners' AF decision making. The relationship between socio-psychological factors (e.g., cultural, demographic, economic, and social variables) and how people make decisions in practicing AF is inseparable, and must be considered if policy makers, extension agents, and agricultural educators hope to influence and improve landowners' AF management.

Although research has not examined the wide range of socio-psychological variables that affect farmers' AF decision making in terms of C sequestration, past research has shown that the demographic, social, economic, and cultural make-up of a land-owner shape how they manage land (Fig. 1). Within each of these broad categories a variety of specific variables can be identified and measured to better explain why and how landowners make land management decisions. For example,

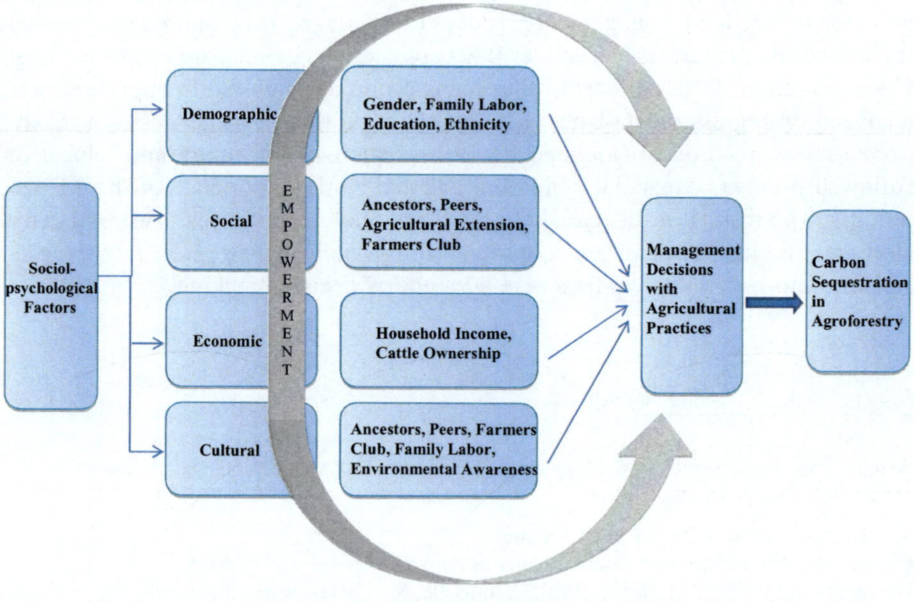

Fig. 1 Interaction between socio-psychological factors and empowerment of farmers in agricultural decisions making that consequently affects carbon sequestration

demographic factors, including age, education, ethnicity, and gender, likely influence what a person plants, how he or she manages and harvests the crops, and how he or she might use those crops. Similarly, economic factors (e.g., household income, sources of income, and availability/ownership of resources), social elements (e.g., connections to ancestors, peers, extension officers, and availability of agricultural networks or organizations), and cultural variables (e.g., influence of ancestors and peers, religious beliefs, women's role in household tasks, family labor and food habits) also influence farmers' beliefs, perceptions, and behavior. Although many of these socio-psychological factors play indirect and hidden roles in AF management, they are extremely important in land management decisions, which in turn affect the sequestration of C in AF systems. As more information is gained on the biophysical aspects of AF and C sequestration, it is time for the social sciences to step in and begin to uncover the role these social variables play. The goal of this chapter is to present a case study of a research project designed to shed light on this important issue.

The extent of C stored (and sequestered) in an agricultural system including AF is influenced by a number of land management practices including tillage, plant residue management, manure usage, and fertilizer application (FAO 2004; Nair et al. 2010). Tillage decreases C stocks in agricultural soils through the destruction of aggregates and exposure of stored C to microbial degradation (Six et al. 2000). Application of manures and other organic materials to soils increases the formation and stabilization of soil macroaggregates (Whalen and Chang 2002), which are beneficial to the storage of soil C (Six et al. 2002). In general, fertilizer application is likely to enhance C sequestration (Lal et al. 1999) because of enhanced biomass production and consequent increase in biomass addition to soil; however, nitrogenous fertilizers are considered to generate nitrous oxide (N_2O), a greenhouse gas (Robertson et al. 2000). The adoption and intensity of these management practices are influenced by a variety of socio-psychological factors, thereby, indirectly affecting the soil C stocking.

Socio-psychological factors such as farmer's economic and educational status, demography, social connections, culture, and resource availability are important to understand why and how farmers select certain management practices (Seabrook et al. 2008). Agricultural decisions made by individuals (or farmers) are often influenced by their economic opportunities (Lambin et al. 2001). Similarly, the positive effects of education on adoption of desirable land management practices have been reported (Anjichi et al. 2007; Matata et al. 2008). Availability of resources such as raw materials, labor, and domestic animals also influence farmers' decision to adopt specific practices (Williams 1999). Social connection (ancestors, peers, extension agents) is another important factor in this context. Aguilar-Støen et al. (2009) reported that connections with peers increased plant diversity in the smallholder farms of Oaxaca, Mexico. Farmers' contacts with extension agents and participation in extension workshops have been reported to promote adoption of improved agricultural technologies such as mixed intercropping in Malawi (Thangata and Alavalapati 2003), and composting technology in Burkina Faso (Somda et al. 2002).

Although the socio-psychological factors have the potential to influence the AF management practices, the magnitude of such influence depends largely on the economic importance of the system. In commercial AF systems, in which management decisions are made with the goal of higher production and profit maximization, the influence of cultural, demographic, and social factors are seldom considered. The management of AF systems, which are practiced predominantly in smallholder farms, however, is influenced by a number of socio-psychological factors other than economics. The homegarden (HG) is one such system, where profit maximization is not generally the main objective and socio-psychological factors have strong influence on farmers' decisions regarding the management practices.

Homegardens are intimate, multistory combinations of various trees and crops, sometimes in association with domestic animals, around the homesteads (Kumar and Nair 2004). These integrated systems are distributed in tropical areas like Asia, Africa, Central America, the Caribbean, and the Pacific Islands (Nair and Kumar 2006). They represent a traditional agroforestry land use that has been practiced in Kerala, India, from time immemorial. The 4.32 million HGs in Kerala covering 1.4 Mha (Kumar 2006) provide an array of products for the immediate consumption by the homegardener (homegardeners are farmers of their own garden and henceforth used synonymously) as well as a supplemental source of income. Homegardens in Kerala have been recognized for their ecological and socioeconomic sustainability values (Nair and Sreedharan 1986; Jose and Shanmugaratnam 1993; Kumar et al. 1994; Peyre et al. 2006) and have high levels of C in the soil (Saha et al. 2009) comparable to natural systems (Saha et al. 2010). It is in this scenario that the present study was undertaken in the HGs in Kerala state with the specific objectives of analyzing the effects of cultural, demographic, economic, and social factors on farmers' decisions with homegarden management practices, which are known to have a relatively high impact on soil C sequestration.

The Conceptual Framework: Theory of Planned Behavior (TPB)

In order to assess how various traditional beliefs of farmers influence their decision making in relation to land management practices, we used the Theory of Planned Behavior (TPB) (Ajzen 1985, 1991) as the conceptual framework (Fig. 2). The TPB assumes that people behave in accordance to their belief, which are based on their experiences, social or peer influence, and availability of resources (Ajzen 1985). The underlying beliefs of TPB are associated with the social factors that influence human behavior (Ajzen 1991) and any cultural, demographic, economic, or social differences among people should, if relevant to the behavior, be reflected in their beliefs (Beedell and Rehman 2000). Although numerous social, psychological, and economic studies have used TPB to understand the human behavior, only few have applied this theory in the field of agriculture to explain farmer behavior (Beedell and Rehman 2000; Burton 2004; Colemont and Van den Broucke 2008).

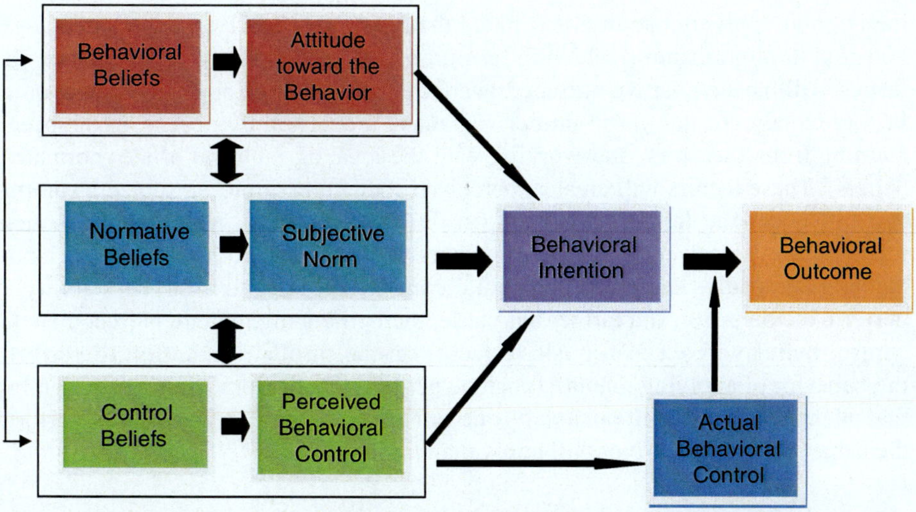

Fig. 2 Schematic diagram of the Theory of Planned Behavior (Source: Adapted from Ajzen 1985, 1991)

Ajzen (1991) identified three basic beliefs that influence the outcome of a behavior: behavioral belief, normative belief, and control belief. Behavioral belief is the belief by an individual that a specific behavior will result in a given outcome (e.g., application of manure will help plants to grow and develop). Behavioral belief in association with subjective values of expected outcomes determine the positivity or negativity of attitude towards the behavior. Factors such as education and experience of the farmer influence his or her behavioral beliefs. Normative belief refers to the perceived behavioral expectations of such important referent individuals or groups (e.g., my father applied manure and my peers apply it, too). It is believed that the normative beliefs in combination with the person's motivation determine the prevailing subjective norm, which is the perceived positive or negative social pressure about a behavior. In this study normative beliefs refer to referent individuals or groups such as ancestors, peers, agricultural extension agents, and experts from the farmers' club. Control beliefs refer to the perceived presence of factors that may facilitate or impede performance of a behavior (e.g., I have household cattle, so I will have supply of manure). It is assumed that these control beliefs, in combination with the perceived power of each control factor determine the prevailing perceived behavioral control, which refers to people's perceptions of their ability to perform a given behavior. Control beliefs in this study refer to the financial condition (e.g., availability of funds to perform an action), availability of family labor, cattle, and raw materials such as plant residue and manure. Actual behavioral control, which is similar to perceived behavioral control, refers to the extent to which a person has the skills, resources, and other prerequisites needed to perform a given behavior.

Considering manure application to a homegarden as an example to explain TPB, the behavioral belief (education/experience) of the farmer will contribute to the intention of applying manure. It is likely that the farmer has learned that plants get nutrition through manure, which in turn promotes growth and yield; therefore, the farmer will have a positive attitude toward the behavior and apply manure if other factors remain constant. The farmer will make a decision by observing the peers, learning from ancestors, and working with the local agricultural office (normative beliefs). These factors will create a perceived social pressure (i.e., subjective norm) that influences the farmer's decision (in this case positive). Here, control beliefs relate to the perceived presence of factors such as availability of funds, labor, manure, and cattle. Perceived behavioral control will refer to a farmer's ability to perform the behavior. If a farmer has cattle, then producing manure is free, then the control belief will act as a positive factor on the farmer's intention to perform the behavior of applying manure (positive perceived behavioral control). Assuming that all three beliefs result positively, one can predict a positive intention to perform the behavior, and the farmer will apply manure to the crop.

Methods

Study Area

The study was conducted in the district of Thrissur in the central part of the State of Kerala (Fig. 3). Three villages (Pandiparambu, Chirakkakode, and Vellanikkara) in Madakkathara *Panchayath* (a *panchayath* is the smallest administrative unit) located at 10º 32′ and 10º 36′N latitudes and 76º 14′ and 76º 18′E longitudes in the northeast part of Thrissur district were chosen randomly for the study. The major land use types of the *panchayath* include rice-paddy (*Oryza sativa* L.) fields, rubber (*Hevea brasiliensis* Kunth. Muell.) plantations, cashew (*Anacardium occidentale* L.) plantations, vegetable fields, banana (*Musa paradisiaca* L.) intercropped with coconuts (*Cocos nucifera* L.), forests, and homegardens. Monocultures of rice-paddy, rubber, cashew, and banana cover 15%, 16%, 4.5%, and 4% of cultivable land, respectively. Mixed plantations with coconut and homegardens cover about 34% of cultivable land (Government of Kerala 2005). The majority of the land holdings are small (< 1 ha) and 45% of the population are involved in farming, of which only 30% depends entirely on agriculture for their livelihood (Government of Kerala 2005). Total population is more than 20,000 in Madakkathara *panchayath* and literacy is about 85.4%.

Focus Group Meeting

In order to get an overview of the agricultural, cultural, demographic, economic, and social information of the area, a focus group meeting was arranged with ten representative homegardeners (eight males, two females) at the agricultural office

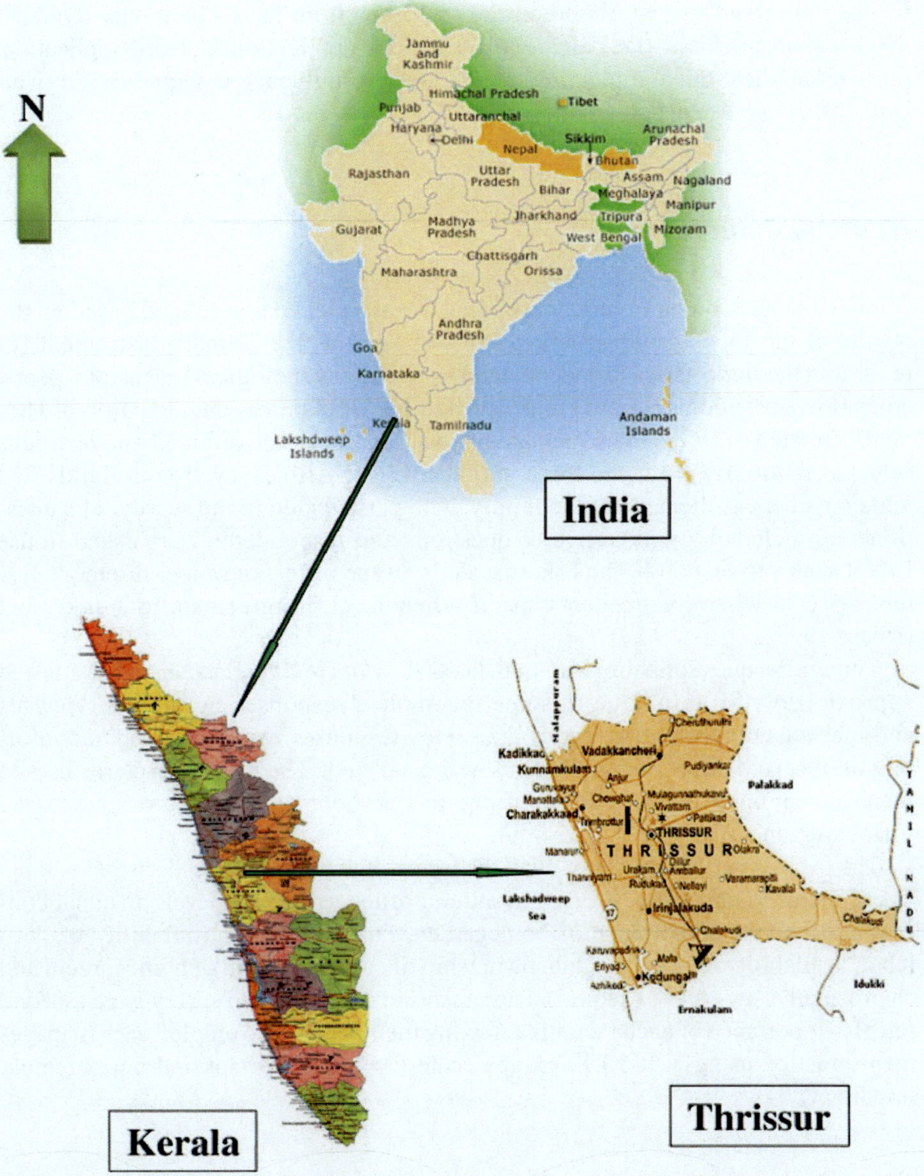

Fig. 3 Geographical location of the study site (Thrissur, Kerala, India) (Source: http://maps.locateindia.com, http://hikerala.googlepages.com. Last accessed October 2010)

of Madakkathara *panchayath*. The demographic information of the participants was collected followed by a discussion session with open ended questions. Questions were asked in local language (*Malayalam*) and the responses were translated to English. Active responses were obtained and the meeting generated information that enriched the knowledge base as well as helped in fine tuning the questionnaire for the household survey. Based on the responses from the focus group, five land management practices (i.e., tillage, plant residue application, manure application, fertilizer application, and planting trees) known to influence C sequestration (Nair et al. 2010) were selected.

Household Survey

A questionnaire for surveying households was developed and refined based on the results of the focus group meeting and was used as the primary instrument for recording the importance farmers placed on socio-psychological (ancestors, peers, education etc.) and situational (e.g., availability of plant residue, location of HG) factors related to HG management decisions. The survey covered 65 homegardens selected randomly from the three study villages. All surveyed households had male members as the head of the family, who participated in the survey. The questionnaire included a wide range of questions and respondents were asked to use Likert scales to respond. The Likert scale is a type of response measurement system or scale where respondents specify their level of agreement to a statement (Likert 1932).

The entire questionnaire was translated to *Malayalam*. Each interview lasted approximately 30 min. Due to some incomplete responses by the interviewees and inadequate information, six of the survey responses were discarded and information received from 59 respondents was analyzed. The HG characteristics and management intensities as gathered from the questionnaire survey are summarized in Tables 1 and 2.

The focus group meeting identified the factors that influence HG decision making. These included ancestors, peers, agricultural office, education level, financial condition, economic importance of homegarden, farmers' clubs, availability of farm labor, availability to plant residue, ownership of cattle, location of homegarden, and environmental awareness (Table 3). Participants in the household survey were asked to rate the importance of each factor in affecting their decision to employ each management practice using a 1–5 Likert-type scale (1 = not important and 5 = extremely important).

Statistical Analysis

Independent factors were the items (e.g., ancestors, peers, education, financial resources etc.) that were identified by the focus group participants as the determining factors in HG management decision making. The dependent factors were the intensity of five management practices (i.e., tillage, plant residue application, manure application, inorganic fertilizer application, and tree planting). A regression analysis was performed to determine the relationship the independent factors had with the

Table 1 Basic information about the household and homegardens (HG) in Thrissur, Kerala, India

Characteristics	Mean	Median	Mode	Lowest	Highest
Age of HG (years)	71.75	60	100	10	200
Area of HG (ha)	0.22	0.12	0.04	0.016	2.4
Total annual income (USD)	679.7	375	250	75	7,500
Annual income from HG (USD)	166.4	125	125	0	2,500
Age of the homegardener (years)	54.37	58	60	24	82
Size of family	4.57	4	4	2	12
Male members in family	2.07	2	2	0	6
Annual plant residue application (kg ha^{-1})	2,306	1,339	1,250	0	10,000
Annual manure application (kg ha^{-1})	1,742	1,250	1,250	0	7,500
Annual fertilizer application (kg ha^{-1})	172.5	93.75	0	0	781.3
Tree cover in HG (%)	48.5	40	40	20	90

Table 2 Homegarden (HG) and homegardener characteristics in Thrissur, Kerala, India

Characteristics	Low	Medium	High
[a]Income from HG	55.2	27.6	12
[b]Education of homegardener	37.3	47.5	11.9
[c]Plant residue application	34.5	44.8	20.7
[d]Manure application	39	44	17
[e]Fertilizer application	45.8	23.7	30.5
[f]Tree population in HG	23.7	49.2	27.1

[a]Low = 1–33, medium = 33–66, high = 66–100% of total income. 5.2% of the surveyed farmers did not have any income from HG

[b]Low = primary, medium = secondary, high = post-secondary education. 3.4% of the surveyed farmers did not have any formal education (below primary)

[c]Low = 0–1,000, medium = 1,001–2,500, high = >2,500 kg ha^{-1} year^{-1}

[d]Low = 0–1,000, medium = 1,001–2,500, high = >2,500 kg ha^{-1} year^{-1}

[e]Low = 0–100, medium = 101–250, high = >250 kg ha^{-1} year^{-1}

[f]Low = 0–30, medium = 31–60, high = 60–100% tree cover in the homegarden

dependent variables. Statistical tests were performed with software package SPSS (ver. 11 2001) and differences were considered significant at p value < 0.1.

Results and Discussion

The socio-psychological and situational factors had significant influence on farmers' decision with all five management practices (Table 4). Ancestors had the most influence on farmers' decisions about HG management practices, followed by education (Fig. 4). Peers, financial condition, and economic importance of homegarden carried equal importance in influencing farmers' decision. Finally, the agricultural office and the farmers' club resulted in comparatively less effects on the farmers' decision regarding the HG management practices (Fig. 4). Using the Theory of Planned Behavior as the underlying theoretical framework, we examined how different belief

Table 3 Socio-psychological and situational factors in the context of homegarden (HG) management practices in Kerala, India

Socio-psychological factors	Belief category (TPB)[a]	Concept	Impact
Ancestors	Normative	In India, farmers generally receive their agricultural education from their ancestors.	Farmers continue the traditional practices from generation to generation.
Peers	Normative	Management decisions of HGs are made by imitating peers that include friends, relatives, and neighbors.	Farmers learn the consequences of a behavior from their peers and make the decision accordingly.
Agricultural office	Normative	Local government agricultural offices and their agents play a major role in controlling the trend of agriculture by supplying inputs.	Farmers visit agricultural offices to resolve their agricultural problems and learn about modern technologies.
Education	Behavioral	Education indicated academic qualifications and knowledge of modern agricultural developments.	Education gives the farmers exposure to a variety of modern agricultural developments.
Financial condition	Control	The financial condition of farmers affects the management decisions (e.g. buying costly inputs or employing external labors).	More financial resources do not necessarily ensure performing all farm management practices more frequently.
Farmers' Club	Normative	Farmers' clubs are informal village organizations, where farmers discuss and learn about different agricultural issues.	Several farming decisions are made based on the suggestions from the experienced farmers at the farmers' club.
Availability of family labor	Control	Availability of family labor may influence the decision on intensity of HG practices such as tillage or applying plant residue.	Economically less well-off farmers depend on family labors to perform the farm management practices (Ali 2005).
Ownership of cattle	Control	Farmers traditionally make composted farmyard manure from cattle dung.	Ownership of cattle ensures supply of the raw materials, which might boost the farmers' decision to apply manure.
Environmental awareness	Behavioral	The level of environmental awareness or the ecological concept may encourage farmers to make eco-friendly decisions.	An environmentally aware farmer may avoid application of chemical fertilizers and adopt/promote organic manures.
Situational factors			
Availability of plant residue	–	Availability to plant residues is a determining factor for its application.	The more plant residue is available, the more likely farmers will apply it.
Economic importance of HG	–	Economic importance of HGs depends on the percentage of the total income generated from the land holding.	Farmers' willingness to invest time and resources depend on the economic importance of the land holding/HG.
Location	–	The location of the HG determines access to resources (e.g. plant residue).	Proximity to forests or plantations may increase access to plant residue.

[a]*TPB* (Theory of Planned Behavior)

Table 4 Influence of socioeconomic and demographic factors on homegarden management decisions in Thrissur, Kerala, India

		Positive effects				
Management practice	R^2	Level 1[a]	Level 2[a]	Level 3[a]	Level 4[a]	Negative effects
Tillage	0.4	ANC ($p=0.001$) (NB)	ECN ($p=0.025$)	PER ($p=0.075$) (NB)	FLB (NS)	EDU ($p=0.082$) (BB)
Plant residue application	0.4	ANC ($p=0.018$) (NB)	PER ($p=0.05$) (NB)	PRD ($p=0.053$) (CB)	EDU (NS)	FIN ($p=0.01$) (CB)
Manure application	0.3	MNR ($p=0.019$) (CB)	CTL ($p=0.022$) (CB)	PER (NS)	ANC (NS)	–
Fertilizer application	0.3	ECN ($p=0.005$)	ANC ($p=0.089$) (NB)	PLT (NS)	FIN (NS)	EDU ($p=0.008$) (BB)
Planting trees	0.3	EDU ($p=0.053$) (BB)	PER ($p=0.088$) (NB)	ENV (NS)	AOF (NS)	–

AOF agricultural office, *ANC* ancestors, *AST* aesthetic sense of farmer, *CTL* cattle availability, *ECN* economic importance of homegarden, *EDU* education, *FIN* financial solvency of the farmer, *FLB* family labor, *PER* peers (friends, relatives, neighbors) *MNR* manure availability, *PLT* plant type, *PRD* plant residue availability, *NS* statistically not significant

NB normative belief, *CB* control belief, and *BB* behavioral belief

[a]The levels 1, 2, 3, and 4 represent the extent of positive influence on the farming practice, level 1 the highest and 4 the lowest. For example, tillage was most influenced by 'Ancestors', followed by ECN followed by PER and FLB in that order. The order of importance was obtained from the SPSS regression analysis. Out of factors analyzed, only top four significant levels are mentioned under each farming practice

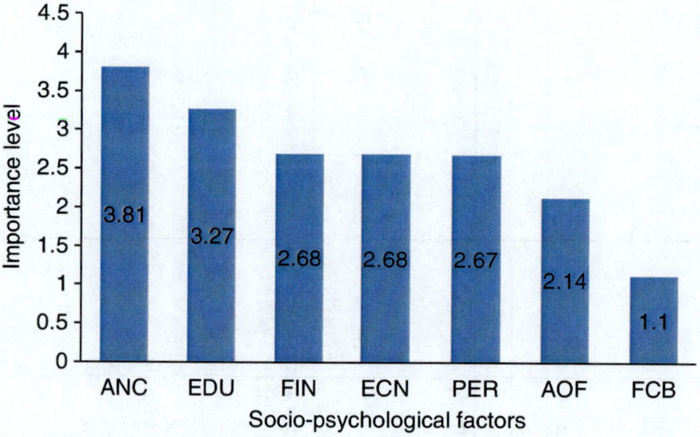

Fig. 4 Importance of socioeconomic and demographic factors in homegarden management decision in Kerala, India. *AOF* agricultural office, *ANC* ancestors, *ECN* economic importance of homegarden, *EDU* education, *FIN* financial solvency of the farmer, *PER* peers, *FCB* farmers' club. The importance values refer to the participants' rating of importance of each factor in affecting their decision to employ each HG management practice using a 1–5 Likert-type scale (1 = Not Important; 5 = Extremely Important)

mechanisms (ancestors, peers, education, financial condition, agricultural office etc.) might have impacted each of the five management practices.

Tillage Operations

Tillage is a popular practice in the homegardens of Kerala with over 86% of respondents saying they till their homegardens. Tillage, however, is known to inversely affect the soil C sequestration (Six et al. 2000). Therefore, factors promoting tillage are in turn negatively influencing the process of soil C sequestration. The decision on whether to practice tillage seemed to be significantly influenced by ancestors ($p=0.001$) and peers ($p=0.075$) (Table 4). These two factors come under normative beliefs and show how social norms, as mentioned in the TPB, influence farmer behavior. Tillage is a traditional practice and the current generation of farmers have observed their ancestors performing it for reasons such as soil aeration, better root growth, and improved plant health. Therefore, it is likely that recommendations from the ancestors would influence the farming decisions today. Similarly, if neighbors/peers practice tillage on their lands, peer pressure would influence the farmer to till his homegarden.

Decision on tilling was also positively influenced by the economic importance of the homegarden ($p=0.025$). The more economically important the HG is, the more likely the farmer will till the land. Most of the homegardeners, who said that they did not till had little income from the HG. Out of eight farmers, who negatively responded about tilling, six had less than 12% of their total income from the homegardens. The decision to till was inversely influenced by education ($p=0.082$) (Table 4). This is supported by the results of Anjichi et al. (2007), who observed that formal education was a critical factor in influencing the efficacy of the farmers' decision to adopt soil conservation measures in Kenya.

Plant Residue Application

Plant residues supply C to the soil through gradual decomposition (Brady and Weil 2008) and contribute to the process of soil C sequestration. Out of the 59 surveyed homegardeners, 95% applied plant residue with varying amounts. Factors such as ancestors ($p=0.018$) and peers ($p=0.05$), which are normative beliefs under TPB, positively influenced the decision of plant residue application (Table 4). Application of plant residue is an age-old practice in the HGs and generally, there is no external purchase of plant residues. Thus, irrespective of the economic importance of the HGs, farmers may apply plant residues.

Availability of plant residues, which falls under the control belief is also a factor significantly influencing the farmers' decision ($p=0.053$) (Table 4). The main source of plant residue is the HG itself, however, due to fragmentation of holdings or sparse tree growth, a HG may not be self-sufficient in producing plant residues. In such cases, farmers collect residues from external sources such as nearby public lands or from neighbors. The amount of plant residues applied is much higher than manure and fertilizers (Table 1), and, on the other hand, there is no organized way of buying or selling plant residues. Therefore, the ease of availability of plant residues is very important in making a decision about the intensity of its application.

Another control belief, financial solvency of the farmer, was observed to negatively influence the decision on plant residue application ($p=0.01$) (Table 4). This could be due to two reasons. First, the resource-rich farmers can buy concentrated organic fertilizers, in turn replacing plant residues as a source of nourishment. Second, resource-rich homegardeners often may have other sources of employment and the income from the HGs may not be substantial. This would discourage them from investing time and labor for maintaining HGs.

Manure Application

Application of manure supplies organic matter, which in turn promotes C sequestration in soil (Whalen and Chang 2002; Six et al. 2002). The results indicated that 93% of the farmers under the survey applied manure to the HGs in the past year; however, the intensity varied. Factors that significantly affected the farmer's decision to apply manure, were the availability of manure ($p=0.019$) and ownership of the household cattle ($p=0.022$) (Table 4). Both these factors fall under the control belief of the TPB, because they directly relate to the ability of the farmers to control this behavior. This finding is consistent with that of Williams (1999), who observed that in semiarid West Africa, the likelihood of applying manure increased with the herd size of animals. Traditionally, manure is produced from raw materials such as animal/leaf litter that are obtained from the HG itself. Manure is also procured from friends, neighbors, and relatives, and sometimes purchased externally.

Inorganic Fertilizer Application

Application of fertilizers promotes biomass production, which consequently may get incorporated in soil and influence the C sequestration process (Lal et al. 1999). Our results indicate that 53% of the homegardeners applied inorganic fertilizers within the last year and factors such as economic importance of HG, ancestors, and education significantly impacted the application. Fertilizer application decision was positively ($p=0.005$) influenced by the economic importance of HGs (Table 4). The more important economic factors were (i.e. more the expectation of monetary returns from the HG), the more the farmer invested in HGs. Unlike plant residues and manure, fertilizer has to be purchased at a price. To meet the recommended dosage of fertilizers for any homegarden crop, the farmer may have to spend a considerable amount of money and most of the farmers would make decisions to earmark funds for a complete package of fertilizers only if he or she is expecting significant economic returns from the HG. This is in agreement with the observations of Lambin et al. (2001) who found that the agricultural decisions made by individuals are mostly influenced by economic opportunities.

Our results also indicate that the ancestor factor, which represents normative belief, has an impact ($p=0.089$) on the farmers' decision to apply fertilizers (Table 4). Farmers have learned and experienced the benefits of fertilizers from their ancestors, which encouraged them to use fertilizers. The decision to apply fertilizer was inversely associated with a behavioral belief, which is education of the farmer ($p=0.008$). This means the higher the farmers' education level (or exposure to new developments), the more likely he or she will apply fertilizers more judiciously than the less educated farmers, as observed by Welch (1970).

Tree Planting

Trees play a major role in sequestering C both above- and belowground (Montagnini and Nair 2004; Nair et al. 2010); therefore, farmers' decisions to plant trees affect the total C sequestration of the HG. Tree cover percentage in HGs in our study varied from 20% to as high as 90%. Our results showed that the decision to plant trees is influenced ($p=0.053$) by the education (behavioral belief) of the farmer (Table 4) and are consistent with the findings of Pichon (1997) and Geoghegan et al. (2001).

In South America, they observed that higher education had a negative effect on the levels of cutting trees. Matata et al. (2008) reported that almost all the participants in new agroforestry activities (planting trees with crops) in Tanzania had primary education. The fact that education influences tree planting means that the more educated the farmer is, the more likely will he or she be inclined to plant (more) trees. It is possible that farmers who are educated and exposed to new developments understand the additional environmental benefits of trees. However, it should not be concluded that less or uneducated farmers are unaware of tree's environmental benefits. The difference, though, is that educated farmers have more access and exposure to external scientific information from various sources, which strengthen their knowledge base and may increase environmental awareness.

Another factor that significantly ($p = 0.088$) influenced the decision to plant trees was peers (normative belief: Table 4). Several management decisions of HG are influenced by friends, relatives, and neighbors. The profitability from a particular tree species does not remain the same forever, and selection of potential new species with a market demand takes place in course of time. Kerala homegardens are highly dynamic and the cropping patterns change depending on the market demand. For example, cacao (*Theobroma cacao* L.) was once a profitable species in Thrissur HGs; but, with time, it was replaced with nutmeg (*Myristica fragrans* Houtt.), although they still coexist in some places. This information on change in market trend and preference to a new species circulates through people and the peer factor influences the farmers. Typically, a few farmers introduce a new species and the rest of the farmers learn from the opportunities and make a similar decision. Thus, when it comes to selecting any type of tree, the farmer is influenced by peers. In addition, farmers learn about the additional environmental benefits of the trees from their educated peers and this might also affect their decision to plant trees.

Summary and Conclusions

The cultural, demographic, economic, and social factors associated with agroforestry have influenced the land management decisions to varying degrees, which in turn have the potential to affect soil C sequestration in the homegarden systems. Factors such as ancestors (normative belief) influenced tillage, plant residue application, and fertilizer application, and had an overall high impact on decision making. This was followed by education (behavioral belief), which influenced the decision on tree planting in a positive manner and affected the decision with tillage and fertilizer application in a negative manner. Other important factors influencing the farmer behavior were peers (normative belief), financial conditions (control belief) of the farmer, and economic importance of the HG.

To put these results in an appropriate context, it must be understood that farmers do not make their decisions on land management practices based on which practice has the most C sequestration potential. Carbon sequestration is a relatively new issue and it is not widely discussed among agricultural extension officers, who usually bring new ideas to farmers, in Kerala and in many areas throughout the world. Therefore, the effect of land management practices on C sequestration is mostly indirect, in the sense that any practice that contributes to higher productivity will indirectly lead to higher C sequestration. Since some management practices (e.g., soil tillage) are known to have a greater bearing on soil C sequestration, we focused on those practices in this study to better explain the factors influencing those practices. The impact of farm management on C sequestration *per se* was not the objective of this study.

An important outcome of this study is that landowners place great importance on socio-psychological factors (i.e., ancestors, peers, and education), which influence how they manage the HGs (i.e., tillage, plant residue application, use of fertilizer, etc.). With an understanding of these connections, policies could incorporate past history and learning from friends and family to encourage farmers to conduct C sequestration friendly (CSF) practices. For example, policies should be enacted to work with social groups to educate on the benefit of C sequestration and how it might be related to traditional land use practices. Educating them on this unexplored virtue could be a way to enhance their confidence in their traditional life-style. Homegarden farming is an essential aspect of such traditional life style, but has been overlooked by modern development paradigms and commodity-oriented chemical agriculture. Efforts leading to a better and deserving appreciation, and even amplification, of the role of traditional knowledge and elaborating the possibility of future financial gains as compensation for C sequestration would be of considerable advantage to the subsistence farming community. These issues will need to be built into the training curricula of extension officers, who are seldom required or encouraged to appreciate and promote such 'out-of-the-box' and 'non-modern' practices.

Further research along these lines is required in other agricultural and social systems and geographic locations to understand the detailed effects of individual socioeconomic and demographic factors on management practices and how TPB can be applied to explain agricultural decision making. Such studies will provide answers to a number of interesting questions such as how traditional knowledge influences the adoption of CSF practices, if more education may lead to promotion of organic agriculture and C sequestration friendly practices, and if strengthening the agricultural extension service and farmers' clubs may promote eco-friendly agriculture. Such information will have implications to Clean Development Mechanism (CDM) in developing emission reduction projects in the context of climate change migration and adaptation through Reduced Emission from Deforestation and Degradation (REDD) of UNFCCC.

Acknowledgements We are grateful to Kerala Agricultural University (Kerala, India) and the university personnel, especially Dr. B. Mohan Kumar, who helped us organize and conduct the study in Thrissur, Kerala. We would like to thank the Kerala Department of Agriculture for their cooperation and all the farmers who participated in the survey and the meetings. Finally, we are thankful to the University of Florida for providing financial support to conduct the study.

References

Aguilar-Støen M, Moe SR, Camargo-Ricalde SL (2009) Homegardens sustain crop diversity and improve farm resilience in Candelaria Loxicha, Oaxaca, Mexico. Hum Ecol 37:55–77

Ajzen I (1985) From intentions to actions: a theory of planned behavior. In: Kuhl J, Beckman J (eds.) Action-control: from cognition to behavior. Springer, Heidelberg, pp 11–39

Ajzen I (1991) The theory of planned behavior. Organ Behav Hum Decis Process 50:179–211

Ali AMS (2005) Homegardens in smallholder farming systems: examples from Bangladesh. Hum Ecol 33:245–270

Anjichi VE, Mauyo LW, Kipsat MJ (2007) The effect of socioeconomic factors on a farmer's decision to adopt farm soil conservation measures: an application of multivariate logistic analysis in Butere/Mumias district Kenya. In: Bationo A, Waswa B, Kihara J, Kimetu J (eds.) Advances in integrated soil fertility management in Sub-Saharan Africa: challenges and opportunities. Springer, Dordrecht, pp 915–920

Beedell J, Rehman T (2000) Using social- psychology models to understand farmers' conservation behavior. J Rural Stud 16:117–127

Brady NC, Weil RR (2008) The nature and properties of soil, 14th edn. Prentice Hall, New York

Burton RJF (2004) Reconceptualising the 'behavioral approach' in agricultural studies: a socio-psychological perspective. J Rural Stud 20:359–371

Colemont A, Van den Broucke S (2008) Measuring determinants of occupational health related behavior in Flemish farmers: an application of the theory of planned behavior. J Saf Res 39:55–64

FAO (2004) Carbon sequestration in dryland soil. In: World Soil Resources Reports 102, FAO, Rome. Available at http://www.fao.org/docrep/007/y5738e/y5738e0d.htm. Accessed Dec 2010

Geoghegan J, Villar SC, Klepeis P, Mendoza PM, Ogneva- Himmelberger Y, Chowdhury RR, Turner BL, Vance C (2001) Modeling tropical deforestation in the southern Yucatan peninsular region: comparing survey and satellite data. Agri Ecosys Environ 85:25–46

Government of Kerala (2005) Soil and land resources of Madakkathara *panchayath*. Report no. 468. Soil Survey Organization, Kerala

Jose D, Shanmugaratnam N (1993) Traditional homegardens of Kerala: a sustainable human ecosystem. Agroforest Syst 24:203–213

Kumar BM (2006) Carbon sequestration potential of tropical homegardens. In: Kumar BM, Nair PKR (eds.) Tropical homegardens: a time-tested example of sustainable agroforestry. Springer, Dordrecht, pp 185–204

Kumar BM, Nair PKR (2004) The enigma of tropical homegardens. Agroforest Syst 61:135–152

Kumar BM, George SJ, Chinnamani S (1994) Diversity structure and standing stock of wood in the homegardens of Kerala in peninsular India. Agroforest Syst 25:243–262

Lal R, Hassan HM, Dumanski J (1999) Desertification control to sequester C and mitigate the greenhouse effect. In: Rosenberg N, Izaurralde RC, Malone EL (eds.) Carbon sequestration in soils: science, monitoring and beyond. Battelle Press, Columbus, pp 83–149

Lambin EF, Turner BL, Geist HJ, Agbola SB, Angelson A, Bruce JW, Coomes OT, Dirzo R, Fischer G, Folke C, George PS, Homewood K, Imbernon J, Leemans R, Li X, Moran EF, Mortimer M, Ramakrishnan PS, Richards JF, Skanes H, Steffen W, Stone GD, Svedin U, Veldkamp TA, Vogel C, Xu J (2001) The causes of land- use and land- cover change: moving beyond the myths. Glob Environ Change 11:261–269

Likert R (1932) A technique for the measurement of attitudes. Arch Psychol 140:1–55

Matata PZ, Ajayil OC, Oduol PA, Agumya A (2008) Socio-economic factors influencing adoption of improved fallow practices among smallholder farmers in western Tanzania. Int NGO J 3:68–73

Montagnini F, Nair PKR (2004) Carbon sequestration: an under-exploited environmental benefit of agroforestry systems. Agroforest Syst 61–62:281–298

Nair PKR, Kumar BM (2006) Introduction. In: Kumar BM, Nair PKR (eds.) Tropical homegardens: a time-tested example of sustainable agroforestry. Springer, Dordrecht, pp 4–5

Nair MA, Sreedharan C (1986) Agroforestry farming systems in the homesteads of Kerala southern India. Agroforest Syst 4:339–363

Nair PKR, Nair VD, Kumar BM, Showalter JM (2010) Carbon sequestration in agroforestry systems. Adv Agron 108:237–307

Peyre A, Guidal A, Wiersum KF, Bongers F (2006) Dynamics of homegarden structure and function in Kerala, India. Agroforest Syst 66:101–115

Pichon FJ (1997) Settler households and land- use patterns in the Amazon frontier: farm-level evidence from Ecuador. World Dev 25:67–91

Robertson GP, Paul EA, Harwood RR (2000) Greenhouse gases in intensive agriculture: contributions of individual gases to the radiative forcing of the atmosphere. Science 289:1922–1925

Saha SK, Nair PKR, Nair VD, Kumar BM (2009) Soil carbon stock in relation to plant diversity of homegardens in Kerala, India. Agroforest Syst 76:53–65

Saha SK, Nair PKR, Nair VD, Kumar BM (2010) Carbon storage in relation to soil size-fractions under some tropical tree-based land-use systems. Plant Soil 328:433–446

Seabrook L, McAlpine C, Fensham RJ (2008) What influences farmers to keep trees? A case study from the Brigalow Belt Queensland Australia. Landscape Urb Plan 84:266–281

Six J, Elliott ET, Paustian K (2000) Soil macroaggregate turnover and microaggregate formation: a mechanism for C sequestration under no-tillage agriculture. Soil Biol Biochem 32:2099–2103

Six J, Conant RT, Paul EA, Paustian K (2002) Stabilization mechanisms of soil organic matter: implications for C-saturation of soils. Plant Soil 241:155–176

Somda J, Nianogo AJ, Nassa S, Sanou S (2002) Soil fertility and socio economic factors in crop-livestock systems in Burkina Faso: a case study of composting technology. Ecol Econ 43:175–183

Statistical Package for the Social Sciences (SPSS) (2001) Statistical package for the social sciences, version 11.0.1. SPSS Inc, Chicago

Thangata PH, Alavalapati JRR (2003) Agroforestry adoption in Southern Malawi: the case of mixed intercropping of *Gliricidia sepium* and maize. Agri Syst 78:57–71

Welch F (1970) Education in production. J Pol Econ 78:35–59

Whalen JK, Chang C (2002) Macroaggregate characteristics in cultivated soils after 25 annual manure applications. Soil Sci Soc Am J 66:1637–1647

Williams TO (1999) Factors influencing manure application by farmers in semi-arid West Africa. Nutr Cycl Agroecosys 55:15–22

Index

B.M. Kumar and P.K.R. Nair (eds.), *Carbon Sequestration Potential of Agroforestry Systems: Opportunities and Challenges*, Advances in Agroforestry 8,
DOI 10.1007/978-94-007-1630-8, © Springer Science+Business Media B.V. 2011